The Evolution and Extinction of the Dinosaurs

SECOND EDITION

This new edition of *The Evolution and Extinction of the Dinosaurs* is a unique, comprehensive treatment of a fascinating group of organisms. It is a detailed survey of dinosaur origins, their diversity, and their eventual extinction.

The book is written as a series of readable, entertaining essays covering important and timely topics in dinsoaur paleontology and natural history. It will appeal to non-specialists and all dinosaur enthusiasts, treating subjects as diverse as birds as "living dinosaurs," the new feathered dinosaurs from China, and "warm-bloodedness." Along the way, the reader learns about dinosaur functional morphology, physiology, and systematics using cladistic methodology – in short, how professional paleontologists and dinosaur experts go about their work, and why they find it so rewarding.

The book is spectacularly illustrated by John Sibbick, a world-famous illustrator of dinosaurs, with pictures commissioned exclusively for this book.

The Evolution and Extinction of the Dinosaurs

SECOND EDITION

David E. Fastovsky
University of Rhode Island

David B. Weishampel
The Johns Hopkins University

With illustrations by
John Sibbick

CAMBRIDGE UNIVERSITY PRESS

CAMBRIDGE UNIVERSITY PRESS

Cambridge, New York, Melbourne, Madrid, Cape Town, Singapore, São Paulo

Cambridge University Press
40 West 20th Street, New York, NY 10011-4211, USA

http://www.cambridge.org
Information on this title: www.cambridge.org/9780521811724

First published 1996
Reprinted 1996, 2001
Second edition 2005

Printed in the United States of America

A catalog record for this book is available from the British Library,

Library of Congress Cataloging in Publication Data

Fastovsky, David E.
The evolution and extinction of the dinosaurs / David E. Fastovsky,
David B. Weishampel. – 2nd ed.
    p.   cm.
Includes bibliographical references and index.
ISBN 0-521-81172-4
1. Dinosaurs – Evolution.   2. Extinction (Biology).   I. Weishampel, David B., 1952–
III. Title.
QE861.6E95F37 2005
567.9—dc22   2004049261

ISBN-13 978 0-521-81172-9 hardback
ISBN-10 0-521-81172-4 hardback

Contents

Sir Richard Owen, the brilliant nineteenth century English anatomist and father of the term "Dinosauria." (Photography courtesy of the American Museum of Natural History.)

Preface to the second edition

The idea behind this book is much as it was for the first edition: to use dinosaurs as an attractive entrée into aspects of natural history and scientific inquiry. Here then is a cohesive take on Dinosauria, with emphasis on the extinct rather than the extant members of the group. We remain passionately convinced that dinosaurs – when properly understood – illuminate not only the past but the present.

Belying their popular persona, dinosaurs are not about binomial Linnaean Latin litinies nor moribund minutiae. Instead, the study of dinosaurs is blessed by all the intellectual and creative opportunities afforded by science. An explicitly phylogenetic approach, as well as the presentation of multiple viewpoints, offers readers a sense of what is (and what is not) possible using scientific inquiry. Interested students and amateurs may use the book as a means of developing sophistication in not only dinosaurs, but also in the logic of scientific discovery.

For us the study of dinosaurs is all about the history of life and of the earth, the nature of nature, and, ultimately, who we are.

Acknowledgements

We have been aided by many generous, yet critical, friends and colleagues. We wish to extend our thanks and appreciation to the reviewers of and contributors to these pages, among them A. K. Behrensmeyer, M. J. Benton, J. A. Cain, K. Carpenter, P. Dodson, L. A. Fastovsky, C. Forster, P. M. Galton, R. E. Heinrich, C. Janis, J. S. McIntosh, M. B. Meers, K. Padian, H.-D. Sues, B. H. Tiffney and L. M. Witmer. Naturally all these individuals are held blameless for any errors that have crept into this book. Those that have are clearly Dave's fault.

Several colleagues, museums, and other institutions provided photographs for use in this book. We greatly appreciate the courtesy of reproducing these photos, acknowledging each throughout the text.

We are grateful to the publishing staff of Cambridge University Press, particularly Dr Ward Cooper and especially Mrs Sandi Irvine, for help and encouragement throughout this project. The work was supported in part by the Department of Geosciences of the University of Rhode Island, the Instituto de Geología, Universidad Nacíonal Autónoma de México, the Department of Functional Anatomy and Evolution of The Johns Hopkins School of Medicine, and by a Fulbright Garcia-Robles Fellowship (to D.E.F.).

PART I

Setting the stage

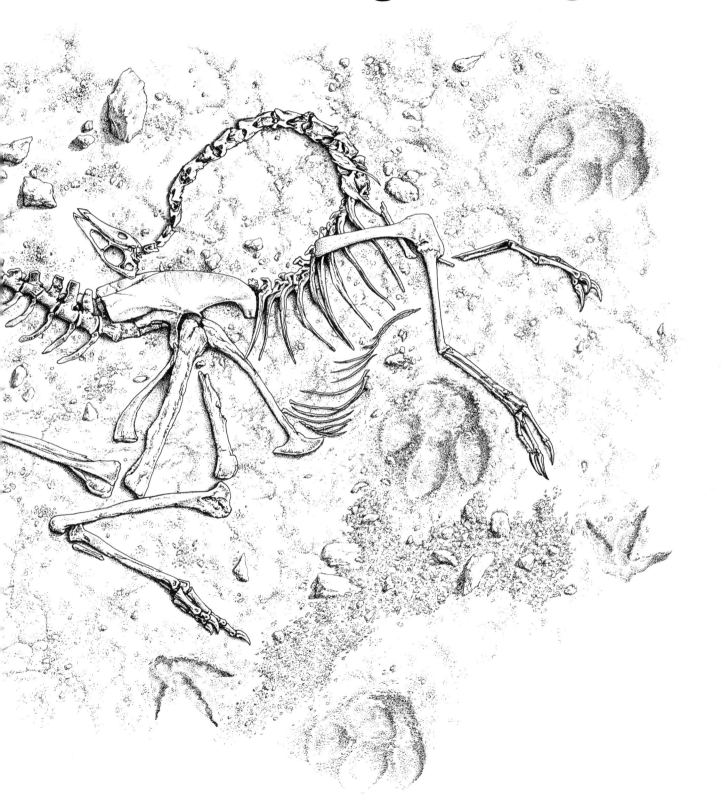

CHAPTER 1

Introduction

We live in dinosaur-crazy times. We watch dinosaur sit-coms and soap operas; we endure "Barney"; we absorb dinosaur documentaries on dinosaur extinctions, dinosaur descendants, dinosaur habits, and all else dinosaurian. We have thrice feasted on the *Jurassic Park* movies. Godzilla – not quite a dinosaur, but then not obviously anything else either – has returned episodically since 1954. We enjoy dino toys, and savor dino candy while surfing a kaleidoscope of dino websites. There are coloring books, erasers, stick-ons, refrigerator magnets, wooden and plastic models, pen covers, clothes … and we're not even discussing the dino paraphernalia available in museum gift shops. Tabloids fill us with the latest dinosaur "research," and new dinosaur books appear almost faster than we can count them. Even a few paleontologists – a profession that in previous generations was stereotyped by mild-mannered bookishness – have become minor media personalities. In this multi-media feeding frenzy of things dinosaurian, the science can become lost.

Here we attempt to present dinosaurs as professional paleontologists view the group. Because dinosaurs have been marveled over (at least by Westerners) since 1819, a good deal is known; by the same token, a 30-year-old revolution in methods of studying them has only in the past 20 years (or less) begun to overturn long-held ideas about them and their 160 million-year tenure on earth. Much of the recent media excitement is a reflection of the changes wrought in our understanding of dinosaurs. We hope that the give-and-take of scientific dialog are well recorded in these pages, for this most accurately reflects the sometimes tortuous path of scientific progress.

Scientists and scientific research are not without social context, a point that is ably made by W. T. J. Mitchell in *The Last Dinosaur*. Hence, what was compelling about dinosaurs in a culture of nationalism and manifest destiny might not be as compelling in a culture in which concern for the family and cultural diversity predominate. Still, the fact of bones and other remains exists whatever the interpretations and, although various aspects of dinosaur biology have resonated popularly at different times, the objective reality of dinosaurs and even the long-standing interpretations of their morphology and behavior are generally not at issue.[1]

The ideas presented here are not the final answer about things dinosaurian because, as we shall stress repeatedly, science is a process and not a static solution. Our field would lose its vibrancy if time and further research didn't modify what we present. These pages contain only our best call on what is known about dinosaurs, and ideas presented here will surely be modified with time. It is ironic that, although the fossil record may be written in stone, its interpretation is not.

So what follows is on one level a tale of dinosaurs: who they were, what they did, and how they did it. But on another and more significant level, it is a tale of natural history. In writing of dinosaurs, we are really developing a much fuller concept of the biosphere. Commonly we think of the biosphere as occurring only in the present; a kind of three-dimensional organismal insulation wrapping the globe. But in fact the biosphere has a 3.8 billion-year history, and we and all the organisms around us are products of that fourth dimension: its history. The history of life is a grand pageant and to be unaware of it is to be unaware of who we are. Part of our goal in this book, therefore, is to explore the relationships of organisms to each other and to the biosphere. Historically, humanity has maintained a distorted sense of its position in the biosphere; with wilderness ever-diminishing, we would do well to refine (or even redefine) our understanding of our relationship to the earth and its inhabitants. Dinosaurs have some significant information to impart in this regard; as we learn who dinosaurs really are, we can better understand who *we* really are.

Finally, ours is a tale of science, itself: we live in an increasingly technical world, and we can no longer afford to ignore science and the way it impinges on lives. But science is often poorly understood. One of us remembers a song from the hopeful post-World War II days when it truly seemed that technology could solve all the world's problems:

> It's a Scientific Fact – A Scientific Fact!
> It has to be Correct! It has to be Exact!
> Because it is, because it is a Sci-en-ti-fic Fact![2]

---

1 The issue of the reality of dinosaurs and dinosaur biology (as opposed to their being social constructs) is addressed in detail by K. M. Parsons in his book, *Drawing Out Leviathan: Dinosaurs and the Science Wars*.

2 From the 1959 children's record *Space Songs* by H. Zarett.

If science were as it is portrayed in *Space Songs*, neither of us would have become scientists, either by disposition or by inclination. Science is essentially a creative enterprise requiring, depending upon the problem, approximately equal parts intuition and art. In the following pages, therefore, we hope to build a sense of the beauty of science and a feel for the meaning of scientific observation.

Like any other discipline, paleontology can be mastered only in a stepwise fashion. It has its own language and its own concepts, quite apart from the related disciplines of geology and biology. So, although our book is written as a series of individual essays on selected topics relating to dinosaurs, the development of concepts and ideas is sequential, and each chapter builds upon the previous ones.

The word "dinosaur" in this book

The term "dinosaur" (*deinos* – terrible; *sauros* – lizard) was established in 1842 by the English anatomist Sir Richard Owen to describe a few fossil bones of large, extinct "reptiles." With modifications (e.g., "large" no longer applies to all members of the group), the name proved resilient. It has become clear in the past 10 years, however, that not all dinosaurs are extinct; most vertebrate paleontologists now agree that *birds are living dinosaurs*. This leaves us with a problem, because much of what we will discuss concerns non-avian dinosaurs; that is all dinosaurs *except* birds. We could use the cumbersome, but technically correct, term "non-avian dinosaurs," but it would be far easier if the term "dinosaurs" is used as a kind of shorthand for "non-avian dinosaurs." The distinction between non-avian dinosaurs and all dinosaurs will be most relevant only when we discuss the origin of birds and their early evolution in Chapters 13 and 14; there, we will take care to avoid confusing terminology.

Fossils

That we even know that there ever were such creatures as dinosaurs is due to plain luck: some members of the group just happened to be preserved in rocks. Exactly what is preserved and how that occurs provides insight into the kinds of information we can expect to learn about these extinct beasts. Dinosaurs last romped on earth 65 million years ago. This means that their soft tissues – muscles, blood vessels, organs, skin, fatty layers, etc. – are long gone. If any vestige remains at all, it is usually hard parts such as bones and teeth. Hard parts are not as easily degraded as the soft tissues that constitute most of the body. Obviously the kinds of change that organic remains can undergo are of great interest to us as paleontologists. The study of those changes, and in fact the study of all of what happens to organisms after death, is called taphonomy.

Taphonomy is extremely important because, by looking at a group of fossils and the sedimentary rocks in which they are encased, we can learn something about the deaths (and lives) of the animals represented. Because all dinosaurs were land-dwelling vertebrates, our primary interest will be in the taphonomic changes that occur in terrestrial settings. Such settings generally consist of deserts, wetlands and lakes, river systems (channels as well as floodplains), and, to a lesser extent, deltas. Each of these settings is an environment in which bones can be preserved.

Taphonomy

Before burial

Consider what happens to a dinosaur – or any land-dwelling vertebrate – when it dies. If it is killed, it can be disarticulated (dismembered), first by the animal that killed it, and then by scavengers. In modern environments, the best known of these scavengers are vultures or hyenas, but there are smaller animals of far greater significance, such as scarily efficient carrion-eating dermestid beetles. Of course most of the heavy lifting in the world of decomposition is done by bacteria that feast on rotting flesh (leaving no doubt that something died). The bones are commonly stripped clean of meat and occasionally left to bleach in the sun. Some bones might get carried off and gnawed somewhere else. Sometimes the disarticulated remains are trampled by herds of animals, breaking and separating them further. The more delicate the bones, the more likely they are to be destroyed. So there sit the sum of all the earthly remains of the animal: a few disarticulated bleached bones lying in the grass.

If the animal isn't killed by some predator but just dies (old age, drowning, and disease all qualify) it may or may not be disarticulated immediately, depending upon which scavengers get to it and when (Figure 1.1). Left intact, it is not uncommon for a carcass to swell up (bacterial decomposition produces gasses that inflate it), eventually deflate (sometimes catastrophically: this can be just a tad grotesque), and then dry out (if not in water), leaving bones, tissues, ligaments, tendons, and skin hard and inflexible. The tissues shrink as they dry, bending the limbs and pulling back the head and lips into a hideous rictus (producing the illusion that the animal died in agony). Under such conditions, the creature is essentially mummified, and the carcass can be exposed for a very long time without further decomposition. Then we get "jerky": dried meat that resists decomposition. Later in this book, we will encounter genuine dino jerky.

Occasionally catastrophic things happen to herding animals. Floods catch herds as they cross swollen rivers and there are mass drownings. The carcasses may bloat and float, eventually to pile up along the edge of a channel, wash up onto a floodplain, or accumulate on the surface of a point bar at bends in the waterway. Left alone, the bones may be stripped, partially disarticulated, and bleach in piles of semi-articulated skeletons of one type of animal.

Burial

Sooner or later bones become either destroyed or buried. If they aren't turned into somebody's lunch, destruction can come from weathering; eventually, the minerals in the bones break down and the bones disintegrate. But the game becomes interesting for paleontologists when

Figure 1.1. A wildebeest carcass, partly submerged in mud and water and on its way to becoming permanently buried and fossilized. If the bones are not protected from scavengers, air, and sunlight, they decompose rapidly and are gone in 10–15 years. Bones destined to become high-quality fossils must be buried soon after the death of the animal. (Photograph courtesy of A. K. Behrensmeyer.)

the bones are buried. At this point, they become fossils (the bones, not the paleontologists). The word "fossil" comes from the latin word *fodere* (to bury), and refers to anything that is buried. There is no implication of how much time the remains have been buried; a dog burying a bone is technically producing a fossil. A body fossil is what is produced when a part of an organism is buried. We distinguish these from trace fossils, which are impressions in the sediment left by an organism.

Burial can take several forms. The simplest type of burial is when a bone or accumulation of bones is covered by sediment. For example, in a desert, burial might occur when a sand dune migrates over another, covering anything that was there before: desert lore resonates with mysteries of shifting sands relentlessly burying all who passed through. Equally inevitably, floodwaters also bury; ask the unhappy homeowner whose basement was silted as the waters receded. A more subtle type of deposition can occur, however, when sediments are reworked, which means that they are actively eroded from wherever they were originally deposited, and redeposited somewhere else.

Rivers are notorious reworkers. They flow, well behaved, within the confines of their channels. But in storms, they can breach their channels, flooding the landscape and eroding the edges of the channel. The eroded material from the channel margin is carried within the channel or spewed out with the floodwaters onto the floodplain. Rivers practice equal-opportunity erosion: if any buried bones are within the eroded part of the channel margin, they too will be swept along wherever the floodwaters see fit.

Here is where a paleontologist can be fooled. Seeing bones in a river channel, he or she might interpret these to be the remains of organisms that died together. But reworking concentrates fossils within the confines of the channel, and a collection of fossils in a channel is not necessarily the remnants of a community that actually lived (and died) together. Instead, it might be a reworked assemblage of bones that includes fossils eroded out of a much older floodplain, mixed with material of the same age as the channel. In the upper Great Plains of North America, 65 million-year-old dinosaur bones have been found jumbled with deposits of 10,000-year-old bison bones: as glaciers melted 10 thousand or so years ago, stream channels of glacial meltwater eroded 65 million-year-old floodplain sediments and mixed the bones of Mesozoic dinosaurs and ice-age mammals.

So paleontological work is not just digging up old bones. We want to know how the bones got the way in which we find them, because that may tell us something of how dinosaurs lived. Different types of concentration of bones can come about through different processes (or can be explained by different scenarios). When the bones are articulated (connected), this suggests that they have not been transported far from where the animal died. The idea here is that none of the destructive forces we described above has had much effect on the fossils that we have found. On the other hand, if we find a collection of disarticulated bones of several types of vertebrates, we can be fairly sure that the deposit has been reworked, and that the bones got there through sedimentary processes sometime after death and initial burial. Then there are the bonebeds: accumulations of bones of many individuals of a very few *kinds* of organism, sometimes articulated and sometimes not. A bonebed with one or two species may represent a herd subjected to a catastrophic event; alternatively, if we could determine that the bonebed accumulated over a long period of time, then it might simply represent a location where many animals of the same type chose to live (and die). Finally, isolated finds – a thigh bone here or a vertebra there – could represent almost any of the possibilities that we described above. In Figure 1.2, two taphonomic sequences, leading to two different results, are shown.

After burial

Bone is made out of calcium (sodium) hydroxyapatite, a mineral that is not stable at temperatures and pressures at or near the surface of the earth. This means that bones can change with time, which in turn means that most no longer have original bone matter present after fossilization.

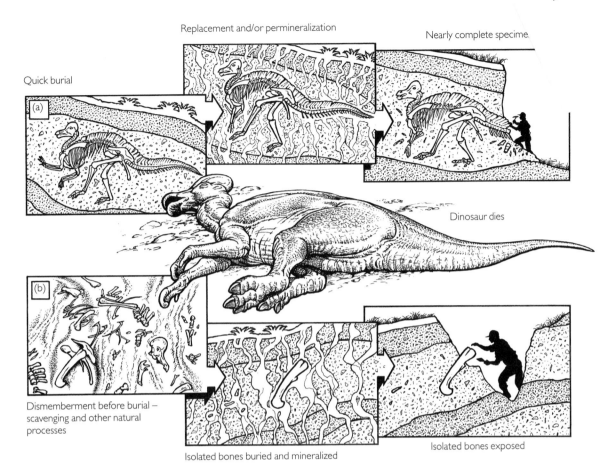

Replacement and/or permineralization

Nearly complete specime.

Quick burial

(a)

Dinosaur dies

(b)

Dismemberment before burial —
scavenging and other natural
processes

Isolated bones exposed

Isolated bones buried and mineralized

Figure 1.2. Two endpoint processes of fossilization. In both cases, the first step is the death of the animal. Some decomposition occurs at the surface. In the upper sequence (a), the animal dies, the carcass undergoes quick burial, followed by bacterial decomposition underground, and permineralization and/or replacement. Finally, perhaps millions of years later, exposure. Under these conditions, when the fossil is exhumed, it is largely complete and the bones articulated. This kind of preservation yields bones in the best condition. In the lower sequence (b), the carcass is dismembered on the surface by scavengers and perhaps trampled and distributed over the region by these organisms. It may then be carried or washed into a river channel and buried, replaced and/or permineralized, eventually to be finally exposed perhaps millions of years later. Under these conditions, when the fossil is exhumed, it is disarticulated, fragmented, and the fossil bones may show water wear and/or the gnaw marks of ancient scavengers. Different conditions of fossil preservation tell us something about what happened to the animals after death.

This is especially likely if the bone is bathed in the variety of fluids that is associated with burial in the earth (e.g., ground water). If, however, no fluids are present throughout the history of burial (from the moment that the bone is buried, to when it is exhumed by paleontologists, a time interval that could be measured in millions of years), the bone could remain unaltered, which is to say that original bone mineralogy remains. This situation is not that common, and is progressively rarer in the case of older and older fossils. Unaltered bone, however, is obviously crucial for studies involving aspects of bone mineralogy (see Chapter 15); for example, the discovery of genuine red blood cells from *Tyrannosaurus* required unaltered bones.

Most bones, as we have suggested, are altered to a greater or lesser degree. Since bones are porous, the spaces in fossil bones fill up with minerals. This situation is called permineralization. More significant than permineralization is recrystallization, where the bone itself (made of hydroxyapatite) is dissolved and reprecipitated, retaining the exact original form of the fossil. Recrystallization can be very obvious – for example, when small crystals are replaced with large ones – but it can also be very subtle, and occur on a microscopic scale. Bones can also be replaced, in which case the original bone minerals are replaced with other minerals.

Nothing is simple, and, in general, fossil bones undergo a combination of replacement, permineralization, and recrystallization. The resultant fossil, therefore, is a magnificent natural forgery: chemically and texturally unlike the original bone matter, although commonly retaining its exact shape and most delicate features. Still, accept no substitutes: fossil bones tend to be much heavier (they're permineralized) than their living counterparts and are more brittle. In virtually every respect, most fossil dinosaur bone is really closer to rock than to the living inorganic and organic mix of materials that we call bone.

In general, the more quickly a bone is buried, the better the chance it has of being preserved. This is because quick burial generally inhibits the weathering processes that would normally break down the bone minerals.[3] In fact, it is not uncommon to find evidence of weathering *before* fossilization. For example, dinosaur bones that appear to have been transported in water show water-wear that is exactly the same as those found in modern bones that have been transported by water: they are rounded and commonly the surface of the bone is partly or completely worn off. In the case of such dinosaur bones, the bone was transported by flowing water shortly after the animal was alive and was buried water-worn. The resultant fossil perfectly preserves the water-wear.

So if the fossils are buried, how is it that we find them? The answer is really luck: if fossil-bearing sedimentary rocks happen to be eroded, and a paleontologist happens to be looking for fossils at the moment that one is sticking out of an actively eroding sedimentary rock, the fossil *may* be observed and *may* be collected. Does this mean that great numbers of fossils lie buried within sedimentary rocks that happen not to be eroding at the earth's surface? Undoubtedly. Have fossils been eroding out of rocks since the very first fossil was formed? No question. Are important fossils currently eroding that will never be collected? No doubt.

---

3 An exception to this is when a bone ends up buried in an active soil; under certain conditions, the bone is then destroyed by biotic and abiotic soil processes.

Paleontology has been pursued in an active and systematic fashion only for the past 150 years; however, as A. Mayor has noted in her fascinating book, *The First Fossil Hunters*,[4] it was not unknown in classical antiquity. This means that, although fossils with hard parts have been produced for 540 or so million years (and trace fossils have been produced for even longer), most of those few that happened to be fossilized in the first place and then happened to be exposed at the earth's surface were never collected: they eroded away with the rest of their host rock.

Dinosaurs first appeared about 228 million years ago, and all but birds went extinct 65 million years ago or earlier. We may be sure that, throughout the 160 million-year existence of dinosaurs on earth, their fossils were constantly eroding out of sediments, being weathered away and lost for eternity.

The odds are stacked against fossilization. And, although many fossils are found, it is clear that most creatures – even those with hard parts – that have lived on earth are not preserved. Some paleontologists estimate that 25% of all the *types* of dinosaurs that ever lived have been found as fossils. This may be true but then again it may not. Who knows what percentage of all the dinosaurs that ever lived we presently know? The fossilization process is fraught with opportunities for fossils not to be found by that uniquely fossil-crazy group of primates (ourselves) that has been systematically seeking them for the past 150 or so years.

Other kinds of fossil

Bones are not all that is left of dinosaurs. Occasionally the fossilized feces of dinosaurs and other vertebrates are found. Called coprolites, these sometimes-impressive relics can give an intestine's-eye view of dinosaurian diets. Likewise, as we shall see later in this book, eggs and skin impressions have also been found.

Still, the single most important type of dinosaur fossil, other than the bones, themselves, are trace fossils. Dinosaur trace fossils (sometimes also called ichnofossils; (*ichnos* – track or trace)) are most commonly isolated footprints or complete trackways. Figure 1.3 shows a track left by a dinosaur. To be produced, the footprint must be made in material that can hold an impression (mud or fine sand). Again, rapid burial is a good way to ensure that tracks and trackways stick around. One finds molds that, in the case of tracks, represent the original impression itself, and casts, which are made up of material filling up the mold. Thus, a cast of a dinosaur footprint is a three-dimensional object that formed inside of the impression.

It would be terrific if we could link tracks with dinosaurs known from bones. But we can't; sometimes we can identify the trackmaker by a broad category like, for example "theropod." For this reason, footprints have their own names, and are classified separately as ichnotaxa, or footprint types. This may seem at first to complicate the number of

---

4 Mayor, A. 2000. *The First Fossil Hunters*. Princeton University Press, Princeton, NJ, 361 pp.

Figure 1.3. Theropod dinosaur footprint from the Early Jurassic Moenave Formation, northeastern Arizona, USA. Human foot for scale.

dinosaurs out there, but in fact it is an important way of keeping different kinds of data distinct.

While historically trackways have tended to be undervalued by vertebrate paleontologists, in the past 15 years their importance has become recognized. Trackways have demonstrated that dinosaurs walked erect, much as mammals and birds do today. Moreover, the trackways have shown the position of the foot: up on the toes in some cases and flat-footed in others. The speeds at which dinosaurs traveled (see Box 15.3) have been calculated using trackways: we can actually assess the possibility of *T. rex* running down a Jurassic Park Jeep. Trackways can give some indication of the faunal composition of a particular locality. Trackways have been used to document a predator (in this case, a theropod) stalking a herd of sauropods. The very idea of dinosaur herds has been supported by trackways. Trackways, therefore, are important clues to dinosaur behavior, and we will have recourse to them throughout the pages of this book.

Collection

The romance of dinosaurs is bound up with dinosaur collection: one travels to far-flung locales, undergoes heroic field conditions and, like Dr Seuss' Gerald McGrew, manfully extracts exotic beasts. Regardless of romance, dinosaur collecting turns out to be non-trivial. The first step is prospecting; that is, hunting for them. The second step is collecting them, which means getting them out of whichever (usually remote) locale they are situated. The final step is making them available for study and/or display in a museum. This last step involves preparing and curating them; that is, getting them ready for viewing and incorporating them into museum collections. These steps involve different skills and commonly different individuals.

How deep is your wallet? Let's say that you decided to bag a *Triceratops*, a moderately large but common (finding it would not be extraordinarily difficult) herbivorous dinosaur from the western part of the USA. Presuming you have elected not just to buy the fossil outright (its ±$1,500,000 price tag might prove a small stumbling block), you should budget 1–1.5 months for prospecting (if you and your crew know what you are doing), about 2.5 months for collecting (because the bones will be spread out all over the ground), and somewhere between 8 and 12 months for preparation and for mounting the specimen for display. More than one person will be involved in all of these steps, and it is likely that, after salaries (for your field crew and preparation staff), equipment, transportation, and preparing the display, your dinosaur would cost you upwards of $300,000 to $500,000. Maybe you should go back to trilobites.

Prospecting

A question that is commonly asked of paleontologists is "How do you know where the dinosaur fossils are?" The simplest answer is, "We don't." There is no secret magic formula for finding dinosaurs, unless long hard hours of persistent searching qualify as a secret magic formula. On the other hand, you can make a well-educated guess about where you might search, and in doing so you can greatly increase your odds of finding dinosaurs. Some collectors fare better than others, and a "feel" is surely involved in finding bone, as well as an experienced eye and plain luck (Figure 1.4).

Some basic criteria constrain the search. These are:

1   the rocks must be sedimentary;
2   the rocks must be of the right age; and
3   the rocks must be terrestrial.

Criterion 1

Of the three major kinds of rock, igneous, metamorphic, and sedimentary, only sedimentary rocks have the potential to preserve fossils to any reasonable degree. This is because igneous rocks are derived from molten material; obviously not a suitable habitat for dinosaurs or the preservation of their bones. Metamorphic rocks are formed by the intense folding and recrystallization of sedimentary rocks; these rocks

Figure 1.4. Typical scene at a dinosaur-producing outcrop somewhere in the badlands of the upper Great Plains, USA. In the foreground, paleontologists are excavating the fossils from the side of a butte, and treating them with liquid hardeners. In the swale behind the paleontologists, a grid is laid out, allowing the scientists to record the exact position of each of the fossil bones. Bags of plaster and burlap strips are scattered around, to be used in jacketing the specimens.

generally are no better. But, of course sedimentary rocks form in, and represent, sedimentary environments, many of them places where dinosaurs lived and died.

Criterion 2
If the rocks you search were not deposited sometime between the Late Triassic and the Late Cretaceous, you're out of luck. This is the window of opportunity for finding dinosaurs; older and younger rocks may yield weird and wonderful vertebrate fossils, but not dinosaurs.

Criterion 3
Dinosaurs were terrestrial beasts through and through, which means that their bones will generally be found in river systems, deserts, and

deltas. Dinosaur remains, however, are known from lake deposits and from near-shore marine deposits. Clearly they lived neither in lakes nor in the ocean. In such cases, if the bones are articulated, the bloated carcasses may have floated out into the water, and eventually sunk and been buried. If not articulated, they may have been washed out of the mouths of rivers into the lakes and oceans.

Many of the richest fossil localities in the world are in areas with considerable rock exposure, such as badlands. Fossil localities are common in deserts; plant cover on the rocks is low, and the dry air slows down the rates of weathering so that, once a fossil is exposed, it isn't chemically destroyed or washed away. Paleontologists, therefore, don't often find themselves in the jungle looking for fossils; the weathering rates are too high and the rocks are poorly exposed. Instead, they commonly toil under the sun (because few plants means little shade) in deserts, or at least fairly dry regions.

This is not to say that all dinosaur material has been found in badlands or in deserts; far from it. As long as the three criteria above are met, there is a possibility of finding dinosaurs, and that's usually reason enough for going in and taking a look.

Once we find rocks that match the criteria above, we simply start searching for bone weathering out of the rock.[5] Despite the promise of high-tech gadgetry such as the ground-penetrating radar used in *Jurassic Park I*, there has not yet been found any substitute for a well-trained eye. You walk, head down, covering as much area as efficiently as possible, looking for exposed bone. If you're lucky and/or have a good eye, you'll spot something. Now what?

Collecting

Collecting is the arena in paleontology in which finesse meets brute force. Delicacy is required in preparing the fossils for transport; raw power is required for lifting blocks of bone and matrix (the rock which surrounds the bone) – commonly weighing many hundreds of kilograms – out of the ground and into a truck or some other means of transportation. So that it can be moved safely out of the field, the fossil is encased in a rigid jacket, or protective covering. Figure 1.5 shows how this is done.[6]

Transport out of the field can be difficult or not, depending upon the size (and weight) of the jackets. A small jacket (soccer ball size) can be carried out easily enough. Many jackets, however, are considerably larger, and there are times when braces, hoists, winches, cranes, flatbed trucks, front-end loaders, and even helicopters are necessary. A rule of

---

5 Notice that the term dinosaur "dig" is a misnomer. Nobody digs into sediment to find bones; bones are found because they are spotted weathering out of sedimentary rocks.

6 Purchasing toilet paper for this purpose provokes raised eyebrows. Consider this often-repeated scene: a paleontologist and her two- or three-member crew walk into a small town in a remote part of the world. They go to the local market and purchase a couple of cases of toilet paper rolls. "How long do you plan to be in the field?" the bewildered shopkeeper asks. "Oh, maybe a couple of weeks," the paleontologists reply. The shopkeeper is very impressed.

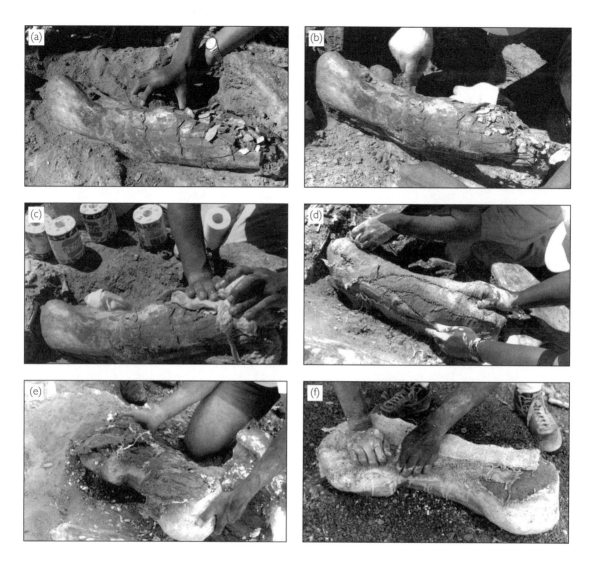

Figure 1.5. Jacketing. (a) A fossil is found sticking out of the ground; now it needs to be cleaned off so that its extent can be assessed. Exposing bone can be done with a variety of tools, from small shovels, to dental picks, to fine brushes. As the bone is exposed, it is "glued"; that is, impregnated with a fluid hardener that soaks into the fossil and then hardens. (b) The pedestal. When the surface of the bone is exposed, the rock around it is then scraped away. For small fossils, this can be quite painless; however, for large fossils, this can mean taking off the face of a small hill. This process continues until the bone (or bones) is sitting on a pedestal, a pillar of matrix underneath the fossil. (c) Toilet paper cushion. Padding is placed around the fossil to cushion it. The most cost-effective cushions are made from wet toilet paper patted onto the fossil. It takes a lot of toilet paper; for example, a 1 m thigh bone (femur) could take upwards of one roll. On the other hand, this is not a step where one should cut corners, because returning from the field with a shattered specimen, or one in which the plaster jacket is stuck firmly to the fossil bone is not so good. (d) Plaster jacket. Jackets are made of strips of burlap cloth soaked in plaster, and then applied to the toilet paper-covered specimen. A bowl of plaster is made up and then precut; rolled strips of burlap are soaked in it and then unrolled onto the specimen and the pedestal. (e) Turning the specimen. After the plaster jacket is hardened, the bottom of the pedestal is undercut, and the specimen is turned; that is, separated at the base of the pedestal from the surrounding rock, and turned over. This is a delicate step in which the quality of the jacket is tested. (f) The top jacket. More plaster and burlap are then applied to the open (former) bottom of the jacket, now its top. At this point, the fossil is fully encased in the plaster-and-burlap jacket and is ready for transport from the field. (Photographs courtesy of D. J. Nichols.)

Figure 1.6. Scenes from a prep lab, in this case, that of the Museum of Northern Arizona in Flagstaff.

(a) Foreground: a large plaster jacket containing the dinosaur *Coelophysis*. Background: sand tables for stabilizing specimen fragments (to be glued), open jackets, storage shelves, and grinding equipment.

(b) The jacket shown in (a). A specimen of the dinosaur *Coelophysis* is visible in the foreground. The arms and hands are to the left; the pelvis, legs, and tail are visible to the right.

(c) A dinosaur skull (*Pentaceratops*) laid out for study. In the background are the large bays in which specimens are stored.

thumb is that the better the fossil, the less conveniently located it is: more than one paleontologist has actually had to cut a road into the middle of nowhere to get a truck to a remote specimen.

Back at the ranch The fossil dinosaur bone is out of the field and back at the museum. The jacket must be cut open, and the fossil prepared, or freed from the matrix. This is done in a laboratory, universally called a prep lab (Figure 1.6). This turns out to be a complex process, for which there is a variety of techniques from the simple to the highly sophisticated. Some fossils require little effort: soft matrix is literally brushed away from the side of the well-preserved bone. In other cases, carbide-tipped needles must be used under a microscope, as the matrix is painstakingly chipped away from the fossil. In still other cases, special air-powered vibrating scribes, called "zip scribes," are used to free the bone. Sometimes an "air dent," a kind of miniature sand-blaster with a very fine spray of baking soda, is used to clear the matrix. A version of this tool is used in some dentists' offices to whiten teeth after cleaning. Finally, in certain circumstances, the matrix can be removed by "acid etching"; that is, dissolution in a bath of weak acid.

It gets tricky when the bone is softer than the matrix, because all those wonderful tools can damage bone more efficiently than they can remove the matrix. A variety of techniques, therefore, is available for hardening the bone. None of these can be too destructive; the preserved fabric and microscopic detail of the bone cannot be destroyed, or the scientific value of the fossil is compromised. Chemically sophisticated low-viscosity hardeners, such as those used in the field, are dripped onto the bone, soak in, and harden. Sometimes the specimen is heated, impregnated with carbowax, cooled, prepared, and then the carbowax removed. There is almost no limit to the ingenuity required for preparation and, in the end, the success of the preparation is measured only by whether the fossil is freed from its matrix undamaged by the process.

Fossils are commonly fragmented and part of the preparation is to put the pieces back together. Sometimes this is rather like a jigsaw puzzle, but sometimes the jacket holds adjacent pieces right next to each other and, by opening the jacket carefully, much of the jigsaw puzzle guesswork is removed. Pieces are stuck together by virtually any glue imaginable (depending upon the requirement): epoxies, superglue, and white, water-based wood glue. Interestingly enough, specimens are not always glued together when broken. This is because broken specimens can permit access to detailed parts of the skeleton – for example, the braincase – that would be very hard to study if the specimen were intact.

We often expect that the preparation process is completed only when a specimen is mounted as a free-standing display in a museum. Although this is the way things were done once, it is no longer the most modern approach to dinosaur fossils. Mounts are admittedly attractive, but using real bone in them presents some serious problems. First, real bone is delicate and needs to be supported. For this reason, a metal frame must be welded to support the bones, a process that is time-consuming, costly, and ultimately destructive to the fossil. In addition, when the specimens are mounted, they cannot be moved around easily and examined for study. Finally, mounted specimens commonly undergo damage over time; slight shifts in the mounts because of the extraordinary weight of the fossil bones, or vibrations in the buildings in which the bones are housed, or museum patrons lifting apparently "insignificant" bits[7] all diminish the quality of mounted specimens.

Most museums, therefore, have begun to cast the bones in fiberglass and other resins and mount the casts. Such mounts are virtually indistinguishable from the originals if done correctly. Casts are lighter and allow for internal frames, and permit spectacular, dynamic, and realistic mounts (Figure 1.7). Finally, mounting casts frees up the actual fossils so that the bone can be optimally protected and studied under the most ideal conditions.

---

7 At the American Museum of Natural History in New York, tail vertebrae of a number of the mounted specimens keep having to be replaced, as some dinosaur enthusiasts have been known to take them home.

Figure 1.7. A spectacular mount of the sauropod *Barosaurus* and the theropod *Allosaurus*. This mount is made of fiberglass and epoxy resin, cast from the bones of the original specimens. A dynamic pose like this would not have been possible using the original fossil bones. (Photograph courtesy of the American Museum of Natural History.)

Some museum curators fear that displaying mounted casts (instead of mounting the fossils themselves) somehow cheats the public of its right to see the originals. A cast, however, is not a poor substitute for the real thing. Leaving the bones disarticulated, properly curated, and available for study maximizes returns on the very substantial investments that are involved in collecting dinosaurs. Paleontology is carried out in large part by public support, and mounted casts give the public the best value for its money.

Important readings

Behrensmeyer, A. K. and Hill, A. P. (eds.) 1980. *Fossils in the Making*. University of Chicago Press, Chicago, IL, 338pp.

Cvancara, A. M. 1990. *Sleuthing Fossils: the Art of Investigating Past Life*. John Wiley and Sons, New York, 203pp.

Gillette, D. D. and Lockley, M. G. (eds.) 1989, *Dinosaur Tracks and Traces*. Cambridge University Press, New York, 454pp.

Kielan-Jaworowska, S. 1969. *Hunting for Dinosaurs*. Maple Press, Pennsylvania, 177pp.

Lessem, D. 1992. *Kings of Creation: How a New Breed of Scientists is Revolutionizing our Understanding of Dinosaurs*. Simon and Schuster, New York, 367pp.

Mitchell, W. J. T. 1998. *The Last Dinosaur Book*. University of Chicago Press, Chicago, IL, 321pp.

Moore, R. C., Lalicker, C. G. and Fischer, A. G. 1952. *Invertebrate Fossils*. McGraw-Hill Book Company, Inc., New York, 766pp.

Parsons. K. M. 2002. *Drawing Out Leviathan: Dinosaurs and the Science Wars*. Indiana University Press, Bloomington, IN, 210pp.

Preston, D. J. 1986. *Dinosaurs in the Attic: an Excursion into the American Museum of Natural History*. St Martin's Press, New York, 244pp.

Sternberg, C. H. 1985. *Hunting Dinosaurs in the Bad Lands of the Red Deer River, Alberta, Canada*. NeWest Press, Edmonton, Alberta 235pp.

Walker, R. G. and James, N. P. (eds.) 1992. *Facies Models: Response to Sea Level Change*. Geological Association of Canada, St Johns, NL, 409pp.

CHAPTER 2

Back to the past: the Mesozoic Era

The Mesozoic Era: it would seem stranger to us than the Land of Oz. The big land animals were dinosaurs, not mammals. You couldn't receive a dozen red roses; roses – or for that matter, any flowers – did not appear until the Mesozoic was more than half-way over. There were no lawns to mow because there was no grass. Until well into the Mesozoic, you couldn't hear birds singing because there were no birds. Indeed, at the beginning of the Mesozoic, the very continents themselves were connected, and they spent much of the ensuing 180 million years getting into more familiar positions.

The Mesozoic would surely seem unfamiliar to a modern visitor, but we can't really understand dinosaurs until we learn something about it. This chapter, therefore, explores a few of the physical qualities of this ancient world. We will answer questions such as "When did the dinosaurs live and how do we know?", "Where were the continents during the time of the dinosaurs?", and "What were climates like during the time of the dinosaurs?" In general, the discussion will be focused on the latter two-thirds of the Mesozoic (the Late Triassic to the end of the Cretaceous Period) because this is when there were dinosaurs. And because dinosaurs roamed continents and not oceans, we will primarily address ourselves to things terrestrial.

When did the dinosaurs live and how do we know?

Geologists dichotomize time and rocks. On the one hand, there is time that passes, regardless of what we do or what happens. On the other hand, there are sedimentary rocks, which constitute the key record on earth of the time that has passed. But sedimentary rocks are imperfect recorders: they preserve only the time during which they were deposited. Without sedimentation, there is virtually no record of time having passed.

Suppose you glance out of your window; you might see lawns, houses, apartments, open fields, something. But almost certainly, the *time* that is going by as you are looking outside your window – indeed, the many years represented by the lawns and houses – is not recorded by the accumulation of sediment in your area. Where, then, is that time – that is, *now* – recorded? Perhaps sedimentary deposition is occurring elsewhere (thereby recording that moment in time there). Perhaps, however, "now" is not represented anywhere on earth by sedimentary deposition. If so, future geologists looking back on this time interval (now) will assume, reasonably enough, that the time must have existed, but that no rocks happened to be preserved to record it. Time passes at a constant rate (because we measure it in consistent units: seconds, minutes, hours, etc.), but the rock record of it is patchy and uneven, recording here an hour and there a week of sedimentation. Nonetheless, geologists assume that time has passed, whether or not there are rocks (or fossils) preserved that represent that time. Geologists are thus affirmatively not existentialists: time is presumed to exist independently, whether or not anything was there to record it.

So the sedimentary record and hence the fossil record (because, of course, fossils are found in rocks) are crude sketches; the amount of time represented by rock deposition is orders of magnitude less than the amount of time that has actually elapsed (Figure 2.1). The challenge is to link together all the sedimentary rock records around the world with the object of developing as complete and precise a record of earth history as possible.

Recall that dinosaur remains are found in layers – or strata – of sedimentary rocks. How old or young are these layers (and the fossils they contain) is the special province of stratigraphy, the practitioners of which are called stratigraphers. Stratigraphy is divided into chronostratigraphy (*chronos* – time) or time stratigraphy, lithostratigraphy (*lithos* – rock) or rock stratigraphy, and biostratigraphy (*bios* – organisms) or stratigraphy as indicated by the presence of fossils.

Chronostratigraphy

Defending religion against the supposed threat of evolution, Matthew Harrison Brady[1] proclaimed, "I am more interested in the Rock of Ages than in the ages of rocks!" Regardless of creed, however, *geologists* are interested in the age of rocks, termed chronostratigraphy. But the ages of rocks are on a scale that is very literally completely out of our experience. Stephen Jay Gould wrote of geological time, "An abstract, intellectual

---

1 The William Jennings Bryan character in the J. Lawrence and R. E. Lee play *Inherit the Wind.*

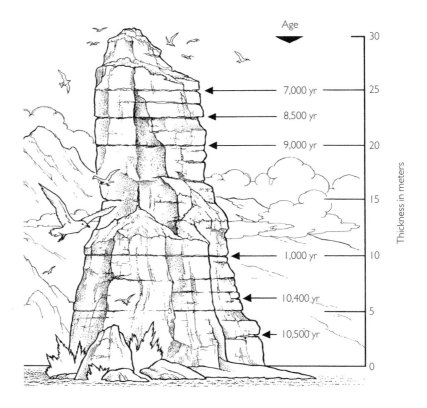

Age

30

7,000 yr ——— 25

8,500 yr

9,000 yr ——— 20

15

Thickness in meters

1,000 yr ——— 10

10,400 yr ——— 5

10,500 yr

0

Figure 2.1. Rocks and time. An outcrop of rock is shown, and the ages of several of the layers are given. Note that the amount of time represented is not equivalent to the thickness of the rock.

understanding of [geological] time comes easily enough – I know how many zeros to place after the 10 when I mean billions. Getting it into the gut is quite another matter."[2] The popular writer, John McPhee, used the expression "deep time," a phrase that is redolent with the antiquity, mystery, richness, and unfathomable extent of earth history.

Geologists generally signify time in two ways: in numbers of years before present, and by reference to blocks of time with special names. For example, we say that earth was formed 4.6 billion *years before present*, meaning that it was formed 4.6 billion *years ago* and is thus 4.6 billion years *old*. Unfortunately, learning the age in years of a particular rock or fossil is not always easy, or even possible. For this reason, time has been broken up into a hierarchy of time intervals, and rocks and fossils can be referred to blocks of time, depending upon how well the age of the rock or fossil can be estimated.

Age of rocks   Geologists are happiest when they can learn the age of a rock or fossil in *years before present*, a determination that is called its absolute age.

2 Gould, S. J. 1987. *Time's Arrow, Time's Cycle*. Harvard University Press, Cambridge, MA. 222pp. (p. 3).

Obtaining an absolute age is not always possible, but it can be accomplished when rocks are found that contain minerals bearing particular radioactive elements.

Absolute age dating

Ages in years before present are reckoned from the decay of unstable isotopes. These spontaneously decay from an energy configuration that is not stable (i.e., that "wants" to change) to one that is more stable (i.e., that will not change, but rather will remain in its present form). The decay of an unstable isotope to a stable one occurs over short or long periods of time, depending upon the particular isotope. The slower the decay process, the longer the amount time that can be deduced from it. The basic decay equation goes as follows:

unstable "parent" isotope $\rightarrow$ stable "daughter" isotope
+ nuclear products
+ heat

Carbon provides a good example. In the unstable isotope of carbon, ^{14}C, a neutron splits into a proton and an electron in the following reaction:

$$^{14}C \rightarrow {}^{14}N + heat$$

Note that the atomic number in the decay reaction changes; it is increased from 6 to 7. Now, with seven protons and seven electrons, the stable daughter is no longer an isotope of carbon but is now nitrogen (see Appendix to this chapter for a quick review of the chemistry underlying these concepts).

The *rate* of the decay reaction is the key to obtaining an absolute age. If we know (1) the original amount of parent isotope at the moment that the rock was formed or the animal died (before becoming a fossil), and we know (2) how much of the parent isotope is left and (3) the rate of the decay of that isotope, we can estimate the amount of time that has elapsed. For example, suppose we know that 100% of an unstable isotope was present when a rock was new, but now only 50% remains. If we know the rate at which the element decays, we can estimate the amount of time that has elapsed since the rock was formed; that is, the age of the rock. This kind of relationship is shown in Figure 2.2. Because radioactive decay is the basis of the absolute age determination, unstable isotopic age estimations are sometimes called radiometric dating methods.

Since the rate of decay is constant (in any given stable isotope), it is convenient to summarize that rate by a single number. That number is called the half-life, which is the amount of time that it takes for 50% of a quantity of an unstable isotope to decay (leaving half as much parent as was present originally). The half-life, then, is an indicator of decay rate, and provides guidance about which isotope is appropriate for which amount of time.

Which unstable isotope is chosen for dating depends upon the probable age of the object in question. To date human remains, not more than several thousand years old, the rubidium/strontium isotopic system

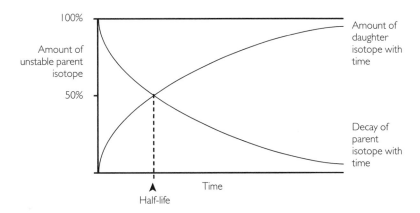

Figure 2.2. As isotopic decay curve. Knowing the amount of unstable isotope that was originally present, as well as the amount of unstable isotope now present and the rate of decay of the unstable isotope, it is possible to determine the age of a rock with that isotope in it. The diagram shows the half-life; that is, the amount of time that it takes for 50% of an unstable isotope to decay.

(^{87}Rb/^{87}Sr), with its half-life of 48.8 billion years, would hardly be the ideal isotopic system. This would be a bit like timing a 100 m dash with a sundial. Likewise, dating dinosaur bones (ages that would be in the hundreds of millions of years) using ^{14}C, with a half-life of 5,730 years, would be like giving your own age in milliseconds.

Radiogenic isotopes are powerful tools but they can't be used directly on dinosaur bone. There has to be a source of radioactive isotopes, which occur commonly in igneous rocks, such as those blasted from a volcano. Then the age of formation of the igneous rock can be obtained. Is it any wonder that artists always seem to depict dinosaurs with active volcanoes in the background? Still, no dinosaur lived *within* an active volcano, and so the challenge is to correlate in time the dinosaur bone with the volcanic event that produced the datable igneous rock. This – the relationships of one body of rock to another – is the province of lithostratigraphy.

Lithostratigraphy

Sedimentation and sedimentary rocks

Sediments – sand, silt, mud, dust, and other less-familiar materials – are deposited in time and in space (geographically). Deposition occurs in strata that can be broad and sheet like or narrow or ribbon shaped, and wedged at the edges, where each layer becomes progressively thinner until it is said to "pinch out." These shapes occur on a scales of meters to hundreds of kilometers, and are the direct result of sedimentation such as flowing water, wind, or explosion from a volcano, to name a few more or less common processes. Virtually every geographical location we can think of – a river, a desert, a lake, an estuary, a mountain, the bottom of the ocean, the pampas – has sedimentary processes peculiar to it that will produce distinctive sediments and, with time and burial, distinctive sedimentary rocks.

Relative age dating

The seventeenth century Danish naturalist Nicholas Steno[3] was the first to recognize that:

1   in any vertical, stacked sequence of sedimentary rocks, the oldest rocks are found at the bottom and successively younger-aged rocks are found above, with the youngest occurring at the top (a observation now termed the "law" of superposition) and

2   all sedimentary rock sequences were originally horizontal (although subsequent geological events may have disrupted their original orientation in space).

That younger sediments are deposited *upon* older sediments seems clearly self-evident, and yet this obvious conclusion is the fundamental basis of all correlations of sedimentary strata in time. If a stratum lies above another (and the rocks have not been subsequently deformed by various geological processes), it is younger than the one below it (Figure 2.3). Ascertaining the relative ages of the two strata is termed relative dating: the type of dating that, while not providing age in years before present, provides the age of one stratum relative to another stratum. In historical terms, we might use the presence of record albums and a record player in a picture to infer that the picture was taken before there were compact discs (CDs) and CD players. Here, then, is part of the solution to dating dinosaur bone. Suppose that a stratum containing a dinosaur bone is

3   An interesting biography of Steno is Cutler, A. 2003. *The Seashell on the Mountaintop: A Story of Science, Sainthood and the Humble Genius who Discovered a New History of the Earth.* Dutton, New York, 228pp.

Figure 2.3. Superposition of strata in Petrified Forest National Park, Arizona, USA. Thick stacks of red mudstones were deposited by rivers 213 Ma to produce the succession of layers visible in this photograph.

sandwiched between two layers of volcanic ash. Ideally, an absolute age date could be obtained from each of the ash layers. We would know that the bone was younger than the lower layer, but older than the upper layer. Depending upon how much time separates the two layers, the bone between them can be dated with greater or lesser accuracy.

Datable ash layers thus truly fall out of the sky, but rarely because stratigraphers need them. More commonly, rocks for which absolute ages can be obtained are widely separated, not only in time, but also geographically. The challenge then is to correlate the strata of known absolute ages with those of unknown age. But how can one tell that two geographically separated deposits were deposited at the same time if absolute ages are unknown? In this, fortunately, stratigraphers are aided by one last, extremely important, tool: biostratigraphy.

Biostratigraphy   Biostratigraphy is a method of relative dating that utilizes the presence of fossil organisms. It is based upon the idea that a particular time interval can be characterized by specific organisms, because different creatures lived at different times. For example, if one knows that dinosaurs lived from 228 to 65 Ma,[4] then any rock containing a

---

4 We use the expression Ma, from the latin *mille annos*, to mean million of years. Thus 65 Ma is 65 million years ago.

dinosaur fragment must fall within that age range. The question is, how precise a date can it really give?

In fact, biostratigraphic correlation – the linking of geographically separated rocks based upon the fossils they contain – can be very precise. Although, like superposition, biostratigraphy cannot provide ages in years before present, the fact that many species of organisms have existed on earth for 1–2 million-year intervals enables them to be used as powerful dating tools. For example, *Tyrannosaurus rex* lived for only about 2 million years, from 67 to 65 Ma. Thus, if we found a *Tyrannosaurus* fossil (a good find, indeed), we would know that, no matter where that dinosaur was found, it would be correlative with *T. rex*-bearing sediments in North America that have been well dated at 67–65 Ma.

Eras and Periods and Epochs, Oh My!

The oldest method of dating sediments is biostratigraphy. Leonardo da Vinci observed marine shells far inland where the ocean clearly was not; he correctly deduced that where he stood had once been covered by an ocean waters. By the early 1800s, French anatomist Georges Cuvier, studying strata around Paris, noted that higher strata had a greater proportion of fossil shells with living counterparts than did lower strata. The increasingly modern aspect of the fauna is due to the fact that the highest rocks are closer in time to the present, and thus the faunas that they contain are the most like those of today.

Within a generation of Cuvier, a remarkable revolution had been wrought in geological thinking. The Phanerozoic (*phaneros* – light, meaning visible; *zoo* – life) time interval, representing that interval of earth history during which there have been organisms with skeletons or hard shells present, was established. Using biostratigraphy as a time indicator, a variety of rock outcrops in northwestern Europe was designated as type sections, or original locations, where a particular interval of time is represented. The names of the largest of the blocks of time within the Phanerozoic, the Eras, came from a description of the life contained within each. These Eras are, from oldest to youngest, the Paleozoic (*paleo* – ancient), the Mesozoic (*meso* – middle), and the Cenozoic (*cenos* – new). Within each are smaller subdivisions (still consisting of 10s of millions of years each) called Periods, and within these, in turn, are yet smaller subdivisions of time called Epochs (consisting of several millions of years each). Figure 2.4 shows the currently

Figure 2.4. The Mesozoic part of the geologic time scale. All ages (except that of the age of the earth, which is given in billions of years) given in millions of years before present. The Mesozoic constitutes only a rather tiny fraction of the expanse of earth time. If you compacted earth time into a single year, from January 1 (the formation of the earth) to December 31 (the past 100,000 years of which, by this way of measuring earth history, would occur in less than a day), then dinosaurs were on earth from about December 11 to December 25. (Dates from Gradstein, F. M., Agterberg, F. P., Ogg, J. C., Hardenbol, J., Van Veen, P., Thierry, J., and Huang, Z., 1995. A Triassic, Jurassic, and Cretaceous time scale. In Berggren, W. A., Kent, D. V., Aubry, M.-P. and Hardenbol, J. (eds.), *Geochronology, Time Scales, and Global Stratigraphic Correlation*. SEPM Special Publication no. 54, pp. 95–126.)

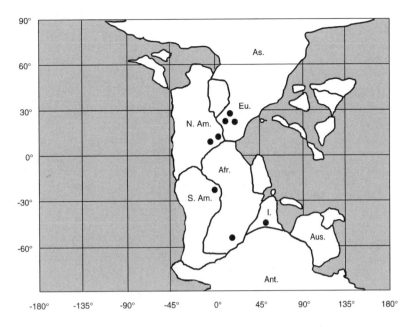

Figure 2.5. The positions of the present-day continents during the Late Triassic. Earth was dominated by the unified landmass Pangaea. Dots indicate locations of major fossil finds. Afr., Africa; Ant., Antartica; As., Asia; Aus., Australia; Eu., Europe; N. Am., North America; I., India; S. Am., South America. (Figures 2.5–2.10: reconstruction by Paleogeographic Information System, M. I. Ross and C. R. Scotese.)

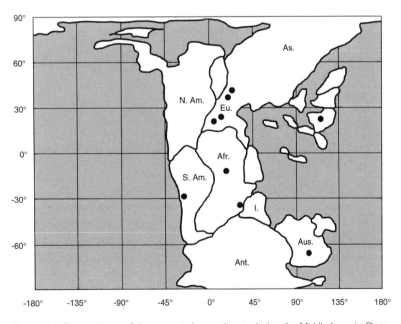

Figure 2.6. The positions of the present-day continents during the Middle Jurassic. Dots indicate locations of major fossil finds. For abbreviations, see caption to Figure 2.5.

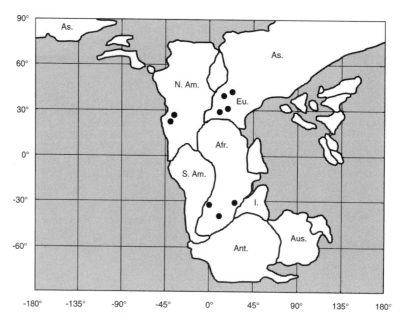

Figure 2.7. The positions of the present-day continents during the Early Jurassic. The dismemberment of Pangaea began, probably, as a rift between the northern continental mass, and the southern supercontinent. Dots indicate locations of major fossil finds. For abbreviations, see caption to Figure 2.5.

Plates, oceans, and seas during the time of dinosaurs

The initial rifting (break-up) of Pangaea took place in the Early Jurassic (Figure 2.6). The effect has been likened to an "unzipping" of the great supercontinent, from south to north. Sediments in the eastern seaboard and Gulf Coast regions of North America and Venezuela, as well as regions in West Africa, record the opening and widening of a seaway.

Also at this time, some of the earliest epicontinental (or "epeiric") seas of the Mesozoic Era first made their appearances. Epicontinental seas are shallow marine waters that cover parts of continents. In the past, epicontinental seas have been considerably more widespread than they are today, because in the past eustatic (or global) sea levels have been higher than they are now.

Eustatic sea level is controlled by many factors, but two of the most common are ice at earth's poles and tectonism. Obviously, if there is a great deal of polar ice (or glaciation, such as took place during the past million years or so), much seawater may be bound up in ice, lowering sea levels worldwide. Likewise, during tectonically active intervals, mid-oceanic spreading centers are topographically elevated, decreasing the volume of the ocean basins and thus displacing ocean water up onto the continents. During the Middle Jurassic, fluctuating epicontinental seas covered large parts of what is now western North America, eastern Greenland, eastern Africa, and Europe, where there developed a complex system of islands and seaways (Figures 2.7 and 2.8).

Although the Middle Jurassic was an important interval of time in dinosaur evolution, terrestrial sediments that record it are rare. Why this is the

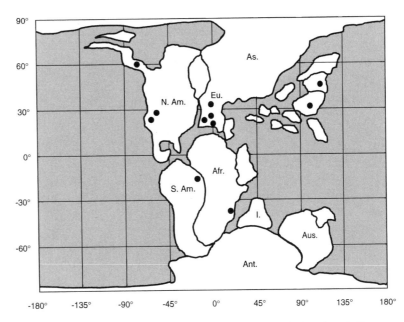

Figure 2.8. The positions of the present-day continents during the Late Jurassic. Dots indicate locations of major fossil finds. For abbreviations, see caption to Figure 2.5.

case is uncertain, but probably it is just preservational "luck of the draw."

In the Late Jurassic (Figure 2.8) and Early Cretaceous (Figure 2.9), continental separation was well underway. A broad seaway, the Tethyan Seaway (after the Greek goddess Tethys, Goddess of the Sea), ran between two supercontinents, one in the north known as "Laurasia" and one in the south called "Gondwana". Eustatic sea level was relatively low, and hence epicontinental seas were not very predominant. Yet, the then-western (present-day northern) coast of Australia seems to have been bathed in a broad, shallow, epeiric sea, as were, periodically, western North America and parts of Europe and Asia.

The mid-Cretaceous was in tectonic terms a swinging time of active mountain-building, sea-floor spreading, high eustatic sea levels, and broad epeiric seas (Figure 2.9). The Tethyan Seaway remained a dominant geographical feature, as rifting between the Europe and North America was initiated from the south to the north. The effects of active tectonism were even more marked in the southern continents. Here, a stable continental merger dating back to the Early Paleozoic Era – the supercontinent of Gondwana – finally underwent rifting involving two of its largest constituents – Africa and South America – as well as two smaller constituents, India and Madagascar. India spent the next 50 million years motoring at breakneck speed (for a continent) across what became the Indian Ocean, to smash head-on into southern Asia and produce the Himalayan Mountains. But this is getting ahead of our story. While India and Madagascar were in the first bloom of unconfinement, Australia and Antarctica remained firmly united (a continental marriage that would not end until 50 Ma), and a land connection remained, as it almost does today, between South America and Antarctica.

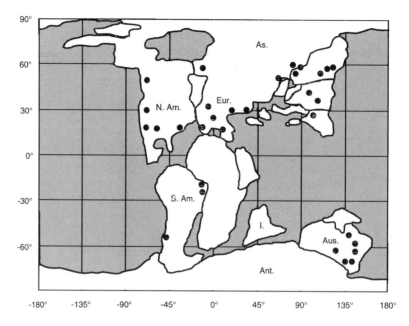

Figure 2.9. The positions of the present day continents during the Early Cretaceous. Note the large number of seas covering continents where today we find dry land. Dots indicate locations of major fossil finds. For abbreviations, see caption to Figure 2.5.

Mid-Cretaceous Europe was an archipelago (group of islands), dissected by small epeiric seaways. Extensive seas developed over both northern Africa and western South America. Finally, a series of North American Western Interior seas bisected the North American craton, deriving their waters from the Arctic Ocean to the north and from the newly formed Gulf of Mexico to the south. These seas came and went, sometimes completely dividing the craton in half, and at other times only partially dividing it.

The global positions of continents during the Late Cretaceous are almost familiar to us (Figure 2.10). North America became nearly isolated, connected only by a newly emergent land bridge across the modern Bering Straits to the eastern Asiatic continent. This land bridge has come and gone several times since the Cretaceous, but it was clearly a significant feature at the time. Africa and South America were fully separated, the former retaining its satellite, Madagascar, and the latter retaining a land bridge to the Antarctica/Australia continent. India was by now well along its way toward southern Asia.

But, if Late Cretaceous global continental positions are somewhat familiar to modern humans, the geography of the Late Cretaceous would still seem strange. The Western Interior Seas that bisected North America remained important geographical features up until the very latest Cretaceous, when they finally receded (only to nostalgically re-emerge briefly, one last time in the early Cenozoic). Europe remained a complex of islands and seaways. Africa is believed to have had extensive internal seaways, as are South America and Asia.

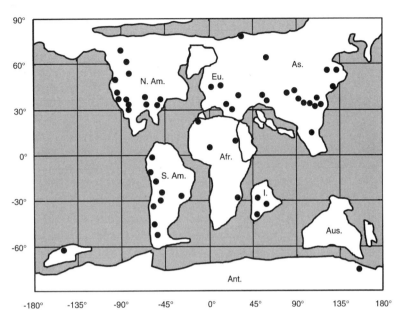

Figure 2.10. The positions of the present day continents during the Late Cretaceous. The positions of the continents did not differ significantly from their present-day distribution. Note the land bridge between Asia and North America, as well as the European archipelago. Dots indicate locations of major fossil finds. For abbreviations, see caption to Figure 2.5.

Just before the end of the Cretaceous, a drop in eustatic sea level caused exposure on all continents.

What were climates like during the time of the dinosaurs?

The flavor of the Mesozoic would be lost without some general sense of Mesozoic paleoclimates or ancient climates. How can we know about past climates? It might seem as if climate is a bit like performance art: if you are not there to experience it, it's gone. Deposition, after all, leaves physical evidence. Climate, on the other hand, might seem to be a more ephemeral quantity. It turns out, though, that just as one can record the sound of a muscial performance, so earth has recorded traces that allow us to infer at least aspects of past climates. The record suggests that, not surprisingly, climates have not remained constant throughout earth history: they, like everything else, have changed through time.

Potential effects of plate motions on climate

Why even suspect that past climates differed from those of today? Distributions of the land masses as well as the amount and distribution of the oceans on the globe drastically modify temperatures, humidity, and precipitation patterns. In the following, we explore this, comparing the extreme case of the continents coalesced into a single landmass (see Figure 2.5) with the equally extreme (but more familiar) case of the continents widely distributed around the globe (Figure 2.10).

Heat retention in continents and oceans

Continents (land) and oceans (bodies of water) respond very differently to heat from the sun. In the height of summer, we cannot walk barefoot on cement, and yet we can cool off in a pool. At night, the cement cools off rapidly; then, the pool seems warm. Because fluids are mobile, heat can be more easily distributed through a larger area in fluids than in solids. While solids quickly become hot to the touch, just a short distance below the surface, temperatures remain cool (a property that keeps cellars cool in the summer). Heat is distributed more evenly through a liquid than through a solid; thus the *entire* liquid needs to be cooled in order for the liquid to feel cool to the touch. In the case of a solid, the exterior can be cool while the interior remains hot; we need cool only the exterior, not the entire solid. Ultimately, this means that oceans are slower to warm and slower to cool than continents.

Consider how these properties of continents and oceans might modify climates. At the dawn of the Mesozoic, the continents were united into the single land mass, Pangaea (see Figure 2.5). Here, continental effects – more rapid warming and cooling of continents than oceans – would have been more intense than today. Pangaea must have experienced wide temperature extremes. It would have heated up quickly and become hotter, and then cooled off rapidly, and become colder, than do modern continents, whose continental effects are mitigated by the broad, temperature-stabilizing expanses of oceans between them.

The post-Late Triassic break-up of Pangaea mitigated the strong continental effects (see Figures 2.6 and 2.7). With the rise in eustatic sea level and supercontinental dismemberment, the effects of the large epeiric seas were superimposed upon the diminishing continental effects. For example, the strong continentality (hot, dry summers; cold winters) observable even today in the interiors of continents would have been decreased by the presence in the Jurassic and Cretaceous of the large continental seaways. These large bodies of water would have stabilized global temperatures, decreasing the magnitude and rapidity of the temperature fluctuations experienced on the continents during times of lower sea level.

Climates through the Mesozoic

Late Triassic–Early Jurassic

The Late Triassic and Early Jurassic were times of heat and aridity. They also were times of marked seasonality; that is, well-defined seasons. Paleontologists studying Late Triassic paleoclimates have recognized three broad climate regimes: an equatorial, year-round dry belt; narrow belts north and south of the dry belt of strongly seasonal rainfall; and middle and upper latitude humid belts. These are based upon the distribution of Upper Triassic deposits of sedimentary rocks indicative of aridity. Such rocks include sand dune deposits (the remains of ancient deserts) and evaporites, rocks that are made up of minerals formed during desiccation. Triassic evaporites and sand dune deposits do not today form continuous belts around the globe;

the continents had to be rotated back into the positions they had during the Late Paleozoic and Early Mesozoic in order to obtain the banding pattern.

Other climatic indicators that are keys to Late Triassic–Early Jurassic paleoclimates have been found. First, continental rocks from this time interval are very commonly red beds, rocks of an orange/red color due to an abundance of iron oxide. Such rocks today generally form in climates that are relatively warm. Secondly, a variety of fossilized soils have been found with caliche (calcium carbonate nodules) in them. Such nodules commonly form today in soils that are located in arid climates. Finally, evidence of warm, dry climatic conditions has been obtained from stable isotopes, isotopes that do not spontaneously decay. It has been shown that the amount of a stable isotope of oxygen, ^{18}O, varies with temperature and salinity. Therefore, by measuring how much ^{18}O is present, one can obtain a direct measure of ancient temperatures and salinities. Obviously, isotopes are an extremely important tool for learning about ancient climates as well as a variety of other subjects, including (as we shall see) warm-bloodedness in dinosaurs (see Box 2.1 and Chapter 15). The Late Triassic–Early Jurassic interval, then, seems to have been a time that was generally warmer and perhaps drier than the present, with strong seasonality.

Middle and Late Jurassic

Although not many sedimentary rocks are preserved from the terrestrial Middle Jurassic, we can still obtain a relatively good idea about Middle and Late Jurassic climates because of a wealth of data available from oceanic sediments. Most importantly, the latter two-thirds of the Jurassic, as well as the entire Cretaceous, are thought to have been without polar ice or glaciers on the northern parts of the the continents. This is quite beyond our own experience; now, glaciers occur at high latitudes at both poles, and the poles themselves are covered in ice. The conclusion that there was no polar ice or glaciers in the Middle and Late Jurassic is based largely upon the presence of warm climate indicators at high latitudes, and upon the absence of any evidence of continental glaciation from that time. Biologically, warm climate indicators include plants and certain fish, whose distributions are thought to have been as high as 75° N and 63° S. This would put them beyond the polar fronts, which in turn suggests that the poles must not have been as cold then as they are today.

The absence of polar ice had an important consequence for climates: water that would have been bound up in ice and glaciers was located in ocean basins. This in turn means higher eustatic sea levels than now, which led to extensive epeiric seas. The increased abundance of water on the continents as well as in the ocean basins had a stabilizing effect on temperatures (because it decreased continental effects), and decreased the amount of seasonality experienced on the continents.

Continental climates are enormously variable and, as we have seen, short distances can encompass huge climatic differences. Indeed, there

BOX 2.1

Stable isotopes, ancient temperatures, and dead oceans

In 1947, Harold C. Urey, a geochemist at the University of Chicago, made an astounding and far-reaching prediction: that stable isotopes could be used to determine ancient temperatures. As it turns out, he had come upon one of the most important tools available to earth historians. Stable isotopes have proven invaluable in reconstructions of ancient oceans, climates, ecosystems, physiologies, and a host of other subjects. Like other scientific techniques, the principle is relatively simple; however, in practice, the methods are complex and require sophisticated techniques and equipment.

Like unstable isotopes, stable isotopes occur naturally. In the case of oxygen, while 99.763% of all oxygen is ^{16}O, 0.191% of all oxygen is the stable isotope ^{18}O. The rest is ^{17}O, another rare stable isotope of oxygen. In the case of carbon, 98.89% of all carbon is the stable isotope ^{12}C; 1.11% of all carbon are the other isotopes of carbon: ^{13}C and ^{14}C.

The critical point is that the minor weight differences in the isotopes cause slight differences in their behavior during chemical reactions, or during natural physical processes such as evaporation, precipitation, or dissolution (the process of being dissolved in a solution). Such differences in behavior are called fractionation, and control how the isotopes separate from each other as a reaction or physical process takes place.

Urey was aware of the fractionation of stable isotopes and observed that, in the case of oxygen, fractionation varied with temperature. For example, in a situation in which calcium carbonate ($CaCO_3$) was precipitated from seawater, if all other variables (e.g., concentration of salts in the solution) were held constant and only the temperature at which precipitation took place was varied, the ratio of ^{18}O:^{16}O would vary predictably. Thus, by studying that ratio, an estimate of the temperature of precipitation could be obtained. The miniscule amounts of stable isotope could be weighed on an instrument called a mass spectrometer, and the ancient temperature could be calculated.

In the past 30 years, a great deal of experimental research with stable isotopes and mass spectrometers has enabled scientists to predict temperatures of precipitation from ^{18}O:^{16}O ratios. The technique is of extreme interest because, for example, stable oxygen isotopes in a $CaCO_3$ shell secreted by a clam that lived on the sea-floor millions of years ago can theoretically be used to provide a clue to temperatures of the water in which the clam was living. Indeed, because clams grow shells throughout the warm seasons, it is possible in certain instances to deduce seasonal temperature fluctuations that occurred millions of years ago: the fossilized shell of the clam records through its stable isotope composition the ancient seasonal temperature fluctuations that it experienced during its life. The method is potentially applicable to any organism that secretes a $CaCO_3$ shell, as well as to vertebrates, which have stable isotopic oxygen incorporated into their bones in the form of phosphate. For example, if indeed dinosaurs were "warm-blooded," their stable isotopic oxygen ratios should show this. Not surprisingly, this subject is revisited in greater detail in Chapter 15, in which dinosaur "warm-bloodedness" is discussed.

In the intervening years since the original stable oxygen isotope fractionation–temperature relationship was uncovered, stable isotopes have been put to a variety of uses. Fluctuations in ^{13}C:^{12}C have been used to record intervals of increased atmospheric CO_2. Also, they have been used to record productivity – the amount of biological activity in an ecosystem – by serving as an indicator of the amount of organic carbon moving through an ecosystem. The flux – or cycling – of organic carbon through an ecosystem is a measure of its activity. Thus it was by studying the ^{13}C:^{12}C ratios from ocean sediments deposited at the very end of the Cretaceous that oceanographers discovered that the oceans went virtually dead at that time: isotopic carbon recorded an astounding plunge in the flux of organic carbon, which was interpreted as a severe drop in the total productivity of the world's oceans. This apocalyptic event, the infamous "Strangelove Ocean," is covered in greater detail in Chapter 18, when the extinction of the dinosaurs is examined.

is some evidence for Late Jurassic aridity in the form of various evaporites deposits. Likewise, Upper Jurassic terrestrial oxidized sediments and caliche deposits in North America suggest that there, to be sure, the Late Jurassic was was marked by at least seasonally arid conditions.

All indications, however, are that the Jurassic was a time of warm equable climates, with higher average global temperatures and less seasonality than we currently experience. It appears that this was in large part a function of high eustatic sea levels and vast flooded areas on the continents.

Cretaceous

Paleoclimates in the Cretaceous are somewhat better understood than those of the preceding periods. During the first half of the Cretaceous at least, global temperatures remained warm and equable. The poles continued to be free from ice, and the first half of the Cretaceous saw far less seasonality than we see today. This means that, although equatorial temperatures were approximately equivalent to modern ones, the temperatures at the poles were somewhat warmer. Temperatures at the Cretaceous poles have been estimated at 0–15 °C, which means that the temperature differerence between the poles and the equator was only between 17 and 26 °C, considerably less than the approximately 41 °C of the modern earth.

More than one culprit bears the responsibility for this climate. The Cretaceous was a time of increased global tectonic activity and associated high volcanic activity. An increase in tectonic activity is associated with increased rates of oceanic spreading, which in turn involves elevated spreading ridges. Raised spreading ridges would have displaced more oceanic water onto the continents and, indeed, there is good evidence for extensive epeiric seas during the Cretaceous. That there was an increase in the atmosphere of carbon dioxide (CO_2) gas during Cretaceous times has been established using ^{13}C. This has been attributed an increase in volcanism related to increased tectonic activity. It turns out that the amount of CO_2 in the atmosphere can be correlated with the amount of ^{13}C isotope present in organic material preserved from the Cretaceous. Increased amounts of CO_2 in the Cretaceous atmosphere meant that the Cretaceous atmosphere tended to absorb more heat (long-wavelength radiation from earth), warming climates globally. These of course are similar to the now-notorious "greenhouse" conditions[5] with which the modern earth is threatened.

So the first half of the Cretaceous was synergistic: tectonism caused increased atmospheric CO_2 and decreased the volume of the ocean basins, which in turn increased the area of epeiric seas. The seas thus stabilized climates already warmed by enhanced absorption of heat in the

---

5 The increase in CO_2 in the modern atmosphere (and consequent global warming) is attributable to anthropogenic (human-originated) combustion of all types, especially automobile exhausts, and not volcanism. Since Earth has *already* undergone an experimental flirtation with greenhouse conditions, the Cretaceous provides insights into how our modern world will respond to such conditions.

atmosphere. A 2.3% increase over today's level of mean global absorbed radiation has been hypothesized. This means that the Cretaceous earth, because of its "greenhouse" atmosphere and abundance of water, retained 2.3% more heat than does the modern earth. And, because of its extensive water masses, heat was not so quickly released during cold seasons; indeed, the seasons themselves were modified. Tropical and subtropical climates have been reconstructed for latitudes as high as 70° S and 45° N.

The last 30 million years of the Cretaceous produced a mild deterioration of these equable conditions of the mid-Cretaceous. A pronounced withdrawal of the seas took place, and evidence exists of more pronounced seasonality. Stable isotopes again provide important evidence of greater fluctuations in temperatures; however, this time they are

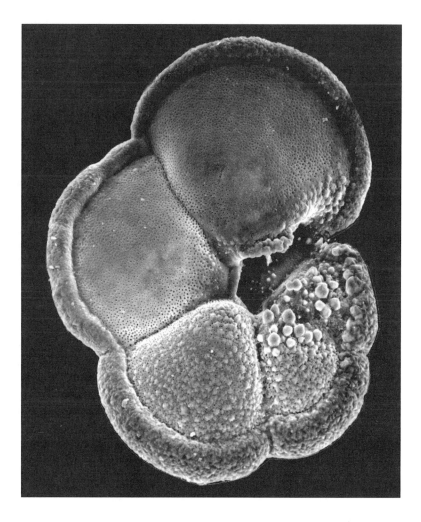

Figure 2.11. The carbonate shell of a modern planktonic (free-swimming) foraminifer, *Globorotalia menardii*. The long dimension is 0.750 mm. (Photograph courtesy of S. L. D'Hondt.)

aided by information from an unexpected source: leaf margin and vein patterns. In the modern world, such patterns can be closely correlated with temperature and moisture. Once this indicator was "calibrated" in the present – that is, once the patterns are correlated with modern temperature and moisture levels – leaf margin and venation patterns could be used to infer previous temperatures and amounts of moisture.

Another important indicator of temperature are single-celled, shell-bearing organisms called foraminifera that live in the oceans (Figure 2.11). The shapes of the shells of foraminifera can be correlated with a relatively narrow range of temperatures, and thus ancient representatives of the group can provide an indication of water temperatures in the past. However foraminifera serve double duty; because their shells and made of calcium carbonate (which contains both carbon and oxygen), the shells are suitable for stable carbon and stable oxygen isotopic analyses.[6]

The Cretaceous was surely a world much different from our own. In its first half, warm, equable climates dominated the period. The second half, however, was marked by well-documented climatic changes, in which seasonality was increased and the equator-to-pole temperature gradient became more like that which we presently experience.

Important readings

Arthur, M. A., Anderson, T. F., Kaplan, I. R., Veizer, J. and Land, L. S. 1983. *Stable Isotopes in Sedimentary Geology*. SEPM Short Course no. 10, 432pp.

Barron, E. J. 1983. A warm, equable Cretaceous: the nature of the problem. *Earth Science Reviews*, **19**, 305–338.

Barron, E. J. 1987. Cretaceous plate tectonic reconstructions. *Palaeogeography, Palaeoclimatology, Palaeoecology*, **59**, 3–29.

Berry, W. B. N. 1987. *Growth of a Prehistoric Time Scale Based on Organic Evolution*. Blackwell Scientific Publications, Boston, MA, 202pp.

Crowley, T. J. and North, G. R. 1992. *Paleoclimatology*. Oxford Monographs in Geology and Geophysics no. 18. Oxford University Press, New York, 339pp.

Dott, R. H., Jr and Batten, R. L. 1988. *Evolution of the Earth*. McGraw-Hill Book Company, New York, 120pp.

Faure, G. 1991. *Principles and Applications of Inorganic Geochemistry*. Macmillan Publishing Company, New York, 626pp.

Frakes, L. A. 1979. *Climates Through Geologic Time*. Elsevier Scientific Publishing Company, New York, 310pp.

Frazier, W. J. and Schwimmer, D. R. 1987. *Regional Stratigraphy of North America*. Plenum Press, New York, 719pp.

Lillegraven, J. A., Kraus, M. J. and Bown, T. M. 1979. Paleogeography of the world of the Mesozoic. In Lillegraven, J. A., Kielan-Jaworoska, Z. and Clemens, W. A., Jr (eds.), *Mesozoic Mammals, the First Two-Thirds of Mammalian History*. University of California Press, Berkeley, CA, pp. 277–308.

---

6 Free-swimming (planktonic) foraminifera first make an appearance in Cretaceous oceans. Because they were (and are) geographically widespread but have relatively short species' durations, they are also superb biostratigraphic indicators for late Mesozoic and Cenozoic marine sediments.

Lutgens, F. K. and Tarbuck, E. J. 1989. *The Atmosphere: an Introduction to Meteorology*. Prentice Hall, Englewood Cliffs, NJ, 491pp.

Robinson, P. L. 1973. Palaeoclimatology and continental drift. In Tarling, D. H. and Runcorn, S. K. (eds.), *Implications of Continental Drift to the Earth Sciences*, vol. I. NATO Advanced Study Institute, Academic Press, New York, pp. 449–474.

Ross, M. I. 1992. *Paleogeographic Information System/Mac Version 1.3*. Paleomap Project Progress Report no. 9. University of Texas at Arlington, 32pp.

Walker, R. G. and James, N. P. 1992. *Facies Models, Response to Sea Level Change*. Geological Association of Canada, St Johns, NL, 409pp.

Wilson, J. T. (ed.) 1970. *Continents Adrift: Readings from Scientific American*. W. H. Freeman Company, San Francisco, 172pp.

APPENDIX

Chemistry quick 'n dirty

Earth is made up of elements. Many of these, such as hydrogen, oxygen, nitrogen, carbon, and iron, are familiar, while others, such as berkelium, iridium, and thorium, are probably not. All elements are made up of atoms, which can be considered to be the smallest particle of any element that still retains the properties of that element. Atoms, in turn, are made up of protons, neutrons, and yet smaller electrons, which are collectively termed subatomic ("smaller-than-atomic") particles. Protons and neutrons reside in the central core, or nucleus of the atom. The electrons are located in a cloud surrounding the nucleus. The electrons are bound within the cloud in a series of energy levels; that is, some electrons are more tightly bound around the nucleus and others are less tightly bound. Those that are less tightly bound are, as one might expect, more easily removed than those that are more tightly bound (Figure A2.1).

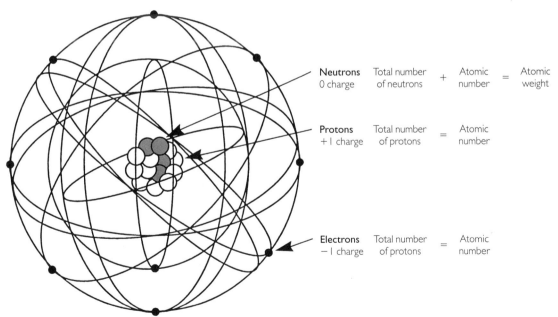

| **Neutrons** 0 charge | Total number of neutrons | + | Atomic number | = | Atomic weight |

| **Protons** +1 charge | Total number of protons | | = | | Atomic number |

| **Electrons** −1 charge | Total number of protons | | = | | Atomic number |

Figure A2.1. Diagram of a carbon atom. In the nucleus are the protons and neutrons. In a cloud around the nucleus are the electrons, whose position relative to the nucleus is governed by their energy state.

Keeping that in mind, let us further consider the subatomic particles. Protons and electrons are electrically charged; electrons have a charge of -1 and protons have a charge of $+1$. Neutrons, as their name implies, are electrically neutral and have no charge. To keep a charge balance in the atom, the number of protons (positively charged) must equal the number of electrons (negatively charged). This number – which is the same for protons and electrons – is called the atomic number of the element, and is conventionally displayed to the lower left of the elemental symbol. For example, the element carbon is identified by the letter C, and it has 6 protons and 6 electrons. Its atomic number is thus 6, and it is written $_6C$.

Along with having an electrical charge, some subatomic particles also have mass. Rather than work with the extremely small mass of a proton (one of them weighs about 6.02×10^{-23} grams!), it is assigned a mass of 1. Neutrons have a mass of 1 as well. Because relative to protons and neutrons, the masses of electrons are negligible, the mass number of an element is composed of the total number of neutrons *plus* the total number of protons. In the case of the element carbon, for example, the mass number equals the total number of neutrons (6) plus the total number of protons (6); that is, 12. This is usually written ^{12}C and is called carbon-12. Note that ^{12}C has 6 protons and therefore must also have 6 neutrons, so its atomic number remains 6, and is written $^{12}_6C$. Because the atomic number is always the same for a particular element, it is commonly not included when the element is discussed. Thus $^{12}_6C$ is usually abbreviated ^{12}C.

Variations in elements exist in nature and those variations that have the same atomic number but different mass numbers are called isotopes. For example, a well-known isotope of carbon-12 (^{12}C) is carbon-14 (^{14}C). Since ^{14}C is an isotope of carbon, it has the same atomic number as ^{12}C (based upon 6 electrons and 6 protons). The change in *mass* number results from additional neutrons. Carbon-14 has 8 neutrons, which, with the 6 protons, increase its atomic mass to 14. Because it is carbon, of course, its atomic number remains 6.

CHAPTER 3

Discovering order in the natural world

Around us there are consistent patterns that obviously constitute order in nature. To cite two simple examples, all plants with flowers have leaves, and all birds have feathers. Indeed, the correlation between birds and feathers is so consistent in our modern world that we might go so far as to identify a bird as such *because* it has feathers. Going further: can we use features such as leaves and feathers to discover underlying patterns of organization among all organisms? In other words, is there some kind of organization that pertains to the reputedly infinite diversity of life?

Hierarchy

Perhaps the most significant pattern applicable to all living organisms is the fact that their attributes – that is, all their features, from eyes, to hair, to chromosomes, to bones – can be organized into a hierarchy. Hierarchy refers to the rank or order of features. Indeed, the hierarchical distribution of features is the most fundamental property of the biota. For this reason, a great deal of attention will be devoted here to the business of hierarchies and to their implications.

Take the group that includes all living organisms. A subset of this group possesses a backbone. We call this subset "vertebrates." Within the vertebrates, some possess fur, while most do not. All members of the group that possesses fur are called "mammals." We choose features that characterize smaller and smaller groups within larger groups, the largest of which is the biota. This arrangement is hierarchical, because those creatures possessing fur are a subset of all animals possessing a backbone, which are in turn a subset of all living organisms (Figure 3.1). Although so far we have limited this discussion to backbones and fur, all features of living organisms, from the possession of DNA – which is ubiquitous – to highly restricted features such as the possession of a brain capable of producing a written record of culture, can be arranged hierarchically.

Although life is commonly referred to as infinitely diverse (indeed, we earlier used such a phrase), this is not correct: life's diversity is most assuredly finite. It is profoundly connected by a hierarchical array of shared features. Diversity actually takes the form of many variations on ultimately the most primitive body plan, with modifications upon modifications that take us to the present. Always, however, unmodified or slightly modified vestiges of the original plan remain, and these provide the keys to revealing the fundamental hierarchical relationships that underpin the history of life.

Characters Identifying the features themselves is a prerequisite to establishing the hierarchy of life's history, so we need to look more closely at what we mean by "features." Features of organisms are termed characters. Characters acquire their meaning not as a single feature on a particular organism, but when their distribution among a selected group of organisms is considered. For example, the group Felidae – cats – is generally linked on basis of distinctive features of the skull. Thus not only is the cartoon character Garfield a felid, but so are cats of all stripes, including bobcats, lions, jaguars, and saber-toothed tigers. And by the same token, if someone told us that some mammal is a felid, we could be confident in the prediction that it has those same unique skull features.

In living organisms, there is a wealth of characters: the macroscopic structure of the organism (skin, feathers, fur, muscles, bones, teeth, organs, etc.), genetic composition (chromosomal structure, DNA and amino acid sequences, aspects of transcription and translation), embryological and developmental stages and patterns, and even ecology and behavior. Many of these features are obviously no longer available to paleontologists, and as a result, we are obliged to rely upon the hard skeletal material provided by the fossils themselves.

Because characters are distributed hierarchically, their position in the hierarchy is obviously a function of the groups they characterize. Consider again the simple example of fur in mammals. Since all mammals have fur, it follows that if one wanted to tell a mammal from a non-mammal (any other organism), he need only observe that the mammal is the one that has the fur. On the other hand, the character "possession of fur" is not useful for distinguishing, say, a bear from a dog; both have fur. To distinguish one mammal from another, characters that identify subsets within mammals must be used.

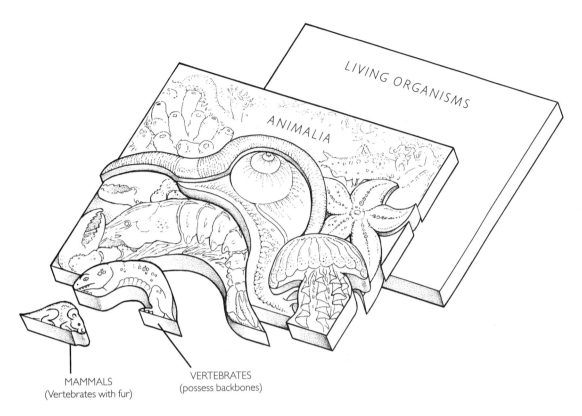

MAMMALS
(Vertebrates with fur)

VERTEBRATES
(possess backbones)

Figure 3.1. The natural hierarchy illustrated as a wooden puzzle. The different organisms represent the larger groups to which they belong. For example, the mouse, representing Mammalia, and the lizard, representing Reptilia, together fit within the puzzle to represent Vertebrata, itself a subset of bilaterally symmetrical organisms (Bilateralia), which would include invertebrates such as the lobster. Bilateralia and other groups constitute the group of organisms that we call Animalia.

These distinctions are extremely important in establishing the hierarchy appropriately, and for this reason, characters may function in two ways: as "general" characters and as "specific" characters. A character is specific when it characterizes (or is diagnostic[1] of) all members of a group, while a character is general when it is non-diagnostic of that group. The same character may be specific in one group but general in a smaller subset of that group (because it is now being applied at a different position in the hierarchy).

Suppose as a description to help you find someone you had never met, you were told, "He has two eyes." This would obviously be of little help, since all humans have two eyes. The character of possession of two eyes is a general character that is found not only among humans but in many other groups of organisms. Indeed, it is a general character among

---

1 The word "diagnostic" has the same meaning here as in medicine. Just as a doctor diagnoses a malady by distinctive and unique properties, so a group of organisms is diagnosed by distinctive and unique characters.

all vertebrates, and the character of two eyes alone would not distinguish a human from a guppy. But at a much deeper level in the hierarchy, the character would be specific: possession of two eyes would distinguish a vertebrate (two eyes) from an earthworm (no eyes) or a spider (four eyes). Likewise, consider yet again the example of fur in mammals. The presence of fur is specific when mammals and non-mammals are considered together (because the presence of fur diagnoses mammals) but is general within mammals (it wouldn't be useful in telling one mammal from another).

Cladograms

Evolutionary biologists, including paleontologists, commonly use a "cladogram" to visualize the hierarchies of characters in the biota. A *cladogram* (pronounced cla-do-gram; *clados* – branch; *gramma* – letter) *is a hierarchical, branching diagram that can be used to depict the hierarchies of shared characters*. Its implications, however, are far greater than those of a mere visual aid, for it and the methods behind it have become the single most important tool for understanding the evolutionary history of organisms.

To understand how a cladogram is constructed, we begin with two things to group; say, a car and a pick-up truck. Notice that anything can be grouped; it does not necessarily apply only to living (and extinct) organisms. Cars and trucks may be linked by any number of characters. The characters must, of course, be observable features of each. For example, "used for hauling lumber" is not appropriate, because hauling lumber is what it *does*, and is not an observable character. Note, though, that a pick-up truck could have characters that make hauling lumber easier than in a car. A cladogram of a car and a truck is shown in Figure 3.2.

Since it is the characters that are distributed hierarchically in the natural world, it is characters that we must choose to diagnose groups.[2] Here, we choose (1) the presence of four wheels, (2) an engine, (3) chassis, (4) seats, and (5) lights. The cladogram simply connects these two separate objects (the car and the pick-up truck) based upon the characters that they share. The features are identified (and itemized) on the cladogram adjacent to the "node," which is a bifurcation (or two-way splitting) point in the diagram. Figure 3.2 shows this relationship.

The issue becomes more complicated (and more interesting) when a third item is added to the group (Figure 3.3). Consider a motorcycle. Now, for the first time, because none of the three items is identical, two of the three items will share more in common with each other than either does with the third. It is in this step that the hierarchy is established. The group that contains all three vehicles is diagnosed by certain features shared by all three. Notice that a subset containing two vehicles has been established. Because the two are linked together on the cladogram, not only do they share the characters pertaining to all three, but above and

---

2 The motor vehicles in this example are not from the natural world, and thus the distribution of their characters may not really be hierarchical. Nonetheless, this example serves as an effective illustration to show how characters function to unite groups on cladograms.

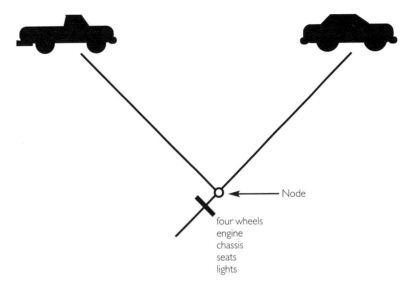

four wheels
engine
chassis
seats
lights

Figure 3.2. A cladogram. The car and pick-up truck are linked by the characters listed at the bar, just below the node. The node itself defines the things to be united; commonly a name is attached to the node that designates the group. Here, such a name might be four-wheeled vehicles.

beyond these they share further characters that link them exclusive of the third vehicle. Lights and seats would be general when one is discussing the subset composed of two vehicles, since those characters are diagnostic only at a higher level in the hierarchy.

It should be clear by now that how these characters, and even the vehicles, are arranged on the cladogram is controlled by the choice of characters. Suppose that instead of "seats" we had specified bucket seats, and instead of "four wheels" we had simply specified "wheels." Bucket seats would then no longer be a general character diagnosing all vehicles, but instead would unite only the car and the truck. Likewise, the presence of wheels would be a general condition pertaining to all three, instead of uniting trucks and cars.

Now suppose that instead of the characters that we listed above, we had chosen wheels, engine, lights, seat, and non-passenger space less than passenger space. These characters produce a cladogram quite different from that in Figure 3.3, in which cars and motorcycles are linked more closely to each other than either is to a pick-up truck (Figure 3.4). This arrangement is counterintuitive, and contradicts the cladogram in Figure 3.3.

How do we choose? We must choose the characters, and order them so that the resultant cladogram doesn't change when other characters are added. Most of the characters that apply to these motor vehicles support the cladogram in Figure 3.3 and not that in Figure 3.4; they suggest that, in its design, a car shares much more in common with a pick-up truck than it does with a motorcycle. Moreover, those features that a car shares with a motorcyle are also present in the pick-up truck;

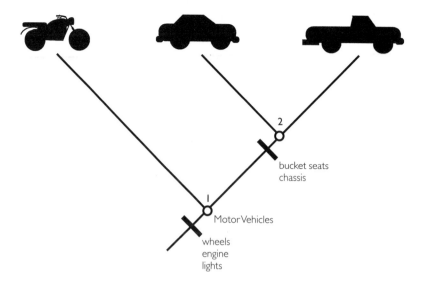

Figure 3.3. One possible distribution of three motor vehicles. Members of the group designated by node **1** are united by the possession of wheels, lights, and an engine; that group could be called Motor Vehicles. Within the group Motor Vehicles is a subset united by possession of bucket seats and a chassis. That subset is designated at node **2**.

they are general features for cars, trucks, and motorcycles, rather than specific features that clearly diagnose a car–motorcycle subset within the three vehicles. Distinguishing among cladograms is an important subject that will be discussed below.

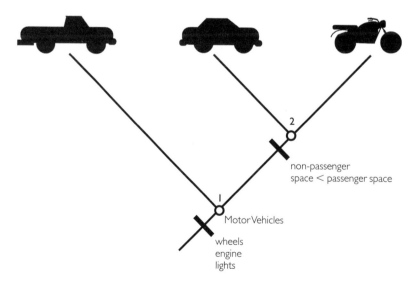

Figure 3.4. An alternative distribution of three motor vehicles. The characters selected suggest that the car and motorcycle share more in common with each other than either does with the pick-up truck.

So far, we have presented the cladogram only as a graphic method of showing hierarchies. Obviously, it must be far more than this, or its relevance to this book would be difficult to fathom. In fact, cladograms are powerful tools for studying the evolutionary relationships among organisms. Their use in the past 25 years has completely revolutionized our understanding of the interrelationships of organisms.[3] Terms as fundamental as "reptile," "dinosaur," and "bird" have startling new meanings as a result of "cladistic analysis," or analysis using a cladogram. For this reason, cladograms play a profound role in this book.

A monkey's uncle

Fundamental to evolutionary biology and paleontology is the recovery of who is related to whom. Before we can understand the great events and rhythms of biotic history, we need a way to discover the pattern of descent of the Earth's creatures. Considered individually, extinct and extant organisms are a myriad of apparently disconnected individuals, but considered as evolving groups (lineages), striking patterns emerge that enrich our view of ourselves and the world around us. It is for this reason that evolution is considered the unifying principle of biology: evolution is the basis of the fundamental genealogical connections among organisms. Accordingly, we are interested in "phylogeny": the history of the descent of organisms. It is in this respect that cladistic analysis makes a key contribution. Using character hierarchies portrayed on cladograms, we can establish "clades" or "monophyletic groups" (to add to the nomenclature, these are sometimes termed "natural groups" as well; here, these terms are all considered to be synonymous). These are groups of organisms that have evolutionary significance because the members of each group are more closely related to each other than they are to any other creature. For example, it is probably no surprise that humans are a monophyletic group: all members of *Homo sapiens* are more closely related to each other than they are to anything else. The idea that a group is monophyletic has a second, more subtle ramification: it implies that all members of that group share a more recent common ancestor with each other than with any other organism.

Evolution

Organic evolution is a fact

By saying organic evolution is a fact, we mean that, if one accepts that the human mind, with its strengths and limitations, is capable of understanding aspects of the natural world, and that scientific

---

3 Cladistic methods were first developed and articulated by an entomologist, Willi Hennig, in *Grundzuge einer Theorie der phylogenetischen Systematik* (1950). Hennig's work had a minor impact on European biologists, but it was not until the 1966 publication of an English translation (entitled simply *Phylogenetic Systematics*) of a revised version of the 1950 work that cladistic methods became relatively well known. During the late 1960s and throughout much of the the 70s and early 80s cladistic methods became a kind of *cause célèbre* as a host of determined advocates foisted it upon a host of equally determined scientists unimpressed by the method (see Box 4.3). The real strengths of the method eventually triumphed, and today, virtually all phylogenetic reconstruction is done by means of cladograms.

methods are an appropriate tool for this type of inquiry, the biota has undergone evolution.[4] Evolution refers to descent with modification: organisms have changed and modified their "morphology" (*morph* – shape; *ology* – the study of) through time, and each new generation is the most recent bearer of the unbroken genetic thread that connects life. In this sense, each new generation is forward looking in that its members potentially contain changes relevant for the future, but is connected to the past by features that they have inherited.

That evolution has occurred is not a particularly new idea; it was articulated by a variety of enlightenment and post-enlightenment philosophers and natural historians. The unique contribution of Charles Darwin and Alfred Russel Wallace (who jointly presented similar ideas at an 1858 meeting of the Linnean Society of London) was that the driving force behind evolution is natural selection. Here, however, we are most concerned with the *record* of evolution – an observable pattern of descent with modification – regardless of the process (natural selection) that may be responsible for it. It is important to intellectually decouple evolution (fundamentally a pattern) itself from natural selection (the process driving the pattern), and our efforts will be directed largely to evolution and not to natural selection.

Evolution amounts to modifications (in morphology, in genetic make-up, in behavior, etc.), so that while some changes are developed in descendants, many of the ancestral features are retained. Clearly implicit in this are the relationships between anatomical structures. For example, we postulate a special relationship between the five "fingers" in the human "hand" and the five "toes" in, say, the front "foot" of, say, a lizard. Here, the English language is confusing; we are really talking about the digits of the forelimbs, a particular feature that happens to have been conserved (or maintained) through time in these two lineages (humans and lizards). In theory, the digits on the forelimbs of lizards and humans can be traced back in time to digits in the forelimb of the common ancestor of humans and lizards. We call these anatomical structures "homologues," and two anatomical structures are said to be "homologous" when they can, at least in theory, be traced back to a single original structure in a common ancestor (Figure 3.5). Thus we infer that the digits in the forelimbs of all mammals are homologous with those of, for example, dinosaurs. That is because these digits can be traced back to the digits in the forelimbs of the common vertebrate ancestor of mammals and dinosaurs. The wings of a fly, however, are not homologous with those of a bird, since they cannot be traced to a single structure on a common ancestor. Because the wings of a fly and the wings of a bird perform in similar fashion (they allow flight to take place), they are considered to be "analogues," and are said to be "analogous" (Figure 3.6). Obviously, the concept of evolution is intimately tied to the concept of homology.

---

4 Scientific debate about the "theory of evolution" is not about whether evolution actually occurred, but rather about the underlying causal mechanisms behind evolution.

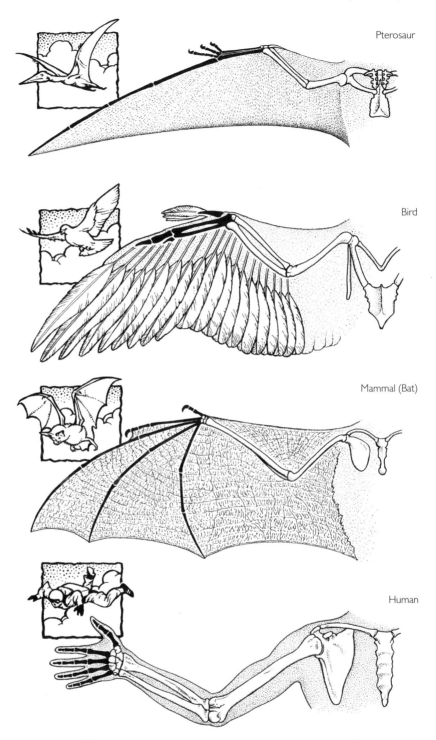

Pterosaur

Bird

Mammal (Bat)

Human

Figure 3.5. Homologues. Homologues are anatomical structures that can, at least theoretically, be traced back to a single structure in a common ancestor. The front limbs of humans, bats, birds, and pterosaurs are all homologous, and retain the same basic structure and bone relationships even though the appearance of these forelimbs may be outwardly different. Homology forms the basis for hypotheses of evolutionary relationships.

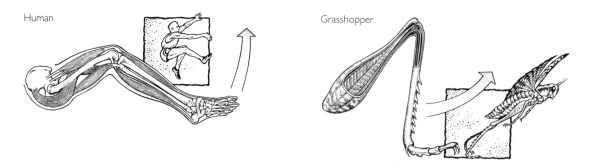

Human

Grasshopper

Figure 3.6. Analogues. Analogues may perform similar functions, and may even look outwardly similar, but internally they can be very different. Here, a human leg is contrasted with that of a grasshopper. Although both have legs, the two structures are different. Human muscles are on the outside of the skeleton, whereas grasshoppers' muscles are on the inside of their skeleton.

An obvious, yet often-ignored clue to the fact that evolution has taken place is the hierarchical distribution of characters in nature. If descent with modification has taken place, what patterns of character distributions might one expect to find? Modification of ancestral body plans through time would produce exactly the distribution of characters that we observe: a hierarchical arrangement in which some homologous characters are present in all organisms, in which other characters are found in somewhat smaller groups, and in which still other characters have a very restricted distribution and are found in only a few organisms.

Chopping down the "tree of life"

Accepting the fact of evolution, there must be a single phylogeny – a single genealogy – that documents the interrelatedness or connectedness of all life. This is not an unfamiliar concept, because most of us have seen "trees" that purport to document who came when and from whom. Such "trees of life" are common in textbooks and museum displays, and deeply influence most people's ideas about the pattern of evolution (Figure 3.7). These trees commonly show an ancestral protoplasmoid rising out of primordial sludge and giving rise to everything else. But how does one make a tree of life? How do we figure out who gave rise to whom? After all, no human was present to observe the appearance of the first dinosaur on the face of the earth. And is it reasonable to suppose that, with the fossil record as incomplete as it is, one fossil that we happen to find serendipitously turns out to be the very ancestor of other fossils that we happened to find? Because of their rarity, the chances of that occurring, especially among vertebrates, are vanishingly small. Thus the oldest hominid fossil known is very unlikely to be the direct ancestor of all subsequent humanity. On the other hand, it is likely to have many features that the real ancestor possessed. In this book, therefore, we avoid trees of life, and instead use cladistic analysis to reconstruct evolutionary events.

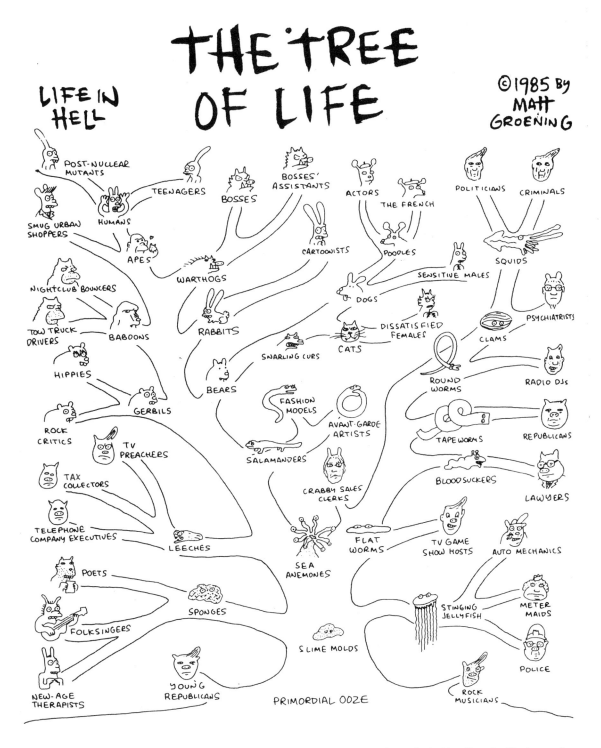

Figure 3.7. A tree of life. This particular one is a satire by Matt Groening. The image of evolution as a tree, however, is completely familiar. (From the *Big Book of Hell* © Matt Groening. All Rights Reserved. Reprinted by permission of Pantheon Books, a division of Random House, Inc., New York.)

Using cladograms to reconstruct phylogeny

To reconstruct phylogeny, we need a way to recognize how closely two creatures are related. Superficially this is very simple: things that are more closely related tend to share specific features. We know this intuitively by simply observing that organisms that we believe are closely related (e.g., a dog and a coyote) share many similarities and because we have seen the results of breeding, in which offspring look, and sometimes act, very much like their parents.

Cladograms were initially described in this chapter without placing them within an evolutionary context. Considered in an evolutionary context, the specific characters that we said characterize groups can be treated as homologous among the groups that they link. Fur in mammals once again (!) provides a convenient example. If all mammals are fur-bearing (and mammals are monophyletic), the implication is that the fur found in bears and that found in horses can in fact be traced back to fur that must have been present in the most recent common ancestor of bears and horses. But this is putting the cart before the horse. It is the distribution of characters that helps us to determine which groups are monophyletic and which are not, and, in the case of mammals, the conclusion that they are monophyletic is in part based upon the fact that mammals all share the specific character of fur (among many other characters). In an evolutionary context, specific characters are termed "derived" or "advanced," and general characters are termed "primitive" or "ancestral." The term "primitive" certainly does *not* mean worse or inferior, and advanced certainly does *not* mean better or superior. These refer instead to the timing of evolutionary change; derived characters evolved later than primitive characters. Only derived characters provide evidence of natural (monophyletic) groups because, as newly evolved features, they are potentially transferable from the first organism to acquire them to all its descendants. Primitive characters – those with a much more ancient history – provide no such evidence of unique natural relationships. To illustrate this, we resort for the last time to mammals and fur. Fur, we said, is among the shared, derived characters that unite the mammals as a monophyletic group. On a cladogram, therefore, we look for characters that unite a bifurcation point in the diagram. All organisms characterized by shared, derived characters are linked by the cladogram in monophyletic groups. The idea is that evolutionary history can be recovered (or reconstructed) using shared, derived characters organized on a cladogram. Box 3.1 exemplifies this for a non-biotic group: watches.

Reflecting the hierarchy of character distributions in nature, the cladogram documents monophyletic groups within monophyletic groups. In Figure 3.8, a small part of the hierarchy is shown: humans (a monophyletic group possessing shared, derived characters) are nested within the mammals (another monophyletic group possessing other shared, derived characters). Notice that the character of warm-bloodedness is primitive for *Homo sapiens*, but derived for mammals. As we have seen, features can be derived or primitive (but not at the same time), all depending upon what part of the hierarchy one is investigating.

BOX 3.1

Wristwatches: when is a watch a watch?

Analogue and digital timepieces are commonly called "watches." Implicit in the term "watches" is some kind of evolutionary relationship: that these instruments have a common heritage beyond merely post-dating a sundial. But is this really so? Here we use cladistic techniques to infer the evolutionary history of watches.

Consider three types of watch: a wind-up watch, a digital watch, and a watch with quartz movement (Figure B3.1.1). Six cladograms are possible for these instruments, but it can be seen

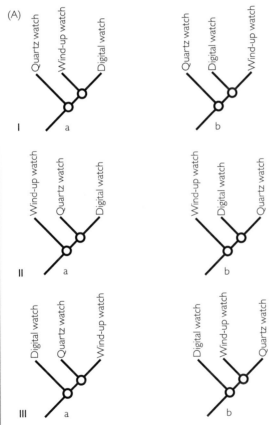

Figure B3.1.1. Six possible arrangements of three timepieces on cladograms. Note that each pair is redundant: the order in which the objects on each "V" is presented is irrelevant. Each pair is said to be "commutatively equivalent."

that, by the definition of a cladogram, a and b for each type are identical. This is because the groups at a node share the characters listed at that node, regardless of order. For this reason, we really have only three cladograms to consider (Figure B3.1.2). One might at first wish to place the digital watch in the smallest subset, in the most derived position (as in types I and II), since it is the most modern, technologically advanced, and sophisticated of the three. Remember, however, how the cladogram is established: on the basis of *shared, derived* characters. Cladogram types I and II say that the digital watch shares the most characters in common with either a wind-up watch (type I) or a quartz watch (type II). A look at the characters themselves suggests that this is not correct: wind-up and quartz watches are both analogue watches (have a dial with moving, mechanical hands) and their internal mechanisms consist of complex gears and cogs to drive the hands at an appropriate speed. The digital watch, on the other hand, consists of microcircuitry and a microchip, with essentially no moving parts. It is apparently something very different, and from its characters, bears little relationship to the other "watches."

What is the digital watch? In an evolutionary sense, it is really a computer masquerading (or functioning) as a timepiece. The computer has been put in a case, and a watchband has been added, but fundamentally this "watch" is really a computer. In our hypothesis of relationship, the watchbands and cases of watches have evolved independently twice (once in computers and once in watches), rather than the guts of the instrument, itself, having evolved two times. That the watchbands and cases evolved twice independently is a more parsimonious hypothesis than arguing that the distinctive and complex internal mechanisms (themselves consisting of many hundreds of characters) of the watches evolved twice independently.

What, then, is a watch? If the term "watch" includes digital watches as well as the other two more conventional varieties, then it should also

(continued on p. 58)

BOX 3.1 (*cont.*)

(B)

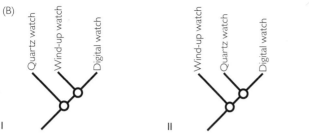

Figure B3.1.2 Because each pair of cladograms in Figure B3.1.1 is commutatively equivalent, there are really only three cladograms under consideration.

include computers, since a digital watch has the shared, derived characters of computers. The cladogram suggests that the term "watch" does not describe an evolutionarily meaningful (monophyletic) group, in the sense that a cladogram that includes digital watches, wind-up watches, and quartz watches must also include computers, as well as a variety of more conventional mechanical timing devices (such as stop watches). Rather, the term "watch" may be thought of as some other kind of category: it describes a particular function (time-keeping) in conjunction with size (relatively small).

In this example, we are fortunate in that, should we so choose, we can test the cladogram-based conclusions by studying the historical record and find out about the evolution of wrist watches, digital watches, and quartz watches. Obviously this is not possible to do with the record of the biota because there is no written or historical record with which to compare our results. The characters of each new fossil find, however, can be added to existing cladograms and the hypothesis of relationship that shows the least complexity will be favored according to the principle of parsimony. In our discussions of the biota, we attempt to establish categories that are evolutionarily significant (monophyletic groups), and avoid groups that have less in common with each other than with anything else.

The cladogram need not depict every organism within a monophyletic group. If we are talking about humans and carnivores, we can put them on a cladogram and show the derived characters that diagnose them, but we might (or might not) include other mammals (e.g., a gorilla). As with the motor vehicles example, if the hierarchical relationships that we have established are valid, the addition of other organisms into the cladogram should not alter the basic hierarchical arrangements established by the cladogram. Figure 3.9 shows the addition of one other group into the cladogram from Figure 3.8. The basic relationships established in Figure 3.8 still obtain even with the new organisms added.

Parsimony   It may be apparent by now that, in an evolutionary context, a cladogram is actually a "hypothesis of relationship," that is, an hypothesis about how closely (or distantly) organisms are related. With a given set of characters, it may be possible to construct several possible cladograms (as we saw that there were in the example of the pick-up truck, the car, and the motorcycle). We can distinguish among different hypotheses of relationship using the principle of "parsimony." Parsimony, a sophisticated philosophical concept first defined by the

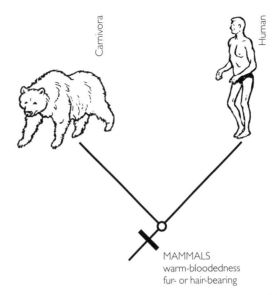

Figure 3.8. A cladogram showing humans within the larger group Mammalia. Mammalia is diagnosed by warm-bloodedness and possession of fur (or hair); many other characters unite the group as well. Carnivora, a group of mammals that includes bears and dogs (among others), is shown to complete the cladogram. Carnivores all uniquely share a special tooth, the carnassial, and humans all uniquely share various gracile features of the skeleton. Note that all mammals (including humans and carnivores) are warm-blooded and have fur (or hair), but only humans have the gracile skeletal features, and only members of Carnivora have the carnassial tooth.

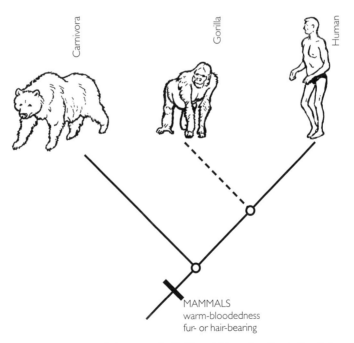

Figure 3.9. Addition of the genus *Gorilla*. The addition of gorillas to the cladogram does not alter the basic relationships outlined on the cladogram shown in Figure 3.8.

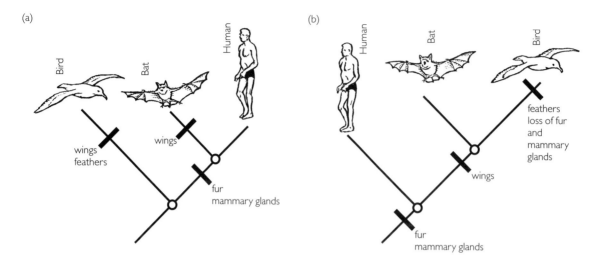

Figure 3.10. Two possible arrangements for the relationships of birds, bats, and humans. Part (a) requires wings to have evolved two times; part (b) requires birds to have lost fur and mammary glands. These as well as many other characters make (a) the more parsimonious of the two cladograms.

fourteenth century English theologian William of Ockham, states that the simplest explanation – that is, one with fewer steps than another – is probably the best. Why resort to complexity when simplicity is equally informative? In other words, why take more steps when fewer can provide the same information?

A bird, a human, and a bat will serve as a simplified example. We will start with the following characters: wings, fur, feathers, and mammary glands. Figure 3.10 shows two cladograms that are possible from these animals and their characters. In the one in which the bird is most closely linked with the bat, the bird has to lose ancestral mammary glands and it has to replace fur with feathers. In the cladogram in which the bat and the human share a most recent common ancestor, wings must be invented by evolution twice. The cladogram in which the human and bat are most closely linked is the simpler of the two because it requires fewer evolutionary events or steps. The cladogram linking the human and bat together remains uncomplicated by the addition of more characters; by contrast, the addition of virtually any other characters that pertain to the creatures in question (e.g., the arrangement, shape, and number of bones, particularly those in the skull and wings, the structure of the teeth, the biochemistry of each organism, etc.) only further complicates the cladogram that most closely links birds and bats. Based upon parsimony, therefore, the cladogram is preferred that shows bats and humans to have more in common with each other than either does with a bird. And indeed, as a hypothesis about the evolution of these vertebrates, it is extremely likely that bats and humans share a more recent common ancestor with each other than either does with a bird (which, obviously, is why they are classified together as mammals).

In this case, the use of shared, derived characters has led us to the most parsimonious conclusion with regard to the evolution of these three creatures.

Science and testing hypotheses

Science is an approach to gaining insight into certain kinds of issue that is rooted in a particular type of logic. In this sense it is nothing more (or less) than a tool for solving a restricted series of problems. Indeed, there is a variety of potentially significant problems that are not particularly amenable to a scientific solution. Examples of such questions are "Is there a God?", "Does she love me?", "Why don't I like hairy men?", and "Is this great music?" Such questions might be answerable, but it will never be by means of science that the answers are discovered.

Other questions, however, are more amenable to scientific inquiry. For example, a scientific hypothesis (although a very simplistic one) is: "The sun will rise tomorrow." This statement can be thought of as a hypothesis with specific predictions. The hypothesis (that the sun will rise tomorrow) is testable; that is, it makes a prediction that can be assessed. The test is relatively straightforward when the right kinds of observation are made: we can wait until tomorrow morning and either the sun will rise or it will not. If the sun does not rise, the statement has been falsified, and the hypothesis can be rejected. If the sun does rise, the statement has not been falsified, and the hypothesis cannot be rejected. For a variety of relatively sophisticated philosophical reasons, scientists do not usually claim that they have proven the statement to be true; rather, the statement has simply been tested and not falsified. *One of the basic tenets of science is that it consists of hypotheses that have predictions which can be tested.* We will see many examples of hypotheses in the coming chapters; all must involve testable predictions. Our ability to test these will determine the importance of these hypotheses as scientific contributions.

In an evolutionary context, cladograms are hypotheses of phylogenetic relationship. They make predictions about the distributions of characters in organisms, both living and extinct. New fossils can test the phylogenetic hypotheses inherent in cladograms because these contain character suites that need to be concordant with the pre-existing cladograms. Living organisms can also test cladograms by the distribution of their characters, including the content of their genetic material (DNA sequences). Cladograms are most "robust" (strongest), if they survive falsification attempts. Parsimony indicates that these are the ones that most closely approximate the course of evolution.

It is now clear how a cladogram is tested. The addition of characters can cause the rejection of a cladogram by demonstrating that it is not the most parsimonious character distribution. In contrast, a tree of life presents more difficulties. Aside from requiring the miniscule probability that ancestors and their direct descendants will be preserved, a tree of life is untestable. How does one identify the actual ancestor and its direct descendant? Given the absurdity of a claim that these have been found, a tree of life is really more of a story or "scenario," than a testable scientific hypothesis. For this reason, here we content ourselves with cladograms, and do not confuse them with trees of life. As will become evident, much can be learned from cladograms that will contribute to our desire to know what occurred in ages long past.

Important readings

Cracraft, J. and Eldredge, N. (eds.) 1981. *Phylogenetic Analysis and Paleontology*. Columbia University Press, New York, 233pp.

Eldredge, N. and Cracraft, J. 1980. *Phylogenetic Patterns and the Evolutionary Process, Method and Theory in Comparative Biology*. Columbia University Press, New York, 349pp.

Hennig, W. 1966. *Phylogenetic Systematics* (translation by D. D. Davis and R. Zangerl). University of Illinois Press, Urbana, IL, 263pp. (Reprinted 1979).

Jepsen, G. L., Simpson, G. G. and Mayr, E. 1949. *Genetics, Paleontology, and Evolution*. Princeton University Press, Princeton, NJ, 445pp.

Nelson, G. and Platnick, N. 1981. *Systematics and Biogeography, Cladistics and Vicariance*. Columbia University Press, New York, 567pp.

Ridley, M. 1992. *Evolution*. Blackwell Scientific Publications, Inc., Cambridge, MA, 670pp.

Stanley, S. M. 1979. *Macroevolution*. W. H. Freeman and Company, San Francisco, 332pp.

Wiley, E. O., Siegel-Causey, D., Brooks, D. and Funk, V. A. 1991. *The Compleat Cladist, A Primer of Phylogenetic Procedures*. University of Kansas Museum of Natural History Special Publication no. 19, 158pp.

CHAPTER 4

Interrelationships of vertebrates

What is a dinosaur and where does it fit in among other vertebrates? The answer to this question uncovers remarkable things not only about dinosaurs but also about many of the vertebrates living around us. Here, we will address questions such as: "How many times has warm-bloodedness evolved in the vertebrates?" (answer: at least two, possibly three times); "How many times has powered flight been invented by vertebrates?" (answer: three independent times), "Is a cow a fish?" (answer: in an evolutionary sense, clearly!), "Did all the dinosaurs become extinct?" (answer: definitely not), and "Which has a closer relationship to a crocodile – a lizard or a bird?" (answer: a bird).

Our approach will be to sequentially construct a series of cladograms, beginning with the most inclusive (largest) group – Chordata. The story will unfold as we systematically encounter each bifurcation in the road, retracing the path of evolution until we reach Dinosauria.

In the beginning

Life is generally understood to be monophyletic. This intuitively comforting conclusion should not be taken for granted, for who can say how many early forms of molecular "life" arose, proliferated, and died out in the primordial oceans of billions years ago? Regardless, all modern life (except for some viruses) is united by the possession of RNA, DNA, cell membranes with distinctive chemical structure, a variety of amino acids (proteins), the metabolic pathways (i.e., chemical reaction steps) for their processing, and the ability to replicate itself (not simply grow). These are all shared derived characters of life.

1 cm

Figure 4.1. *Pikaia graciliens*, a presumed chordate from the Middle Cambrian Burgess Shale.

Jumping to chordates

It is certainly possible to construct a cladogram for all life, but this would require us to blithely encapsulate (given the most recent estimates) about 3.8 billion years of organic evolution. Instead, we'll zip forward to Middle Cambrian time, about 520 Ma, where we first find the diminuitive *Pikaia gracilens* (Figure 4.1), a creature that seems to give tantalizing insights into the ancestry of vertebrates. *Pikaia* harkens from the Burgess Shale, a forbidding windswept outcrop in the Canadian Rockies that was once located at the edge of a tropical, equatorial reef teeming with life, 520 million years ago. Rubble and mud periodically fell from this reef, burying thousands of small invertebrate creatures. These ancient animals are today beautifully preserved (if a bit squashed) and indicate that the Middle Cambrian was a time of remarkable diversity.

Pikaia is about 5 cm in length and, in its flattened condition, looks a bit like a miniature anchovy fillet. It was initially described in 1911 as a "worm," but, on closer examination, *Pikaia* seems to reveal characters that are diagnostic of the chordate body plan (see Box 4.1): a nerve cord running down the length of its back, a stiffening rod (the notochord) that gives the nerve cord support, and V-shaped muscles (composed of an upper and a lower part) with repeated segments, a character that is familiar to most of us because it is present in modern fish. We – and the dinosaurs – would appear to have chordate relatives as far back as the Cambrian.[1]

Although *Pikaia* provides an inkling about our distant relatives, what we know about the early evolution of vertebrates and their forebears among Chordata comes principally from living organisms, with sometimes a goodly mixture of information from other relevant fossils.

The chordates consist of *Pikaia* from the Burgess Shale, urochordates, cephalochordates and, most importantly for our story, vertebrates, all of which can be united on the basis of (1) features of the throat (pharyngeal gill slits); (2) the presence of a notochord at some stage in their life histories; and (3) the presence of a dorsal, hollow nerve cord. Above the notochord in chordates is the nerve cord, encompassed within a distinct tail region behind the gut. This distinctive suite of characters – pharyngeal gills, notochord, and nerve cord – appears to have evolved only once, thus uniting these animals as a monophyletic group (Figure 4.2).

Urochordates, commonly called "sea squirts," have a sessile, shapeless adult form, but evidence of their chordate ancestry is found in their free-swimming larvae (in which the notochord is evident). The larvae eventually give up their roving ways, park themselves on their noses, and develop a filter to trap food particles from water that they pump through their bodies (Figure 4.3).

---

1 The claim that *Pikaia* is some kind of chordate has been contested by paleontologist N. Butterfield. Butterfield believes that chordate tissues cannot be preserved in the way those of *Pikaia* are. Therefore, he is unwilling to place *Pikaia* in any modern group.

BOX 4.1

Body plans

All organisms are subject to design constraints. Organisms live in fluid media (air or water), they are acted upon by gravity, and their ancestry limits the structures that they can evolve. For example, you'll never find a propeller on the nose of a bird (even if that were the most efficient way to propel the animal): the evolutionary process works by descent with modification (of existing structures), not the wholesale invention of new ones. In the Linnaean biological classification (see Box 4.2), the term "phylum" is a grouping of organisms whose make-up is supposed to connote a basic level of organization that is shared by all of its members. The idea is that the members of a phylum may modify aspects of their morphology via evolution, but the fundamental organization – or body plan – of the members of the phylum remains constant. For example, a whale is a rather different creature from a salamander, but few would deny the basic shared similarities of their body plans.

There are many types of body plan out there. But, because organisms are subject to design constraints, many similarities are shared by different body plans. These structural repetitions do not occur as a result of a single evolutionary event. Rather, design constraints are such that different lineages of organisms reinvent each structure. When the reinvention of a structure takes place separately in two lineages, the evolution is said to be convergent (e.g., the characters that have evolved separately converge on each other in form). This means that the structures are analogous rather than homologous (see Chapter 3).

High levels of activity dictate a number of convergent, analogous structures that are repeated in one form or another throughout a variety of different organisms. Because muscles can only contract, opposing muscle masses – termed antagonistic – are the means by which most animals accomplish most movement. In vertebrates, antagonistic muscles are distributed around the rigid support provided by a jointed, internal skeleton. This is quite the opposite case in arthropods (the group that includes spiders, crabs, and insects) in which antagonistic muscle masses are enclosed within a jointed, external skeleton (see Figure 3.6).

Complex movements generally require complex musculature, and complex musculature requires sophisticated coordination. This is usually accomplished by a centralized cluster of nerves and neural material, called a brain. In highly active creatures, regardless of origin, the brain is usually located at the front of the animal in a head that is distinct from the rest of the body. The condition of an anterior head region with brain is called encephalization. With motility and encephalization comes bilateral symmetry, in which the right and left halves of the body are mirror images of each other.

Segmentation is another example of convergence in body plans. Segmentation simply involves the division of the body into repeating units, which in turn permits the isolation of parts of the body. Sequential, coordinated motion becomes possible, because one part of the body can respond independently of another. Moreover, the segments themselves can be modified through evolution into specialized organs for a variety of functions, most commonly for locomotion and/or for obtaining food. We all know that arthropods are segmented animals; less well known is that vertebrates are segmented animals, too. The segmentation of the vertebrate body plan is still seen in the repeated structure of the vertebrae, in the ribs, and in muscles.

All of these features – antagonistic muscle masses working on a skeleton, encephalization, bilateral symmetry, and segmentation – are shared by arthropods and vertebrates, but of course this does not mean that arthropods and vertebrates are phylogenetically close. All evidence suggests that the development of these features was convergent in the two groups.

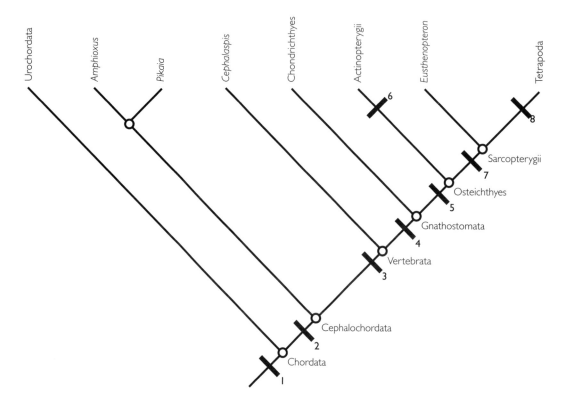

Figure 4.2. Cladogram of the Chordata. Because this is a book about dinosaurs (and not all chordates), we have provided diagnoses for only some of the groups on the cladogram. Bars denote the shared, derived characters of the groups. The characters are the following: **1** pharyngeal gill slits, a notochord, and a nerve cord running above the notochord along its length; **2** segmentation of the muscles of the body wall, separation of upper and lower nerve and blood vessel branches, and new hormone and enzyme systems; **3** bone organized into elements, neural crest cells, the differentiation of the cranial nerves, the development of eyes, the presence of kidneys, new hormonal systems, and mouthparts; **4** true jaws; **5** bone in the endochondral skeleton; **6** ray fins; **7** distinctive arrangement of bones in fleshy pectoral and pelvic fins (see Figure 4.4); **8** skeletal features relating to mobility on land – in particular, four limbs. Consistent with a cladistic approach, only monophyletic groups are presented on the cladogram. Some of the groups may not be familiar; for example, *Cephalaspis* and *Eusthenopteron* are not discussed in the text. *Cephalaspis* was a primitive, jawless, bottom-dwelling, swimming vertebrate, and *Eusthenopteron* was a predatory lobe-fin, bearing many characters present in the earliest tetrapods. *Cephalaspis* and *Eusthenopteron* are included here to complete the cladogram as monophyletic representatives of jawless vertebrates (the non-monophyletic "Agnatha") and lobe-finned fishes, respectively.

Cephalochordates, the closest relative to the vertebrates, are best known to us in the form of the small, water-dwelling lancet (*Amphioxus* or *Branchiostoma*; Figure 4.3). Again, known primarily from present-day forms (one specimen was recovered from the middle of the Paleozoic Era), these creatures share with the vertebrates a host of derived features, including segmentation of the muscles of the body wall, separation of

Figure **4.3.** Two primitive chordates. (a) *Amphioxus (Branchiostoma)*; (b) an adult sea squirt (*Ciona intestinalis*); and (c) the larval form of the sea squirt. Note that the juvenile form is free-swimming and has a notochord running down its tail. It metamorphoses into into the stationary adult by planting itself on its nose and rearranging its internal and external structures.

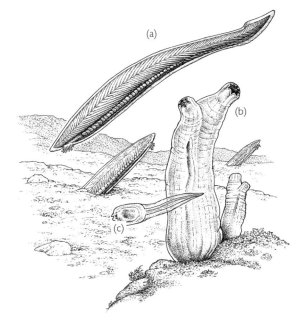

upper and lower nerve and blood vessel branches, and many newly evolved hormone and enzyme systems. As our first cousins among chordates, cephalochordates themselves are united by important derived characters of the head region. Figure 4.2 takes us through the major groups of chordates to Tetrapoda.

Vertebrata   We have now reached Vertebrata. The features that unite vertebrates, with one exception, include calcified skeletal tissue (i.e., bone) divided into discrete parts called elements, and a variety of other characters (see Figure 4.2).

Gnathostomata   What we might unthinkingly regard as the body plan for all vertebrates is really the plan for a subset of the vertebrates called Gnathostomata (*gnathos* – jaw; *stome* – mouth). Gnathostomes are vertebrates with true jaws and a variety of other features (see Figure 4.2). In fact, the absence of jaws is primitive for vertebrates, and the evolution of jaws was an important innovation. So, who are these gnathostomes? Well, *we* certainly are, as are sharks (starring in the movie *Jaws*), ray-finned fishes (familiar in our aquaria and filleted with lemon and french fries on our dinner plates), and lobe-finned fish (fleshy-finned creatures that tinkered with muscular fins and, possibly, air breathing). Other, less familiar, extinct gnathostome fishes that do not concern us here also cruised the ancient seas.

Sharks and their relatives belong to a clade within Gnathostomata called Chondrichthyes (*chondros* – cartilage; *ichthys* – fish). The remaining fishes are grouped together within Osteichthyes (*os* – bone), bony fishes that include the ray-finned and lobe-finned gnathostome groups. The ray-finned branch of osteichthyans is called Actinopterygii (*actino* – ray; *ptero* – wing) and is one of the most diverse groups (more than 20,000 living species) ever to have evolved among vertebrates (see Figure 4.2). The other

BOX 4.2

Biological classification: what's in a name?

Organisms are commonly classified according to the biological classification system, the codified series of categories that have been dutifully memorized by generations of nascent biologists. The classification was first developed by the Swedish naturalist Carolus Linnaeus (1707–1778) who, a full 100 years before Darwin first published *The Origin of Species*, created a formalized hierarchical nomenclature to be applied to the biota. His system of increasingly small categories has been used ever since. The categories that he established – now a memorized mantra – are (in order of decreasing size), kingdom, phylum, class, order, family, genus, species.

Thus a cat would be classified in the following way:

Kingdom: Animalia (a cat is an animal, and not a plant, or a fungus, or some kind of bacterium)

Phylum: Chordata (the embryo has a notochord);

Class: Mammalia (it has fur and mammary glands)

Order: Carnivora (it bears a distinctive pair of shearing teeth, its carnassials)

Family: Felidae (it has certain features in the ear)

Genus: *Felis* (it has distinctive soft anatomy in its nose)

Species: *domesticus* (it's a house cat)

Individuals are generally referred to by italicized generic (genus) and specific (species) names: in the case of a cat, *Felis domesticus*. Each of these names – indeed, any name in the hierarchy – is considered to be a taxon (plural: taxa).

Organisms have names applied to them by the scientists who first recognize them as new or distinct. Hence paleontologists have created names ranging from *Deinonychus* ("terrible claw," to acknowledge the formidable claws on the feet of this carnivorous dinosaur) to *Cuttysarchus*, a small lizard whose name celebrates a well-known scotch whiskey.

All classifications have a purpose, and the biological classification is no exception. We classify by many things; for example, our movies are classified both by subject (Drama, Horror, Comedy, etc.) as well as by suitability for viewing (PG 13; R, etc.). In the case of the biota, implicit in the classification is the degree of relatedness. Thus all members of a single species are said to be more closely related to each other than anyone is to anything else. At any level in the classification, it is assumed that all members of a group are more closely related to each other than they are to members of an equivalent, but different, group. In the case of the order Carnivora, therefore, it is assumed that all carnivores are more closely related to each other than they are to taxa in any other order. In this chapter, we see how this gives a new meaning to the word "Reptilia."

Here we will avoid using most Linnaean terms except for generic and specific ones. The reason for this is that the level in the hierarchy that is designated by any particular Linnaean term is arbitrary: there is no absolute quantity of morphological disparity that is equated with a particular Linnaean term. For example, some people might designate the dinosaur group Ornithischia an order, but some might consider it a suborder. Since there is no quantified disparity embodied in the level "order," neither viewpoint is right and, what's worse, we have learned nothing more about Ornithischia, itself. For us, the important facts about Ornithischia are its shared, derived characters and where (relative to other groups) it is situated in the hierarchy. Moreover, if a formal name were introduced to designate every rank or monophyletic group revealed by the cladogram, this book, already filled with a formidable nomenclature, would be nothing but an endless list of formal names. The cladogram plainly indicates hierarchical relationships without resorting to formalized Linnaean designations.

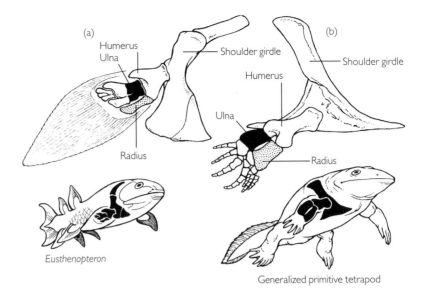

Figure 4.4. Some homologous features between lobe-finned fishes and tetrapods.
(a) the shoulder girdle of *Eusthenopteron*, an extinct sarcopterygian; (b) the shoulder
girdle in early tetrapods. Because aspects of the forelimb in different early tetrapods are
incomplete, the forelimb shown here is a composite prepared from two early tetrapods,
Acanthostega and *Ichthyostega*. Key homologous bones are labeled in both drawings.

osteichthyan branch, the lobe-fins or Sarcopterygii (*sarco* – flesh), includes
lungfish (of which only three types are alive today); coelocanths (of which
only one type is alive today); extinct barracuda-like carnivorous forms;
and, surprisingly, tetrapods (which share the derived characters of the
group; see Figure 4.2 and Box 4.2). Indeed, bones homologous with those
of the limbs, pelvis, vertebral column, and skull of tetrapods are all found
within non-tetrapod members of the lobe-finned clade, strongly uniting
the tetrapods to other members of this group (Figure 4.4). Those homologies – and many others – indicate that it is here, among the lobe-fins, that
the ancestry of Dinosauria – as well as our own ancestry – is to be found.

Tetrapoda

Those gnathostomes central to our story are tetrapods. Tetrapoda (*tetra*
– four; *pod* – foot) connotes the appearance of limbs, an adaptation that
is strongly associated with land. According to the conventional
Linnaean classification (Box 4.2), there are four classes of tetrapods:

Tetrapoda
1 Amphibia,
2 Reptilia,
3 Aves, and
4 Mammalia

The living amphibians are frogs, salamanders (and newts), and a group
of rare tropical, limbless creatures called caeclians. The living reptiles
are crocodiles, turtles, snakes, and lizards, and the tuatara, an unusual

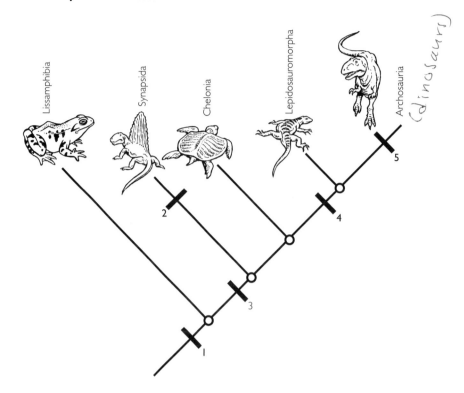

Figure 4.5. Cladogram of Tetrapoda. Diagnostic characters include: at **1** the tetrapod skeleton (see Figures 4.7 and 4.8); at **2** a lower temporal fenestra (see Figure 4.10); at **3** presence of an amnion (see Figure 4.9); at **4** lower and upper temporal fenestrae (see Figure 4.10); at **5** an antorbital fenestra (see Figure 4.13). Lepidosauromorpha is a monophyletyic group, the living members of which are snakes, lizards, and the tuatara. Chelonia – turtles – are reptiles whose primitive, completely roofed skulls place them near the base of Reptilia.

lizard-like creature that lives only in New Zealand. Mammals and living birds (Aves) are common forms familiar to all of us.

The traditional classification is actually a very inadequate way to reflect the interrelationships of the tetrapods (Box 4.3). Figure 4.5 is a cladogram showing the major groups of tetrapods. It shows their phylogenetic relations and leads to a very different understanding of vertebrate interrelationships from that implied by the traditional classification. Even the apparently monophyletic Aves (for who cannot diagnose a bird by its feathers?) is most accurately viewed as an artifact of our post-Mesozoic perspective. As this chapter unfolds, we will address many issues relating to tetrapod relationships.

The tetrapod skeleton made easy

Backbone

The tetrapod skeleton is a modification of the skeletal component of the fundamental vertebrate body plan. We shall see in succeeding chapters how, through evolution, dinosaurs have modified this basic skeleton in a variety of ways. The backbone is composed of distinct, repeated structures (the vertebrae), which consist of a lower spool (the

BOX 4.3

Fish and chips

As 1978 turned to 1979, a provocative and entertaining letter and reply were published in the scientific journal *Nature*, discussing the relationships of three gnathostomes: the salmon, the cow, and the lungfish.[1] English paleontologist L. B. Halstead argued that, obviously, the two fish must be more closely related to each other than either is to a cow. After all, he pointed out, they're both *fish*! A coalition of European cladists disagreed, pointing out that, in an evolutionary sense, a lungfish is more closely related to a cow than to a salmon. In their view, if the lungfish and the salmon are both to be called "fish," then the cow must also be a fish. Can a cow be a fish?

The vast majority of vertebrates are what we call "fishes." They all make a living in either salt or fresh water and, consequently, have many features in common that relate to the business of getting around, feeding, and reproducing in a fluid environment more viscous than air. But as it turns out, even if "fishes" describes creatures with gills and scales that swim, "fishes" is not an evolutionarily meaningful term because there are no shared, derived characters that unite all fishes that cannot also be applied to all non-fish gnathostomes. The characters that pertain to fishes are either characters present in all gnathostomes (i.e., primitive in gnathostomes) or characters that evolved independently.

The cladogram in Figure B4.3.1 is universally regarded as correct for the salmon, the cow, and the lungfish. In light of what we have discussed, this cladogram might look more familiar using groups to which these creatures belong: salmon are ray-finned fishes, cows are tetrapods, and lungfishes are lobe-finned fishes. Clearly, lobe-finned fishes share more derived characters in common with tetrapods than they do with ray-finned fishes. Thus there are two clades on the cladogram:

1 lobe-finned fishes + tetrapods; and
2 lobe-finned fishes + tetrapods + ray-finned fishes.

Clade 1 is familiar as Sarcopterygii. Clade 2 occurs at the level of Osteichthyes and looks like part of

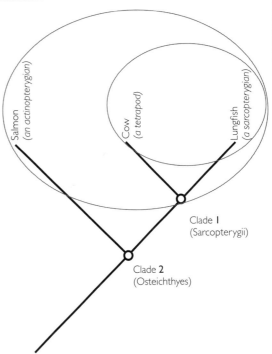

Figure B4.3.1. The cladistic relationships of a salmon, a cow, and a lungfish. The lungfish and the cow are more closely related to each other than either is to the salmon.

the cladogram presented in Figure 4.2 for gnathostome relationships. If only the organisms in question are considered, the only two monophyletic groups on the cladogram must be 1 lungfish + cow; and 2 lungfish + cow + salmon (i.e., representatives of the sarcopterygians and Osteichthyes, respectively).

Which are the "fishes?" Clearly the lungfish and the salmon. But the lungfish and the salmon do not in themselves form a monophyletic group unless the cow is also included. The cladogram is telling us that the term "fishes" has *phylogenetic* significance only at the level of Osteichthyes (or even below). But we can and do use the term "fishes" informally. "Fish and chips" will never be a "burger and fries."

1 See Gardiner, B. G., Janvier, P., Patterson, C., Fortey, P. L., Greenwood, P. H., Mills, R. S. and Jeffries, R. P. S. 1979. The Salmon, the cow, and the lungfish: a reply. *Nature*, **277**, 175–176. Halstead, L. B. 1978. The cladistic revolution – can it make the grade? *Nature*, **276**, 759–760.

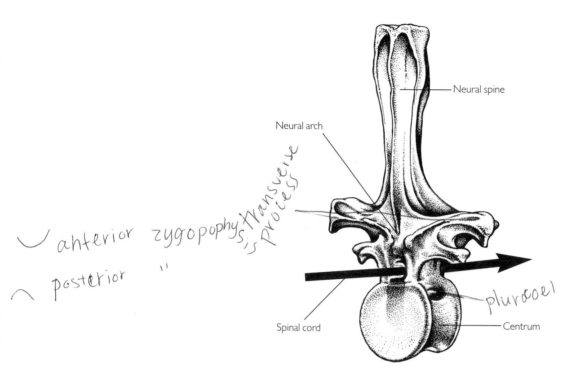

anterior zygopophysis *Transverse Process*

posterior "

pleurocoel

Figure 4.6. A vertebra from *Apatosaurus*, a sauropod dinosaur. The nerve cord (indicated by arrow) lies in a groove at the top of the centrum and is straddled by the neural arch.

centrum), above which, in a groove, lies the spinal cord (Figure 4.6). This relationship was first developed in gnathostomes and modified in tetrapods. Planted on the centrum and straddling the spinal cord is a vertically oriented splint of bone called the neural arch. Various processes, parts of bone that are commonly ridge, knob, or blade shaped, may stick out from each vertebra. These can be for muscle and/or ligament attachment, or they can be sites against which the ends of ribs can abut. The repetition of vertebral structures, a relic of the segmented condition that is primitive for vertebrates, allows flexibility in the backbone. In general, however, the tetrapod backbone is considerably more complex than the backbones that preceded it, because in tetrapods the backbone acts not only to facilitate locomotion, but also to support the body out of the water.

Pelvic and shoulder girdles.
Sandwiching the backbone are the pelvic and pectoral girdles (Figure 4.7). These are each sheets of bone (or bones) against which the limbs attach for the support of the body. The pelvic girdle – which includes a block of vertebrae called the sacrum – is the attachment site of the hindlimbs; the pectoral girdle is the attachment site of the forelimbs. Each side of the pelvic girdle is made up of three bones: a flat sheet of bone, called the ilium, that is fused onto processes from the sacrum; a

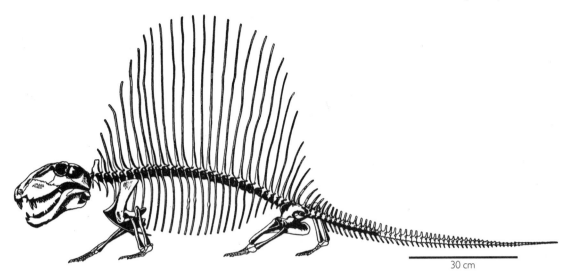

30 cm

Figure 4.11. *Dimetrodon grandis,* a fin-backed synapsid from the late Paleozoic of eastern Texas, USA. (From Romer, A. S. and Price, L. I. 1940. *Review of the Pelycosauria.* Geological Society of America Special Paper no. 28.)

happened to synapsids in the Late Triassic and Early Jurassic remains a mystery (see Chapters 5 and 15).

Reptilia

The other great clade of amniotes is Reptilia (*reptere* – to creep; see Figure 4.5). The living exemplars include about 15,000 total species comprising turtles, snakes, lizards, crocodiles, the tuatara, and birds, but – and this knowledge is a prerequisite for admission to Kindergarten – Reptilia also includes dinosaurs (as well as their close relatives, pterosaurs, as well as many other forms). With so many extinct vertebrates, nobody really knows how many members of this clade have come and gone.

Reptilia is diagnosed by a braincase and skull roof that are uniquely constructed and by distinctive features of the neck vertebrae. Figure 4.8 shows the typical reptilian arrangement of bones in the skull roof and braincase.

The inclusion (above) of birds among the living members of Reptilia is counter to the conventional way of classifying birds, but more accurately reflects their phylogenetic relationships. Clearly we have a decidedly different Reptilia from the traditional motley crew of crawling, scaly, non-mammalian, non-bird, non-amphibian creatures that were once tossed together as reptiles. If it is true that crocodiles and birds are more closely related to each other than either is to snakes and lizards, a mono-phyletic group that includes snakes, lizards, and crocodiles *must* also include birds. The implication of calling a bird a reptile is that birds share the derived characters of Reptilia, as well as having unique characters of their own. These arguments are developed in Chapter 13.

Anapsida

Within Reptilia are two equally important clades: Anapsida and Diapsida. The first, Anapsida (*a* – without), consists of Chelonia (turtles) and some extinct, bulky quadrupeds that do not concern us here.

Legendary stalwarts of the world, turtles are unique: these venerable creatures with their portable houses, in existence since the Late Triassic (210 Ma), will surely survive another 200 million years if we let them.

Diapsida

for Muscle space

Diapsida (*di* – two) is united by a suite of shared, derived features including having two temporal openings in the skull roof, an upper (or supra) and a lower (or infra) temporal fenestra. The upper and lower temporal openings are thought to have provided space for the bulging of contracted jaw muscles, as well as increased the surface area for the attachment of these muscles. There are two major clades of diapsid reptiles. The first, Lepidosauromorpha (*lepido* – scaly; *sauros* – lizard; *morpho* – shape; note that the suffix *sauros* means "lizard" but is commonly used to denote anything "reptilian") is composed of snakes and lizards and of the tuatara (among the living), as well as a number of extinct lizard-like diapsids.[4]

Archosauromorpha

Finally we come to the other clade of diapsids, the archosauromorphs (*archo* – ruling). Archosauromorpha is supported by many important, shared, derived characters that are included on the cladogram in Figure 4.12. Within archosauromorphs are a series of basal members that are known mostly from the Triassic. Some bear a superficial resemblance to large lizards (remember, however, that they cannot be true lizards, which are lepidosauromorphs), whereas others look like reptilian pigs (see Figure 16.5).

The last of the aforementioned – prolacertiforms – possess a number of significant evolutionary innovations (Figure 4.12), most notably an opening on the side of the snout, just ahead of the eye, called the antorbital fenestra (Figure 4.13). These are the characters that unite Archosauria, the group that contains crocodilians, birds, and dinosaurs. Crocodilians and their close relatives belong to a clade called Crurotarsi[5] (*cruro* – shank; *tarsus* – ankle); birds and their close relatives constitute a clade called Ornithodira (*ornith* – bird; *dira* – neck).

Modern crocodiles are but an echo of what preceded them and in the past there have been sea-going crocodiles with flippers instead of legs, crocodiles with teeth that look more mammalian than crocodilian, and crocodiles that appear to have been well adapted for running on land. Other crurotarsans included a variety of carnivorous (Figure 4.14), piscivorous (fish-eating), and herbivorous forms.[6]

Ornithodira brings us quite close to the ancestry of dinosaurs. This group is composed of two major clades, Dinosauria (*deinos* – terrible) and

---

4 Two marine groups, ichthyosaurs and plesiosaurs (see Figure 18.8) have also been placed among the diapsids, but these are not germane to our story.

5 Some paleontologists prefer J. A. Gauthier's (1986) use of the term Pseudosuchia for this group. Membership in Pseudosuchia is similar to but not identical with that of Crurotarsi.

6 Historically, basal archosauromorphs have all been jumbled together under the name Thecodontia (*theco* – socket; *dont* – tooth; see Chapter 13) because their teeth are set in sockets (much as our own are). The teeth in sockets applies to all archosauromorphs (Figure 4.12); how, therefore, can this character be used to distinguish one archosauromorph from another?

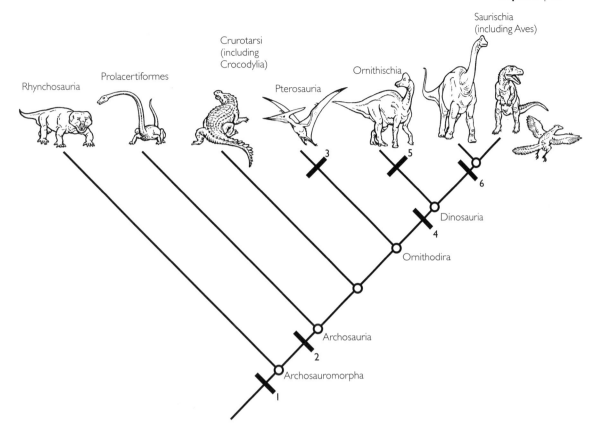

Figure 4.12. Cladogram of Archosauromorpha. Diagnostic characters include: at **1** teeth in sockets, elongate nostril, high skull, and vertebrae not showing evidence of embryonic notochord; at **2** antorbital fenestra (see Figure 4.13); loss of teeth on palate and new shape of articulating surface of ankle (calcaneum); at **3** a variety of extraordinary specializations for flight, including an elongate digit IV; at **4** erect stance (shaft of femur is perpendicular to head; the ankle has a modified mesotarsal joint) perforate acetabulum (see Chapter 5 for greater detail); at **5** predentary and rearward projection of pubic processes (see introductory text for Part II: Ornithischia); and at **6** asymmetrical hand with distinctive thumb, elongation of neck vertebrae, and changes in chewing musculature (see introductory text for Part III: Saurischia).

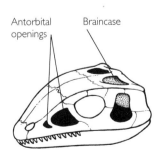

Figure 4.13. An archosaur skull with the diagnostic antorbital opening.

Pterosauria (*ptero* – winged). Pterosaurs, the brainy, impressive "flying reptiles" from the Mesozoic, are highly modified ornithodirans, with as many as 40 derived features that unite them as a natural group. Their smallest members were sparrow sized, and their largest members had wingspans as large as 15 m, making them the largest flying organisms that have ever lived (for reference, the wingspan of the two-person Piper Cub (airplane) is about 12 m).[7] That they are unapologetically Mesozoic and utterly extinct has led to their being called "dinosaurs," but in fact they are something utterly different from either birds or dinosaurs. They are pterosaurs.

_____

7 These extraordinary Mesozoic beasts demand a detailed treatment unfortunately not possible here (see Wellenhfer, 1996).

50 cm

Figure 4.14. A reconstruction of the carnivorous crurotarsan *Euparkeria*.

Dinosaurs This leaves us at long last with the subject of our book, Dinosauria. Dinosaurs can be diagnosed by a host of shared, derived characters, many of which are elaborated in Chapter 5 (Figures 5.4 and 5.5). Most strikingly, dinosaurs are united by the fact that, within archosaurs, they possess an erect, or parasagittal stance; that is, a posture in which the plane of the legs is perpendicular to the plane of the torso and is tucked under the body (Figure 4.15; see also Box 4.4).

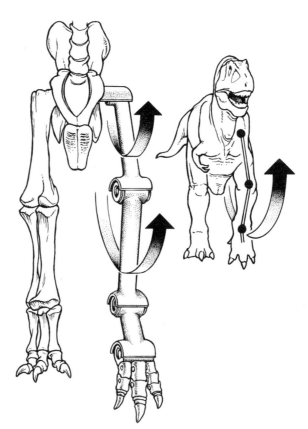

Figure 4.15. The fully erect stance in dinosaurs. Unlike in, for example, a human, the bones of the leg permit motion forward and backward in only one plane.

BOX 4.4

Stance: it's both who you are and what you do

Tetrapods that are most highly adapted for land locomotion tend to have an erect stance. This clearly maximizes the efficiency of the animal's movements on land, and it is not surprising that, for example, all mammals are characterized by an erect stance. Tetrapods such as salamanders (which are adapted for aquatic life) display a sprawling stance, in which the legs splay out from the body nearly horizontally. The sprawling stance seems to have been inherited from the original position of the limbs in early tetrapods, whose sinuous trunk movements (presumably inherited from swimming locomotion) aided the limbs in land locomotion.

Some tetrapods, such as crocodiles, have a semi-erect stance, in which the legs are directed at something like 45° downwards from horizontal (Figure B4.4.1). Does this mean that the semi-erect stance is an adaptation for a combined aquatic and terrestrial existence? Clearly not, because a semi-erect stance is present in the large, fully terrestrial monitor lizards of Australia (goanna) and Indonesia (Komodo dragon). If adaptation is the only factor driving the evolution of features, why don't completely terrestrial lizards have a fully erect stance, and why don't aquatic crocodiles have a fully sprawling stance? The issue is more complex and is best understood through adaptation to a particular environment or behavior, as well as through inheritance.

If we consider stance simply in terms of ancestral and derived characters, the ancestral condition in tetrapods is sprawling. An erect stance represents the most highly derived state of this character, but are animals with sprawling stances not as well designed as those with erect stances? In 1987, D. R. Carrier of Brown University hypothesized that the adoption of an erect stance represents the commitment to an entirely different mode of respiration (breathing) as well as locomotion (see Chapter 15 on "warm-bloodedness" in dinosaurs). Those organisms that possess a semi-erect stance may reflect the modification of a primitive character (sprawling) for greater efficiency on land, but they may also retain the less-derived type of respiration. Dinosaurs (see Figure 4.15) and mammals both have fully erect stances, which represent a full commitment to a terrestrial existence as well as to a more derived type of respiration. The designs of all these organisms are thus compromises among inheritance, habits, and mode of respiration. Who can say what other influences are controlling morphology?

Interestingly, the cladogram (see Figure 4.5) shows that the most recent common ancestor of dinosaurs and mammals – some primitive amniote – was itself an organism with a sprawling stance. Because dinosaurs and mammals (or their precursors) have been evolving independently since their most recent common ancestor, an erect stance must have evolved twice in Amniota: once among the synapsids and once in dinosaurs.

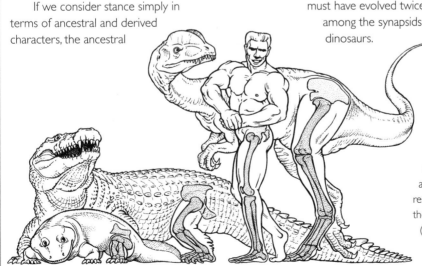

Figure B4.4.1. Stance in four vertebrates. To the left, the primitive amphibian and crocodile (behind) have sprawling and semi-erect stances, respectively. To the right, the human and dinosaur (behind) both have fully erect stances.

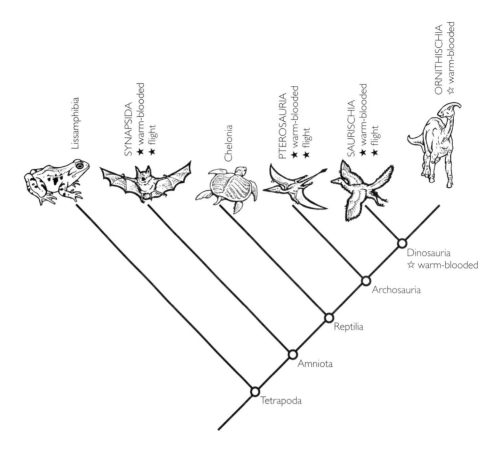

Figure 4.16. The distribution of warm-bloodedness and flight in Tetrapoda, presented on a cladogram constructed from Figures 4.5 and 4.12. Filled stars denote known instances of warm-bloodedness and flight; unfilled stars and italics denote the possibility that warm-bloodedness and/or flight characterized a group. The cladogram shows that warm-bloodedness and flight evolved at least three times in Tetrapoda. Within Synapsida, warm-bloodedness occurs in all mammals, and flight occurs in bats. Within Archosauria, warm-bloodedness and flight occurred in pterosaurs (which are now known to have been insulated), and in birds (Saurischia). Warm-bloodedness has been proposed for Ornithischia, suggesting to some that all Dinosauria might be characterized by warm-bloodedness (see Chapter 15). Warm-bloodedness and flight, however, are not fundamental characteristics of Archosauria; Figure 4.12 shows that archosaurs (e.g., crocodiles) were certainly primitively non-flying, and almost certainly cold-blooded animals. The cladogram thus shows that warm-bloodedness and flight evolved independently three times: once in bats, once in pterosaurs, and once in birds (or their dinosaurian near-relatives).

In dinosaurs, an erect stance consists of a suite of anatomical features with important behavioral implications. In particular, the head of the femur (thigh bone) is distinctly offset from the shaft. The head of the femur itself is barrel shaped (unlike the familiar ball seen in a human femur), so that motion in the thigh is largely restricted to forward and back, within a plane parallel to that of the body. The ankle joint is modified to become a single, linear articulation. This type of joint, termed a modified mesotarsal joint, allows movement of the foot only in a plane parallel to that of the body (forward and back). Note that again this situation differs from that in humans, in which the foot is capable of rotating movement. The upshot of these adaptations of stance is that all dinosaurs are highly specialized for cursorial (i.e., running, as in the "cursor" on a computer screen) locomotion. Dinosaurs are quintessentially terrestrial beasts (see Box 4.4).

This, then, completes our long trek through Vertebrata to find Dinosauria. We can now answer more completely the questions set forth at the beginning of this chapter:

- "How many times has warm-bloodedness evolved in the vertebrates?" It would appear that "warm-bloodedness" (an unfortunate term that we gladly abandon in Chapter 15) has evolved as many as three times: once in synapsids (in mammals or their near ancestors) and twice in ornithodirans (in birds and in pterosaurs). It is possible, however, that it evolved only twice and that the basal ornithodiran – that is, the organism that was the ancestor of all ornithodirans – was itself "warm-blooded." If so, then all ornithodirans were primitively "warm-blooded," which means that all dinosaurs must have been "warm-blooded," too. The evidence for and against this possibility will be explored in Chapter 15.
- "How many times has powered flight been invented by vertebrates?" We know that flight occurred once in the synapsids (in bats) and twice in Ornithodira (in birds and in pterosaurs). If flight had evolved in the ancestor of all ornithodirans, then flight would be primitive among the ornithodirans, and *all* ornithodirans should be flying creatures (at least primitively). Obviously, many among Dinosauria were not (a flying *Stegosaurus* or *T. rex* strains credulity), and so we can be relatively certain that powered flight evolved independently in three lineages of vertebrates. These relationships are shown in Figure 4.16.
- "Is a cow a fish?" This is discussed in Box 4.3, and the bottom line is that many of the features that we might intuitively use to group organisms are primitive characters and not suitable for recognizing the pattern of evolution.
- "Did all dinosaurs become extinct?" The negative answer to this question requires the fuller elaboration provided in Chapters 13, 14, 17 and 18. Finally, it should now be clear why a bird is closer to a crocodile than to a lizard. Birds and crocodiles are archosaurs, whereas lizards are not. The common ancestor of birds and crocodiles was some early archosaur, and thus it is only at the level of Diapsida that lizards, crocodiles, and birds are related. This, as noted previously, bodes ill for the traditional Reptilia: crocodiles, snakes, lizards, turtles, and the tuatara do not form a monophyletic group, unless birds are also included.

Important readings
Benton, M. J. (ed.) 1988. *The Phylogeny and Classification of the Tetrapods*. Vol. I. Systematic Association Special Volume 35A. Oxford, 377pp.

Butterfield, N. J. 1990. Organic preservation of non-mineralizing organisms and the taphonomy of the Burgess Shale. *Paleobiology* **16**, 272–286.

Carpenter, K. and Currie, P. J. (eds.) 1990. *Dinosaur Systematics.* Cambridge University Press, New York, 318pp.

Carrier, D. R. 1987. The evolution of locomotor stamina in tetrapods: circumventing a mechanical constraint. *Paleobiology* **13**, 326–341.

Gauthier, J. A. 1986. Saurischian monophyly and the origin of birds in Padian, K. (ed.) *The Origin of Birds and the Evolution of Flight.* Memoirs of the California Academy of Sciences no. 8, pp. 1–56.

Gauthier, J. A., Kluge, A. G. and Rowe, T. 1988. Amniote phylogeny and the importance of fossils. *Cladistics* **4**, 105–209.

Gould, S. J. 1989. *Wonderful Life: The Burgess Shale and the Nature of History.* W. W. Norton, New York, 347pp.

Gould, S. J. (ed.) 1993. *The Book of Life.* Ebury–Hutchinson, London, 256pp.

LaPorte, L. F. (ed.) 1974. *Evolution and the Fossil Record.* Scientific American Offprint Series. W. H. Freeman and Company, San Francisco, 222pp.

Moy-Thomas, J. A. and Miles, R. S. 1971. *Palaeozoic Fishes.* W. B. Saunders Company, Philadelphia, 259pp.

Padian, K. and Chure, D. J. (eds.) 1989. *The Age of Dinosaurs.* Short Courses in Paleontology no. 2. The Paleontological Society, University of Tennessee Press, Knoxville, TN, 210pp.

Prothero, D. R. and Schoch, R. M. (eds.) 1994. *Major Features of Vertebrate Evolution.* Short Courses in Paleontology no. 7. The Paleontological Society, University of Tennessee Press, Knoxville, TN, 270pp.

Schopf, J. W. 1983. *The Earth's Earliest Biosphere: Its Origin and Evolution.* Princeton University Press, Princeton, NJ, 543pp.

Stahl, B. J. 1985. *Vertebrate History: Problems in Evolution.* Dover Publications, New York, 604pp.

Wellenhfer, P. 1996. *The Illustrated Encyclopedia of Prehistoric Flying Reptiles.* Salamander Books, London, 192pp.

CHAPTER 5

The origin of the Dinosauria

No doubt we all think we know what a dinosaur is – just by saying the word "dinosaur" we imagine *Tyrannosaurus*, *Apatosaurus*, or *Triceratops*, or any of the large land-lubbers from the Mesozoic Era. Among the uninitiated, an ichthyosaur, a plesiosaur, or one of the other sea-going reptiles of the Mesozoic bestiary (none of them are dinosaurs) may come to mind. Worse yet, a few people envision as dinosaurs mere youngsters, such as the 100,000-year-old woolly mammoth.

It turns out that the question "Who are the dinosaurs?" is one and the same as "Where did dinosaurs come from?" because evolution is the driving force behind the history of life. Consequently, our goal in this chapter will be to consider precisely how dinosaurs fit into the history of life.

History of the Dinosauria

What dinosaurs are and how they came to be are questions pondered since the creation of the name by Sir Richard Owen just over 150 years ago. As we learned earlier in this book, Owen created the "Dinosauria" to encompass a group of exceedingly large, pachyderm-like reptiles from what he referred to as the Secondary Age (the Mesozoic Era).[1] *Iguanodon* (an ornithopod), *Megalosaurus* (a theropod), and *Hylaeosaurus* (an ankylosaur) were its first members.

From 1842 onward, membership in Owen's Dinosauria grew by leaps and bounds. Alongside activity in the field and laboratory, paleontologists struggled to figure out their relationships. Much of this was done at

Figure 5.1. The pelvis of the hadrosaurid *Prosaurolophus*. A splint of the pubis lies along the base of the ischium, exemplifying the ornithischian condition. (Photograph courtesy of the Royal Ontario Museum.)

the low and uncontroversial level of genera and families. For example, a scientist might ponder: "What is this creature? A new genus? A species of an existing genus? Maybe even a new family?"

Though the discovery and naming of new beasts might be superficially appealing, such activities are intellectually unadventurous. They do not take by the horns more significant questions about patterns of, and processes driving, the great ebb and flow of the biota through time. Two paleontologists who did, however, were Harry Govier Seeley, a vertebrate paleontologist at Cambridge University in England, and Friedrich von Huene, Dean of German dinosaur paleontology at the University of Tübingen, Germany.

Dinosaurs divided   In 1887, Seeley first recognized a fundamental division among dinosaurs. Ornithischia (*ornith* – bird; *ischia* – hip) were all those dinosaurs that had a bird-like pelvis, in which at least a part of the pubis runs posteriorly, along the lower rim of the ischium (Figure 5.1). Saurischia (*saur* – lizard) were those that had a pelvis more like a lizard, in which the pubis is directed anteriorly and slightly downward (Figure 5.2). This

---

1   The term "Secondary" came from a now outdated concept of how rocks were formed. In 1759, G. Arduino, an Italian mining engineer, developed a history of the rocks in northern Italy. He viewed these rocks as having been deposited by a retreating ocean. For him, the oldest rocks were designated Primary, and, as the seas receded, Secondary and Tertiary rocks were consecutively deposited. Thus Primary, Secondary, and Tertiary strata represented sequentially younger rocks. The scheme was further developed and elaborated by the eminent German naturalist A. G. Werner, who in 1787 published a highly influential history of the earth in which these terms were applied. Although, in both schemes, rock types and age relations were embodied in the terms, the term Tertiary (and another, Quaternary) remain with us today as age designations (the notion of a characteristic rock type has been abandoned, because most rock types can be produced and deposited at any time).

Figure 5.2. The pelvis of the ornithomimosaur *Ornithomimus*. The pubis is directed forward only, exemplifying the saurischian condition. (Photograph courtesy of the Royal Ontario Museum.)

pelvic distinction has held sway ever since, and has been bound up in the debate of the origin of birds (see Chapter 13).

That dinosaurs had one or the other kind of pelvis was of great importance in understanding the shape of the evolutionary tree of these animals, but in Seeley's hands it went considerably further. For it implied to him that the ancestry of Ornithischia and Saurischia was to be found separately and more deeply embedded within what was then called Thecodontia (a name also coined by Owen in 1859 and one that we met in Chapter 4 and will meet again in Chapter 13), a heterogeneous group of non-dinosaurian animals essentially linked by the fact that they were all considered to be primitive archosaurs. Therefore, Dinosauria to

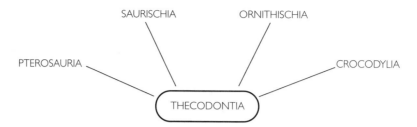

Figure 5.3. H. G. Seeley's evolutionary scenario of the origin of dinosaurs.

Seeley was not a natural group (Figure 5.3).

This major theme, that dinosaurs were diphyletic (i.e., having two separate origins), was continued well into the twentieth century, principally by Friedrich von Huene. Throughout his remarkably long career (he published actively from the early 1900s to the 1960s), von Huene's studies came to epitomize independent dinosaur origins. According to him and many who came after, dinosaurs had at least two and more probably three or four, separate origins from different stem archosaurs, generally called "thecodonts" (see Chapter 13). Certainly, saurischians and ornithischians had separate origins; after all, their hip structure was different. And among saurischians, surely sauropods and theropods had separate origins; after all, they *look* so different. Finally, among ornithischians, ankylosaur ancestry was also often sought separately within some thecodontian group.

Despite a wealth of dinosaur remains, North America never produced workers intent on the Big Picture of dinosaur phylogeny. Most of the ideas that dominated the history of dinosaur studies throughout much of the twentieth century came from studies of European dinosaur material. Following in the tradition of von Huene, British paleontologist A. J. Charig was the chief purveyor of dinosaur systematics in the 1960s and 70s. His work centered on primitive archosaurian taxa (he called them "thecodonts"), among them the semi-aquatic, carnivorous, and crocodile-like proterosuchids and erythrosuchids, as well as a number of early non-dinosaur groups. It was through these studies that Charig developed a detailed scheme of dinosaur phylogeny in which he traced the various dinosaur lineages to their separate origins among the various "thecodont" archosaurs. Dinosaurs to Charig and to many of his co-workers remained an unnatural group.

Dinosaurs united

The first inkling that things were changing as regards dinosaur relationships began in 1974 with a short publication in the British science journal *Nature*. In it, R. T. Bakker and P. M. Galton attempted to resurrect Dinosauria as a monophyletic taxon, using a number of skeletal features and speculations about dinosaurian physiology. Their analyses – which reunited not only saurischians and ornithischians, but also linked Aves within Dinosauria – met a great deal of resistance and, in a few cases, open hostility (see Chapter 13). Similar resistance was met by J. F. Bonaparte, who also speculated in 1976 that Dinosauria is a true clade.

Starting in 1984, however, cladistic analysis exploded onto the dinosaurian systematic scene. The thrust was four-fold: the disbanding of "Thecodontia," the origin of dinosaurs, the internal pattern of relationships within ornithischian and saurischian clades, and the relationships of birds to dinosaurs. In all four aspects, the changes wrought by cladistic analyses in our understanding of archosaurs in general and dinosaurs in particular were nothing short of revolutionary.

Of the four foci above, we discuss birds as dinosaurs in Chapter 13. Dinosaurian interrelationships will be treated within each of the successive taxonomic chapters (see chapters in Parts II and III). As for "Thecodontia," it has been fully dismembered. There are no unifying diagnostic features that are uniquely shared by all members of the group. For this reason, "Thecodontia" is not monophyletic and we have abandoned it.

Without "Thecodontia," what can be said about dinosaur origins? The multiple roots of Dinosauria might still exist and in fact may be more obvious now that the cover of "Thecodontia" has been blown. Are there particular archosaurian taxa that share a close relationship with one or the other (but not all) dinosaur groups? To cut to the chase, the short answer is "no."

How we know this to be the case is through the elegant 1986 cladistic work of J. A. Gauthier, now at Yale University. His research provided ample corroboration of a monophyletic Dinosauria, identifying upwards of 10 derived features uniting all dinosaurs with each other. Since then, numerous cladistic analyses of both new and old taxa have confirmed that dinosaurs share a single, most recent common ancestor, itself a dinosaur.

Dinosaurian monophyly

Dinosauria can be defined as consisting of all of the descendants of the most recent common ancestor of Saurischia and Ornithischia (Figure 5.4). Such a definition of Dinosauria is quite close to that of Owen, even

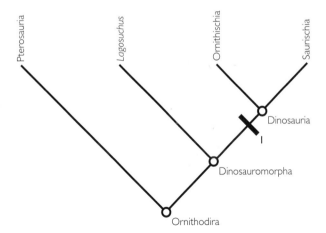

Figure 5.4. Cladogram of Ornithodira, showing the monophyly of Dinosauria. Derived characters include: at **1** loss of postfrontal, elongate deltopectoral crest on humerus, brevis shelf on ventral surface of postacetabular part of ilium, extensively perforated acetabulum, tibia with transversely expanded subrectangular distal end, ascending astragalar process on front surface of tibia.

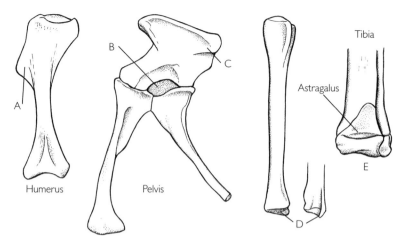

Figure 5.5. Some of the derived characters uniting Dinosauria. (A) Elongate deltopectoral crest on humerus; (B) brevis shelf on ventral surface of postacetabular part of ilium; (C) extensively perforated acetabulum; (D) tibia with transversely expanded subrectangular distal end; and (E) ascending astragalar process on front surface of tibia.

if Sir Richard was not necessarily thinking along such lines. As it turns out, dinosaurian monophyly is upheld by a host of derived features (Figure 5.5), so many that it is now hard to think of some members of the group as having origins elsewhere within Archosauria. These now include (among others) loss of a skull roofing bone – the postfrontal – that lies on the top of the head along the front margin of the upper temporal opening, an elongate deltopectoral crest on the humerus, an extensively perforated acetabulum, a tibia with a transversely expanded subrectangular lower end, and an ascending process of the astragalus on the front surface of the tibia.

The constituent members of Dinosauria – Ornithischia and Saurischia – each have a monophyletic origin (see introductory text to Parts II: Ornithischia and III: Saurischia) and the pattern of internal relationships of taxa within these clades is also reasonably well understood. Still, there are places, as we shall see, where controversy – and therefore intense research activity – still reigns.

Origins

How does one find the ancestor of a clade? Simply put, the hierarchy of characters in the cladogram specifies for us what features ought to be present in an ancestor. It is then simply a question of finding an organism that most closely matches the expected combinations of characters and character states. As we have seen, the likelihood of the very progenitor of a lineage being fossilized is nil; however, we can commonly find representatives of closely related lineages that embody most of the features of the hypothetical ancestor.

So far, we haven't yet identified who within Archosauria might have the closest relationship to Dinosauria, in part because the answer is not yet clear. According to Gauthier and K. Padian, pterosaurs – otherwise

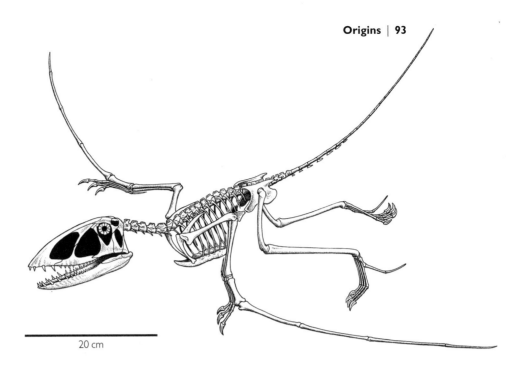

Figure 5.6. A candidate for closest relative to Dinosauria: Pterosauria as represented by *Dimorphodon*.

20 cm

highly modified for flight (Figure 5.6) – may be the closest archosaurian relatives to dinosaurs, together sharing four derived features (see Chapter 4). The clade of pterosaurs + dinosaurs then shares close relationship with a slender, long-limbed animal from the Middle Triassic of Argentina called *Lagosuchus* (*lago* – rabbit; *suchus* – crocodile) to form what Gauthier termed Ornithosuchia. *Lagosuchus* (Figure 5.7) was a very small (less than 1 m), long-legged (hence the name) bipedal carnivore or insectivore. The head of this creature is very poorly known. Nevertheless, few paleontologists would disagree that this relatively tiny creature embodies many of the features that were ancestral for all Dinosauria; the diminutive *Lagosuchus* is probably close to the ancestry of all the spectacular vertebrates encompassed within Dinosauria.

P. C. Sereno, in contrast, places *Lagosuchus*, as well as several other small, contemporary archosaurs (*Lagerpeton* (*erpet* – a creeper), *Pseudolagosuchus* (*pseudo* – false)), and *Marasuchus* (refers to the mara, a rabbit-like rodent that presently lives in Patagonia) as the closest dinosaurian relatives. This clade of dinosaurs and *Lagosuchus*, called Dinosauromorpha, shares a sigmoidal vertebral column in the neck region, a shortening of the forelimb, and several modifications of the ankle bones and of the metatarsals. On the strength of these features, it appears that Dinosauria shares closest relationship with archosaurs such as *Marasuchus*, *Lagosuchus*, *Lagerpeton*, and *Pseudolagosuchus*. More far-flung relationships of these dinosauromorphs are with pterosaurs, which, together with a few other taxa, make up Ornithodira – the bird-necks (see Chapter 4). It is only thereafter that these forms join with Crurotarsi to form Archosauria.

There is an interesting and perhaps surprising consequence of this phylogeny. With archosaurs like *Lagosuchus* closest to dinosaurian ancestry, apparently dinosaurs were primitively obligate bipeds. This

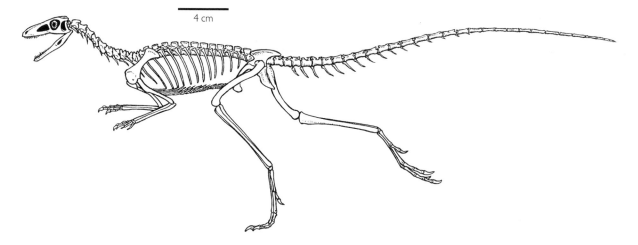

4 cm

Figure 5.7. Another candidate for closest relative to Dinosauria: *Lagosuchus*.

means that the earliest dinosaurs were creatures that were completely and irrevocably bipedal. Because the primitive stance for archosaurs is quadrupedal, and because Dinosauria is monophyletic, it follows that creatures such as *Triceratops*, *Ankylosaurus*, and *Stegosaurus* – in fact, all quadrupedal dinosaurs – must have evolved secondarily (or re-evolved) their quadrupedal stance. They must have (phylogenetically) got back down on four legs, as it were, after having been up on two. In fact, you can see the remnant of bipedal ancestry when you look at a stegosaur or a ceratopsian, in which the back legs are quite a bit longer than those at the front.

The rise of dinosaurs: superiority or luck?

We have now gone through the host of features providing the evidence for the phylogenetic relationships of dinosaurs to other archosaurs closest to them. What, if anything, does this tell us about the success of dinosaurs? That is, did these features confer any advantage to dinosaurs, making them superior to their less well-equipped contemporaries?

Before turning to how features might affect evolutionary success, let's first set the stage for the emergence of dinosaurs in the Triassic. From its outset, some 245 Ma, the Triassic was dominated on land by therapsids. Among these, the sleek, dog-like cynodonts were the chief predators, while the rotund, beaked and tusked dicynodonts were the most abundant and diverse of herbivores. From the middle and toward the end of the Triassic, these therapsids shared the scene with squat, plant-eating, and swine-like archosauromorphs called rhynchosaurs and a few carnivorous crocodile-like archosaurs. Yet toward the tail end of the Triassic, approximately 225 Ma, there was a great change in the fortunes of these animals. The majority of therapsids went extinct (one highly evolved group of therapsids, the mammals, of course survived), as did the rhynchosaurs, while only the dinosaurs and a few other taxa among archosaurs survived. And it was the dinosaurs that somehow rose to become the dominant terrestrial vertebrates, by which it is meant that they became the most abundant, diverse, and probably visible group of tetrapods.

Many of the character changes found higher and higher in the archosaurian evolutionary tree are those related to limb posture and locomotion. It is generally said that archosaurs began as sprawlers and ended up with either semi-erect stance (crocodilians) or fully erect posture (pterosaurs and dinosaurs (including birds)). A. J. Charig envisioned much of this change in limb posture happening along the lineages that produced the dinosaurs.[2] Thus changes at the hip, knee, and ankle enabled a fully erect, parasagittal posture in which the legs acted not only as support pillars when standing but also provided for longer strides and more effective walking and running ability (see Chapter 15). The archosaurs that had these new, "improved" features were then able, he supposed, to outcompete contemporary predatory therapsids for their food sources, the herbivorous therapsids and rhynchosaurs, both groups that lacked such parasagittal limb posture. The immediate descendants of these flashy new archosaurs – indeed some of the parasagittal-limbed forms themselves – were the dinosaurs. The inevitable consequence of such progressive improvements in limb posture, Charig argued, was the gradually changing pattern of faunal succession at the end of the Triassic. Therapsids lose, dinosaurs win – all by virtue of having better designed limbs and thereby more efficient terrestrial locomotion. We can call this and any other evolutionary advantage a competitive edge. The pattern of waxing and waning dominance (as one group supersedes another in evolutionary time) is called the wedge (Figure 5.8).

At nearly the same time, R. T. Bakker was making similar arguments about the competitive superiority of warm-bloodedness – endothermy – in dinosaurs (see Chapter 15 for an elaboration of Bakker's views on this subject). As we shall see, he believed that, instead of limbs, it was the achievement of internally produced heat that gave dinosaurs (or their immediate ancestors) a competitive edge over contemporary and supposedly cold-blooded therapsids and rhynchosaurs. The same conclusions apply: dinosaurs win, therapsids lose. And the truth of the competitive superiority of endotherms over ectotherms can be read directly from the pattern of faunal succession at the end of the Triassic. Again, we have a hypothesis that comes down to the competitive edge producing the wedge. But must the wedge be produced by the edge?

"No," says M. J. Benton (University of Bristol, England). Not that he's adverse to competitive edges and wedges, when and if they can be documented. It's just that they don't appear in the particular fossil record in question; that is, the Middle to Late Triassic fossil record of the earliest dinosaurs and their predecessors. In order for edges to lead to wedges, all

---

2 Things are complicated by the fact that Charig also viewed dinosaurs as being at least diphyletic; nevertheless the scenario we present is reasonably close to his general views on the evolution of dinosaur locomotion.

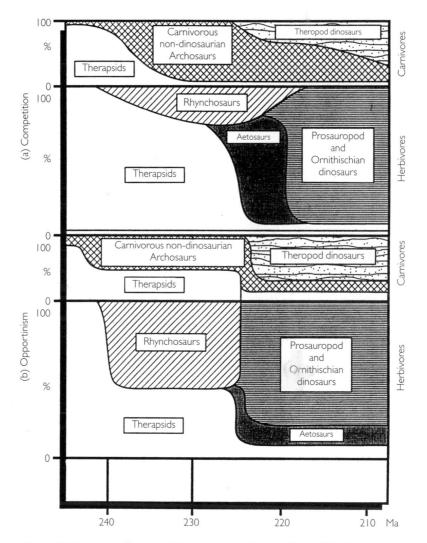

Figure 5.8. Two views of the rise of dinosaurs during the Late Triassic (Tr). (a) Gradual competitive replacement of synapsids, primitive archosaurs, and rhynchosaurs (both herbivores and carnivores) by herbivorous and carnivorous dinosaurs. (b) Rapid opportunistic replacement mediated by extinction.

of the players in the game have to be present to interact with each other. And according to Benton, they were not (note Figure 5.8). Instead, he suggests that the fossil record of the last part of the Triassic is marked by not one, but two mass extinctions. The first appears to have been the more extreme and ultimately most relevant to the rise of dinosaurs. This earlier Late Triassic extinction completely decimated rhynchosaurs and nearly obliterated dicynodont and cynodont therapsids, as well as several major groups of predatory archosaurs. Likewise, there is a major extinction in the plant realm. The important seed-fern floras (the so-called *Dicroidium* flora, which contained not only seed-ferns, but also horsetails, ferns, cycadophytes, ginkgoes, and conifers; see Chapter 16) all but went extinct as well, to be replaced by other conifers and bennettitaleans (large cycad-like plants). Dinosaurs appeared as the dominant land verte-brates only after this great disappearance of therapsids, archosaurs, and rhynchosaurs. Thus the initial radiation of dinosaurs, according to Benton, was done in an ecological near-vacuum, with mass extinction followed by opportunistic replacement. No competitive edge, because there was no competition.

That there was at least one, and more than likely two, mass extinc-tions at the end of the Triassic Period is uncontroversial; most researchers working on this part of earth history are now providing us with a better, more detailed picture of these extinctions. Naturally, one of the key questions is what might have caused these extinctions. Benton has suggested that the Late Triassic extinctions may be linked with climatic changes – the regions first inhabited by dinosaurs appear to have been hotter and more arid, a change from the more moist and equable – and thence to alterations in terrestrial floras and faunas. The abrupt extinction of the *Dicroidium* flora may have caused the extinction of herbivores specialized on them and thereby the predators feeding on the herbivores. According to Benton, far from being a long-term *competi-tive* takeover, this rapid loss of the dominant land-living vertebrates set the stage for the *opportunistic* evolution of dinosaurs.

The end-Triassic extinctions may have been driven by climatic shifts, but Columbia University's P. E. Olsen and colleagues believe that they have identified another, more dramatic forcing factor: asteroid impact. Like the end of the Cretaceous (see Chapter 18), which was marked by severe and abrupt extinctions of the earth's biota, the extinctions at the end of the Triassic also rank in the Extinction Hall of Fame. And like the mass extinction at the end of the Cretaceous, those at the end of the Triassic may also be allied with a "smoking gun": an impact crater close in age to the first of the Triassic extinctions. This impact structure, the Manicouagan crater in northern Quebec, Canada, is 70 km in diameter, large enough, Olsen believes, to have accommodated an asteroid with enough force to have done the job.

Changing climates may produce extinctions, but for catastro-phic events they've got nothing on asteroid impacts. Geologists are coming to believe that these must be among the worst and most wide-ranging disasters that can be suffered by global ecosystems. If it is true that an asteroid contributed to the first of the double extinctions that dev-astated the earth's biota at the end of the Triassic, then perhaps the archosaurian predecessors of dinosaurs may have just squeaked by – survivors, not because they were somehow superior to the presumed competition but because they happened to inherit a deserted earth. Instead of survival having been something intrinsic to dinosaur superiori-ty, it may have been that they simply had better luck. Ironically (as we shall see in Chapter 18), 160 million years later the tables again turned, and mammals inherited an earth *this* time deserted by the very dinosaurs who, by one means or another, had taken it from them 160 million years earlier.

Important readings

Bakker, R. T. 1975. Dinosaur renaissance. *Scientific American*, **232**, 58–78.

Bakker, R. T. and Galton, P. M. 1974. Dinosaur monophyly and a new class of vertebrates. *Nature*, **248**, 168–172.

Benton, M. J. 1983. Dinosaur success in the Triassic: a noncompetitive ecological model. *Quarterly Review of Biology*, **58**, 29–55.

Benton, M. J. 1984. Dinosaurs' lucky break. *Natural History*, **6** (84), 54–59.

Benton, M. J. 2004. Origin and relationships of Dinosauria. In Weishampel, D. B., Dodson, P. and Osmólska, H. (eds.), *The Dinosauria*, 2nd edn. University of California Press, Berkeley, pp. 7–20.

Charig, A. J. 1972. The evolution of the archosaur pelvis and hindlimb: an explanation in functional terms. In Joysey, K. A. and Kemp, T. S. (eds.), *Studies in Vertebrate Evolution*. Winchester, New York, pp. 121–155.

Charig, A. J. 1976. "Dinosaur monophyly and a new class of vertebrates": a critical review. In Bellairs, A. d'A. and Cox, C. B. (eds.), *Morphology and Biology of the Reptiles*. Academic Press, New York, pp. 65–104.

Gauthier, J. A. 1986. Saurischian monophyly and the origin of birds. *Memoirs of the Californian Academy of Science*, **8**, 1–55.

Olsen, P. E., Shubin, N. H. and Anders, M. H. 1987. New Early Jurassic tetrapod assemblages constrain Triassic–Jurassic tetrapod extinction event. *Science*, **237**, 1025–1029.

Sereno, P. C. 1991. Basal archosaurs: phylogenetic relationships and functional implications. *Journal of Vertebrate Paleontology*, **11** (suppl.), 1–53.

Sereno, P. C., Forster, C. A., Rogers, R. R. and Monetta, A. F. 1993. Primitive dinosaur skeleton from Argentina and the early evolution of Dinosauria. *Nature*, **361**, 64–66.

PART II

Ornithischia: armored, horned, and duck-billed dinosaurs

Ornithischia, one of the two major clades of dinosaurs, was first recognized by Harry Govier Seeley of Cambridge University, England, in 1887, but little could he have guessed at that time that ornithischians were such a diverse and anatomically wide-ranging group of closely related dinosaurs. Since then, we have learned an immense amount not only about the existence of new ornithischian taxa (e.g., Pachycephalosauria, Heterodontosauridae) but also about the detailed anatomy and evolutionary diversity of both earlier-known and newly discovered groups. Nonetheless, all the diversity and anatomical details do not cloud the issue of ornithischian monophyly: Ornithischia is monophyletic.

Diagnostic features for the entire clade abound. As we have already learned and as is clear from the name Ornithischia, the pelvis is reminiscent of that found in birds.[1] That is, at least a part of the pubis has rotated backward to lie close to and parallel with the ischium; this is called the opisthopubic condition (Figure II.1). The other landmark condition of ornithischians is the presence of a bone called the predentary, an unpaired, commonly scoop-shaped bone that caps the front of the lower jaws (Figure II.2) and is found nowhere else.

Although these are the *sine qua non* of ornithischians, there are numerous other derived features shared by these dinosaurs, including a toothless and roughened front tip of the snout, a narrow bone

Figure II.1. Left lateral view of the ornithischian pelvis as exemplified by *Stegosaurus*. Note that the pubic bone is rotated backwards to lie under the ischium in what is known as the opisthopubic condition. (Photograph courtesy of the Royal Ontario Museum.)

---

1 Ornithischian dinosaurs are confusingly called "bird-hipped"; but birds themselves belong to the "lizard-hipped" clade (Saurischia) of dinosaurs (see Chapter 13).

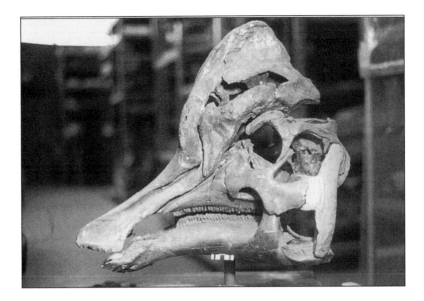

Figure II.2. Left lateral view of the skull of the lambeosaurine hadrosaurid *Corythosaurus*. Note the predentary bone capping the front of the lower jaw. (Photograph courtesy of D. B. Weishampel.)

(the palpebral) that crossed the outside of the eye socket, a jaw joint set below the level of the upper tooth row, cheek teeth with low crowns somewhat triangular in shape, at least five sacral vertebrae, and ossified tendons above the sacral region (and probably further along the vertebral column as well) for stiffening the backbone at the pelvis, among a host of others.

The basal split of Ornithischia is into the lone form *Lesothosaurus* and a clade termed Genasauria (*gena* – cheek) by P. C. Sereno (Figure II.3). The small, long-limbed herbivore *Lesothosaurus* (named for Lesotho, South Africa, where this dinosaur was discovered) was first christened by P. M. Galton in 1978 (Figure II.4). This Early Jurassic form had earlier been grouped with Ornithopoda, principally on the basis of primitive characters. With more recent cladistic analyses, however, it now appears to be fully ensconced as the most basal of known ornithischians.

In contrast, all remaining ornithischians – Genasauria – share the derived characters of muscular cheeks (as indicated by the deep-set position of the tooth rows, away from the sides of the face), a spout-shaped front to the mandibles, and reduction in the size of the opening on the outside of the lower jaw (the external mandibular foramen), among others.

Genasaurs subsequently split into Thyreophora and Cerapoda. Taking each in turn, Thyreophora (*thyreo* – shield; *phora* – bearer; a reference to the fact that these animals have dermal armor) – a name originally proposed by F. Nopcsa in 1915 – consist of those genasaurs in which the jugal (one of the cheek bones) has a transversely broad process behind the eye and there are parallel rows of keeled dermal armor scutes

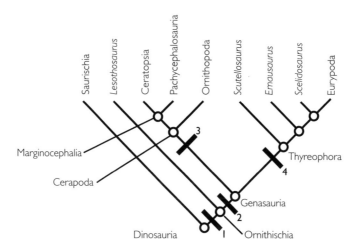

Figure II.3. Cladogram of Dinosauria, emphasizing relationships within Ornithischia; in particular those of *Lesothosaurus*, cerapodans, and basal thyreophorans. Derived characters include: at **1** opisthopubic pelvis, predentary bone, toothless and roughened tip of snout, reduced antorbital opening, palpebral bone, jaw joint set below level of the upper tooth row, cheek teeth with low subtriangular crowns, at least five sacral vertebrae, ossified tendons above the sacral region, small prepubic process along the pubis, long and thin preacetabular process on the ilium; at **2** emarginated dentition (indicating large cheek cavities), and reduction in the size of the opening on the outside of the lower jaw (the external mandibular foramen); at **3** gap between the teeth of the premaxilla and maxilla, five or fewer premaxillary teeth, finger-like anterior trochanter; at **4** transversely broad postorbital process of the jugal, parallel rows of keeled scutes on the back surface of the body.

on the back surface of the body. The most familiar thyreophorans are stegosaurs and ankylosaurs, but there are some more primitive, yet no less important, thyreophorans. At the base of the cladogram is *Scutellosaurus* (*scutellum* – small shield), which was described from Lower Jurassic rocks of Arizona by E. H. Colbert in 1981. *Scutellosaurus* was a

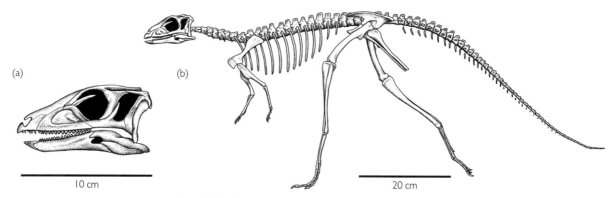

Figure II.4. Left lateral view of the skull (a) and skeleton (b) of the basal ornithischian *Lesothosaurus*.

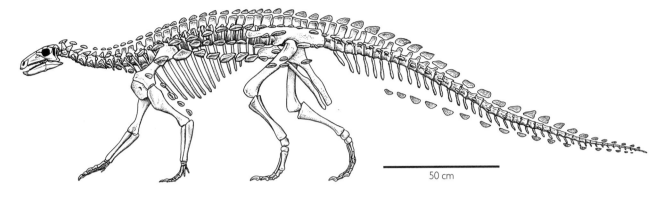

Figure II.5. Left lateral view of the skull and skeleton of *Scelidosaurus*.

small, gracile bipedal herbivore with a back covered by small, oval-shaped plates of dermal armor. Slightly higher up within Thyreophora, we come to one of the newest of these armor-bearers to be discovered, *Emausaurus* (EMAU is the abbreviation for Ernst-Moritz-Arndt-Universität), also from Lower Jurassic strata (but from Germany) by H. Haubold in 1991. Finally, we have another Early Jurassic thyreophoran, *Scelidosaurus* (*skelis* – limb). This primitive ornithischian comes from the southern coast of England. *Scelidosaurus* is a moderate-sized, heavily built herbivore whose limb morphology suggests that it may have been at times a quadruped and at others a biped (Figure II.5). First described by R. Owen in 1860, *Scelidosaurus* is the closest relative to the crowning thyreophoran clade, a taxon called Eurypoda (*eury* – wide; *pod* – foot).

Eurypodans consist of both stegosaurs (Chapter 6) and ankylosaurs (Chapter 7) and share as many as 20 important features not found in any of the more basal thyreophorans. These characters include short and stocky metacarpal and metatarsal bones, reduction in the large process on the shaft of the femur called the fourth trochanter, shortened post-acetabular process of the ilium (the part of the ilium behind the hip socket), and loss of a phalanx in digit IV of the foot.

As earlier mentioned, Thyreophora has as its sister-taxon Cerapoda (*cera* – horn), which are those genasaurs that share a diastema (or gap) between the teeth of the premaxilla and maxilla, five or fewer pre-maxillary teeth, and finger-like lesser trochanter (a process at the top of the femur), among other derived characters. Within this large group, we encounter Ornithopoda, who we will meet in Chapter 10, and Marginocephalia. Marginocephalians (*margin* – margin; *cephal* – head) – a group united by having a narrow shelf formed from both the parietal and squamosal bones that extends over the back of the skull, a reduced contribution of the premaxillary bone to the palate (this bone primitively forms an important part of the front of the palate), and a relatively short pubis – is formed of two well-known ornithischian taxa, pachycephalosaurs (Chapter 8) and ceratopsians (Chapter 9).

Important readings

Haubold, H. 1991. Ein neuer Dinosaurier (Ornithischia, Thyreophora) aus dem unteren Jura des nördlichen Mitteleuropa. *Revue de Paleobiologie*. **9**, 149–177.

Norman, D. B., Witmer, L. M. and Weishampel, D. B. 2004. Basal Ornithischia. In Weishampel, D. B., Dodson, P. and Osmólska, H. (eds.), *The Dinosauria*, 2nd edn. University of California Press, Berkeley, pp. 325–334.

Norman, D. B., Witmer, L. M. and Weishampel, D. B. 2004. Basal Thyreophora. In Weishampel, D. B., Dodson, P. (eds.), *The Dinosauria*, 2nd edn. University of California Press, Berkeley, pp. 335–342.

Sereno, P. C. 1986. Phylogeny of the bird-hipped dinosaurs (Order Ornithischia). *National Geographic Research*, **2**, 234–256.

CHAPTER 6

Stegosauria:
hot plates

Dumb as a dodo. Dumb as a dinosaur. Both of these expressions relate small brain size and unredeemable stupidity to an inexorable march towards extinction. We know extinction took those dim-witted unsophisticates, the dodos and dinosaurs, long before clever life forms like ourselves appeared. And in the pantheon of dumb and dumber, what is more celebrated than *Stegosaurus*? For not only did it boast a notorious "walnut-sized" brain, but allusions have been made to a second, larger brain in its hips.

In this chapter, we will take up the matter of stegosaur intelligence, or at least its measurable surrogate, brain size, along with several other aspects of stegosaur paleobiology and evolution (among them, phylogeny, thermoregulation, and display). But let us here lay to rest the nonsense of lack of evolutionary success among stegosaurs. These dinosaurs did very well. They rose in diversity from their origin sometime in or before the Middle Jurassic to an acme of at least seven species in the Late Jurassic. From then on, stegosaurs declined to one or two species in the Early Cretaceous, and to one by the early Late Cretaceous. During their time on earth, stegosaurs spawned upwards of a dozen species and stood their ground both behaviorally and evolutionarily, often in a unique and complex manner.

Stegosaurs rank today among the most familiar of all dinosaurs (Figure 6.1). As their name implies, stegosaurs (*stego* – roof; *saurus* – lizard) all have rows of special bones called osteoderms (*osteo* – bone; *derm* – skin), which develop into spines and plates along the neck, back, and tail. Other osteoderms – in the form of spines – develop over the shoulder blades. Once thought to have been positioned over the hips and called "parasacral," these spines are now referred to as a *parascapular* spines.

Beyond such features, stegosaurs can be further characterized by their quadrupedal limb posture, and long, thin, yet relatively small heads (containing those infamously small brains) with simple teeth that

Figure 6.1. (see p. 107) *Tuojiangosaurus*, a stegosaur from the Late Jurassic of Sichuan Province, China.

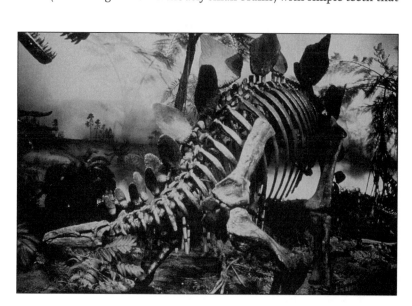

Figure 6.2. The best known of all plated dinosaurs, the North American *Stegosaurus* from the Late Jurassic. (Photograph courtesy of the Royal Ontario Museum.)

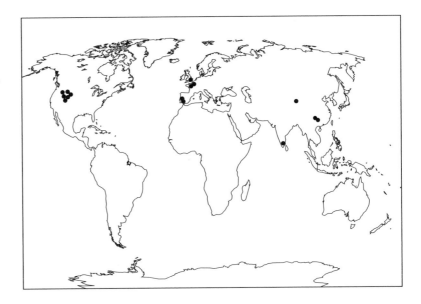

Figure 6.3. Global distribution of Stegosauria.

suggest a herbivorous feeding style. The forelimbs in all stegosaurs are short and massive, while the hindlimbs are long and columnar. The toes of both fore- and hindlimbs ended in broad hooves. The striking disparity in limb proportions gave these animals a profile that sloped strongly forward and downward toward the ground when the animal walked (Figure 6.2).

Stegosaurs spanned 3–9 m in length and weighed in at an estimated 300–1500 kg. By just about any standard, these were large animals, but still only of modest size when compared with many other dinosaur groups, particularly their herbivorous contemporaries, the gigantic sauropods.

The earliest of these so-called roofed or plated dinosaurs – *Huayangosaurus* (after Hua Yang Guo Zhi (the Chin Dynasty name for Sichuan)) – is known from the Middle Jurassic, some 170 Ma, of Sichuan in the People's Republic of China. Thereafter, the record of stegosaurs has its ebb and flow. Although never particularly diverse at any one time, the stegosaur clade nevertheless takes on a somewhat cosmopolitan distribution (Figure 6.3). For example, such forms as *Dacentrurus* (*da* – very; *kentron* – spiny; *oura* – tail) and *Lexovisaurus* (after the Lexovii people of Europe) are known throughout western Europe during the Middle and Late Jurassic (165–150 Ma), *Stegosaurus* and *Hesperosaurus* come from the western USA (Late Jurassic; 150 Ma), *Kentrosaurus* (*kentro* – prickly, referring to its abundant spines) has been found in Tanzania (also from the Late Jurassic), and *Paranthodon* (*para* – around; *antho* – flower; *odon* – tooth, referring to the shape of the teeth) hails from South Africa (from the Late Jurassic–Early Cretaceous transition). To these are added as many as four more Chinese forms (*Chialingosaurus* (from Chialing, China), *Chungkingosaurus* (from Chungking, China), *Tuojiangosaurus* (from

Tuojiang), *Wuerhosaurus* (from Wuerho)) and a final taxon, *Dravidosaurus* (from Dravidandu (the southern part of India)), the last-known stegosaur (early Late Cretaceous). None of these animals ever saw the debacle, whatever it was, that befell the dinosaurs at the very end of the Mesozoic.

Stegosaur lives and lifestyles

It's obvious that there is nothing like a stegosaur alive today, but their occurrence in the fossil record and their individual skeletons give us a few clues about their "wildlife biology."

Stance and gait

The general body plan of any of the stegosaur taxa gives the impression that these animals were more inclined toward the plodding than the sprinting end of the getting-around spectrum (Figures 6.1, 6.2, and 6.4). In all but *Huayangosaurus*, the forelimbs were short and stocky and the hindlimbs, elongate as they were, had very short lower limb segments (the tibia, fibula, and foot) compared with the length of the femur. All of these features conspire to suggest that these animals were not built for great speed. Instead, they give a sense of great stability and strength, almost as elephants do today.

As we have already noted, stegosaurids have much longer hindlimbs than they do forelimbs. Indeed, this disparity in limb length is a shared, derived feature of that group of stegosaurs that excludes the more basal *Huayangosaurus*. Not surprisingly, it has some important biomechanical consequences for stegosaur locomotion, which have been studied by R. A. Thulborn of the University of Queensland. As pointed out by Thulborn, the maximal stride for the short forelimbs is limited to the arc available from all the (short) limb elements. In the same way, the long hindlimbs ensure that hindlimb stride was also proportionately long. As a consequence, stegosaurs had a problem: at the same cadence (the rate of feet hitting the ground), the hindlimbs would have been able to outrun the forelimbs. At high speeds, therefore, the rear end of the animal would have overtaken its head: not a likely scenario.

Thulborn observed that the problem could be avoided in two ways: (1) running stegosaurs could have drawn up the forelimbs up from the ground (i.e., temporarily been bipedal while running) or (2) stegosaur locomotion may have been limited to a slow walking gait. The shear bulk of a stegosaur makes the first option unlikely (although see below for discussion of bipedal stegosaurs in a feeding context) and the second option far more tenable. So far, we have only anatomical observations by which to judge walking or running speeds in these animals; no footprints or trackways are yet known. Nevertheless, it appears likely that the pace of stegosaur life was leisurely, on the order of 6–7 km/h maximal speed. Among contemporaries, stegosaurs may have been slower than the multi-tonne sauropods or such large ornithopods as *Camptosaurus*. More critical to health and longevity, stegosaur running speeds were much less than those of both the small and large theropods that must have hunted – and devoured – them. Indeed, small theropods such as *Ornitholestes* could have run circles around stegosaurs (what *Ornitholestes* would have done when the stegosaur was caught is quite another matter; see Chapter 12).

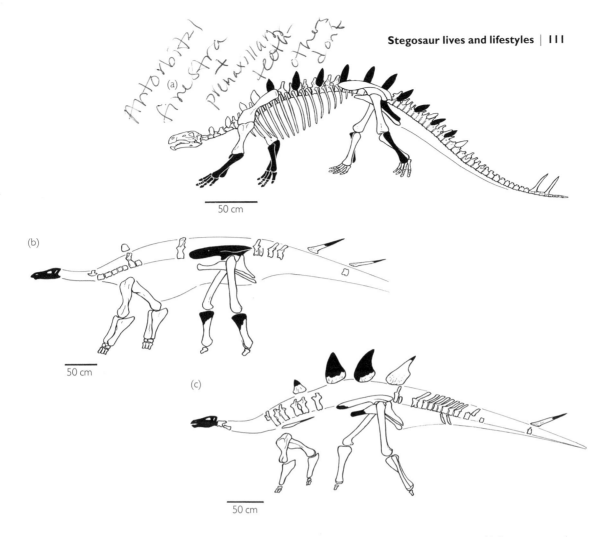

[handwritten annotations: Antorbital fenestra + premaxillary teeth other dont]

50 cm

50 cm

50 cm

Figure 6.4. Left lateral views of the skeletons of (a) *Huayangosaurus,* (b) *Dacentrurus,* and (c) *Lexovisaurus.*

Dealing with mealing

Obviously even these running speeds were not of much importance to a hungry stegosaur, for these animals were herbivorous. Their general body form (large abdominal region for a capacious gut), small head, toothless snout, and simple blunt teeth send a clear message that these animals ate plants. Echoing these features is the thought that stegosaurs had muscular cheeks and strong chewing muscles, both of which help enormously in the way feeding on foliage is conducted, whether you're a dinosaur or a mammal.

For a stegosaur – as for all vertebrates with jaws – the business end of feeding begins at the front of the jaws. Therefore, it might be possible to learn a great deal about how these animals procured their food (Figure 6.5). Here a horn-covered beak or rhamphotheca (*rhampho* – beak; *theca* – cup or sheath) covered the fronts of both the upper jaw (the premaxilla) and the predentary bone of the lower jaw, similarly to rhamphothecae seen in modern turtles and birds. The rhamphothecae of the upper and lower jaws were probably sharp edged, but not hooked like those of a bird-of-prey or a snapping turtle. By bringing these sharp edges together,

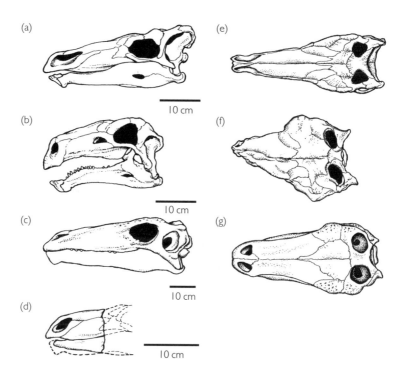

Figure 6.5. Left lateral views of the skull of (a) *Stegosaurus*, (b) *Huayangosaurus*, (c) *Tuojiangosaurus*, and (d) *Chunkingosaurus*. Dorsal views of the skull of (e) *Stegosaurus*, (f) *Huayangosaurus*, and (g) *Tuojiangosaurus*.

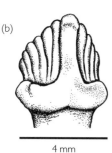

Figure 6.6. Inner views of an upper tooth of (a) *Stegosaurus* and (b) *Paranthodon*.

the rhamphotheca became quite an effective apparatus for cropping or stripping foliage from plants. Only in *Huayangosaurus* – where the upper rhamphotheca was relatively small – was this cropping ability aided by the premaxillary teeth (remember, in those other stegosaurs where we have some evidence, these teeth are absent).

Once enough food was in the mouth, chewing began. But how this was accomplished is hard to tell. The chewing (i.e., cheek) teeth of stegosaurs were relatively small, simple, triangular (Figure 6.6), and not tightly appressed. These teeth apparently did very little grinding, because they lack regularly placed, well-developed wear surfaces on their crowns. Furthermore, the jaw musculature doesn't seem to have been particularly powerful: the coronoid process on the lower jaw was low, lending little mechanical advantage to the jaw musculature.

In view of the small teeth and weak oral musculature, one might look elsewhere for the means by which stegosaurs broke up their food. Where to look would be in the stomach, for it is known that some animals (birds and crocodilians, as well as sauropodomorphs and psittaco-saurs; see Chapters 9 and 11) use stones (gastroliths) to fragment food within the muscular part of the stomach. While it might seem logical that stegosaurs, with their seemingly inadequate chewing apparatus, might have had such gastroliths, skeletons of these animals have never been found with associated gastroliths.

Still, to judge from the inset position of the teeth, stegosaurs probably possessed cheeks. Here is an animal which has relatively weak adaptations for handling food at the mouth but is equipped with cheeks, an adaptation normally associated with sophisticated oral food-processing. Did these animals process food extensively in their mouths, or did they not? This apparent contradiction in stegosaurs has yet to be fully explained. The coexistence in stegosaurs of simple, irregularly worn teeth, a large gut capacity, cropping rhamphothecae, and cheeks all conspire to make stegosaurs unique, yet still poorly understood herbivores.

Even if we cannot understand all of the "hows and whys" of stegosaur food processing, perhaps we can get a glimmer of some more obvious aspects of feeding styles in these plant-eaters. First, recall that the snout is quite narrow in all stegosaurs, suggesting a fair degree of selectivity in the food these animals were seeking. Secondly, remember that all but *Huayangosaurus* have a great disparity in size between the forelimb and hindlimb, yielding the characteristic downturn of the vertebral column in the shoulder and neck region. From these observations, it is obvious that the head must have been brought very close to the ground, most likely near the 1 m level. And if the head was naturally positioned low to the ground, it is highly likely that these stegosaurs were principally low-browsing animals, consuming in great quantities the leaves and perhaps even succulent fruits and seeds of such ground-level plants as ferns, cycads, and other herbaceous gymnosperms.

The Mesozoic world of the low-browsers was not filled just with stegosaurs. It is very likely that these plant-eaters competed with a variety of other forms, principally any number of the smaller, agile, bipedal euornithopods. Although also confined to feeding on low-level vegetation, the latter may have fed on more fibrous, chewable leaves and other portions of the plants.

But were stegosaurs confined to such low-browsing, highly competitive levels? Not necessarily so, suggest some paleontologists. For example, R. T. Bakker has argued that some forms, particularly *Stegosaurus*, were able to rear up on their hindlimbs in order to forage at higher levels, perhaps even into the crowns of tree. This kind of posture is similar to that generally assumed by kangaroos and occasionally by elephants. But stegosaurs may have gone one better on the elephant. First, with the center of gravity near the hips, the hindlimbs would already be supporting nearly 80% of the weight of the body. In addition, the strong, flexible tail would have been able to act as a third "leg" to form a tripod under the animal as it attempted to rear up. Should this kind of behavior have been utilized by stegosaurs, then these animal may have been able to feed on leaves quite high up in the trees (perhaps as much as 3 m for *Kentrosaurus*, and 6 m for *Stegosaurus*). Perhaps, like elephants, which assume a rearing, bipedal stance to feed on very high leaves for only very short periods of time, stegosaurs also foraged in the crowns of contemporary trees, but they may have done so for longer periods of time thanks to the aid of their strong, supporting tails.

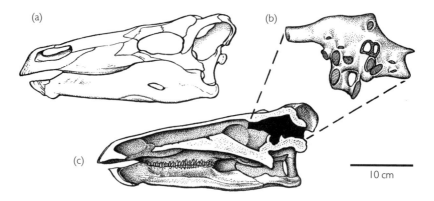

Figure 6.7. Mold of the inside of the braincase of *Stegosaurus* and its silhouette imposed on the skull. (a) Left profile; (b) braincast; (c) section through skull with braincase *in situ*.

Brains

Whatever the aspirations of the dinosaurs to deep thought, stegosaurs cannot have been ranked among their crowning intellects. With very small brains (an estimated 0.001% body weight), it is certainly well justified to consider stegosaurs to be near the bottom of the gray-matter scale. Only sauropods and perhaps ankylosaurs had proportion-ately smaller brains (see Chapters 7 and 11).

The means by which these measures are obtained and their meanings interpreted are exceedingly important and worth describing. How, after all, do we know the size and shape of a dinosaur brain? The brain is soft tissue, and commonly decomposes long before the process of fossilization can begin. As many workers have shown, however, casts can be obtained of the interior of the braincases of fossil vertebrates. Latex is painted onto the inside of a well-preserved, three-dimensional braincase (one that was not crushed during fossilization). When the latex has dried (and is flexible), it can be peeled off the inside of the braincase, and pulled through the foramen magnum ("big hole"), the opening through which the spinal cord enters the skull . The result is a three-dimensional cast of the region occupied by the brain (Figures 6.7 and 15.6). Such casts give some inkling about the shapes and sizes of brains. Dinosaur brains can be no larger than the volume of these casts, but they were probably significantly smaller. Observations made of the brains of living lizards, snakes, and crocodilians show that these brains take up less room within the braincase than do those of mammals or birds. Researchers have long thought that the brains of non-avian dinosaurs should similarly be smaller than the entire volume of the braincase and have provided cor-rection factors, which are reflected in calculations of dinosaurian gray matter.

Starting with general studies of fossil "brains" by Harry Jerison of the University of California at Los Angeles and, further, more detailed research by J. A. Hopson of the University of Chicago (see Box 15.4), it is now clear that brain size in vertebrates scales negatively allometrically with body size (technically, to the 0.67 power). What this means is that, as animals become bigger, either ontogenetically or phylogenetically, their brains also become large, but at a rate not equal to their size. Even for large-brained mammals such as ourselves, as we reach maturity and stop growing, our brains have grown less than have our bodies. The same pattern applies to shrews and elephants, as well as to lizards and crocodilians. And presumably to extinct dinosaurs as well, for we know that the same pattern is found in the living variety: birds.

Hopson used this relationship between estimated dinosaurian brain size (calculated from the expected brain size of lizards, snakes, and crocodilians scaled up to dinosaur size) and dinosaurian body size to make comparisons with measured brain size in actual dinosaurs. Among stegosaurs, casts of the braincase are available only for *Kentrosaurus* and *Stegosaurus*, both of which are reasonably similar in shape and size. The brains of these two stegosaurs are relatively long, slightly flexed, and above all else *small*. The only aspect of the stegosaur brain that appears to have been somewhat large is its extraordinary olfactory bulbs, those portions of the brain that provide the animal with its sense of smell. Why this region should have been so large, we can only guess: perhaps sense of smell was raised to a high art in stegosaurs. Otherwise, the mental and sensory faculties of stegosaurs – as we have already mentioned – were not particularly well developed. Clearly for stegosaurs, animals that assuredly had a relatively unhurried life style and possibly a relatively uncomplicated range of behaviors, these brains were adequate.

Yet the small-brained stegosaur defied even those nay-sayers who spoke about dinosaurs as dullards, because stegosaurs apparently looked elsewhere in order to embellish their anomalous nervous systems. It was O. C. Marsh who, over a century ago, observed the very much enlarged canal – upwards of 20-times the volume of the brain – that accommodated the spinal cord in the hip region. Here began the legend of the dinosaur with two brains: a diminutive one in the head, which monitored and controlled not quite enough; and another in the sacral region, which handled the balance of neural functions. The subject of poetry (Box 6.1), this sacral enlargement has caused a great deal of controversy. Some suggested, as did Marsh, that the enlarged spinal cord "covered" for what the standard brain could not manage, while others hypothesized that it was the logical enlargement for the hindlimb and tail region in an animal that emphasized both, for reasons of locomotion (hindlimb domination) and protection (the tail). While this latter explanation is more in keeping with comparative neuroanatomy, neither has garnered much acceptance in the paleontological literature. Thus a new idea about the stegosaur sacral enlargement has created much excitement.

In 1990, E. B. Buchholz (formerly Giffin) of Wellesley College re-examined the issue of enlargements of the neural canals in sacral vertebrae in dinosaurs. She did so with an eye for how large the neural canals should be, given the size of the spinal cord, in much the same way that Hopson and earlier Jerison did with brain size. She found out that the sacral enlargements in stegosaurs (also in sauropods, as it turns out; see Chapter 11) were more than large enough to accommodate the

BOX 6.1

The poetry of dinosaurs

Dinosaurs have been subjects of numerous poems, limericks, and other bits of doggerel, virtually since the time of their earliest discovery. Most have centered around their enormity and putative lack of brain power and social graces, with few recent efforts to balance such dismal views.

The most famous dinosaurian poem celebrates the mental claims of *Stegosaurus*, in particular the cerebral gymnastics supplied by its double brains, the standard issue and the one in its rear end. The piece, by Bert L. Taylor, a columnist in the 1930s and 40s for the *Chicago Tribune* goes like this:

> Behold the mighty dinosaur,
> Famous in prehistoric lore,
> Not only for his power and strength
> But for his intellectual length.
> You will observe by these remains
> The creature had two sets of brains –
> One in his head (the usual place),
> The other at his spinal base.
> Thus he could reason *a priori*
> As well as *a posteriori*
> No problem bothered him a bit
> He made both head and tail of it.
> So wise was he, so wise and solemn,
> Each thought filled just a spinal column.
> If one brain found the pressure strong
> It passed a few ideas along.
> If something slipped his forward mind
> 'Twas rescued by the one behind.
> And if in error he was caught
> He had a saving afterthought.
> As he thought twice before he spoke
> He had no judgement to revoke.
> Thus he could think without congestion
> Upon both sides of every question.
> Oh, gaze upon this model beast,
> Defunct ten million years at least.

As a poetic counterpoint to the range of Mesozoic intelligensia, we also provide some thoughts by John Maynard Smith, evolutionary biologist *extraordinaire* at the University of Sussex in Brighton, England, entitled *The Danger of Being too Clever.*

> The Dinosaurs, or so we're told
> Were far too imbecile to hold
> Their own against mammalian brains;
> Today not one of them remains.
> There is another school of thought,
> Which says they suffered from a sort

> Of constipation from the loss
> Of adequate supplies of moss.

> But Science now can put before us
> The reason true why Brontosaurus
> Became extinct in the Cretaceous.
> A beast incredibly sagacious
> Lived & loved & ate his fill;
> Long were his legs, & sharp his bill,
> Cunning its hands, to steal the eggs
> Of beasts as clumsy in the legs
> As *Proto-* & *Triceratops*
> And run, like gangster from the cops,

> To some safe vantage-point from which
> It could enjoy its plunder rich.
> Cleverer far than any fox
> Or STANLEY in the witness box
> It was a VERY GREAT SUCCESS.
> No egg was safe from it unless
> Retained within its mother's womb
> And so the reptiles met their doom.

> The Dinosaurs were most put out
> And bitterly complained about
> The way their eggs, of giant size,
> Were eaten up before their eyes,
> Before they had a chance to hatch,
> By a beast they couldn't catch.

> This awful carnage could not last;
> The age of Archosaurs was past.
> They went as broody as a hen
> When all their eggs were pinched by men.
> Older they grew, and sadder yet,
> But still no offspring could they get.
> Until at last the fearful time, as
> Yet unguessed by *Struthiomimus*
> Arrived, when no more eggs were laid,
> And then at last he was afraid.
> He could not learn to climb with ease
> To reach the birds' nests in the trees,
> And though he followed round and round
> Some funny furry things he found,
> They never laid an egg – not once.
> It made him feel an awful dunce.
> So, thin beyond all recognition,
> He died at last of inanition.

> MORAL
> This story has a simple moral
> With which the wise will hardly quarrel;
> Remember that it scarcely ever
> Pays to be too bloody clever.

normal expansion of the nerve cord that passes through this region, giving off nerves to the hindlimbs and continuing backwards to the tail. While this situation was to be expected, the canal – especially at the front of the sacrum – was proportionately larger than might have been expected in order merely to control the legs and tail; that is, it was much too large for just accommodating nerves. Only in living birds is this extra-enlarged situation known to occur, and in these forms there is a structure called a glycogen body that takes up the extra space. The function of the glycogen body is somewhat enigmatic, but it is thought to supply glycogen (one of the carbohydrate reserves of the body) to the nervous system, where it is used in the synthesis of specialized kinds of nervous tissue. By comparison, then, it is possible that the inordinately large expansion of the neural canal of the sacrum in stegosaurs also housed a glycogen body, much as in all modern birds. It did not function as a second brain and in all likelihood did not provide space for more than the usual amount of nerves to the hindlimbs and tail. A glycogen body seems to be the most reasonable interpretation for the enlarged canal in the sacrum of stegosaurs, but what of the function of this body? Time and more research on those living animals that have them – birds – will surely answer this on-going question.

Social lives of the enigmatic

With so much anatomical detail known, albeit some of it somewhat controversial, it is perhaps surprising that the social behavior, reproduction, and growth and development of stegosaurs remains enigmatic. Simply put, we don't have much of an idea about the social behavior of stegosaurs, nor much about their life histories. For example, no nests, isolated eggs, eggshell fragments, nor hatchling material is yet known for any stegosaur. In fact, only a few juvenile and adolescent stegosaur specimens are thus far known for *Dacentrurus*, *Kentrosaurus*, *Lexovisaurus*, and one of the *Stegosaurus* species. Among fully adult individuals, it appears that there is some sexual dimorphism; that is, differences between the sexes. This shows up in, of all places, the number of ribs that contribute to the formation of the sacrum. Whether it is the male or the female that has the greater number of ribs is anybody's guess. Other ways in which the differentiation of the sexes may have been manifested are unknown, but it might be that sexual dimorphism would be found in the size and shape of the spines and/or plates, if only we had better samples. Within species, little is known about the degree of sociality among stegosaurs. The mass accumulation of disarticulated, yet associated *Kentrosaurus* material from Tendaguru (see Box 11.1) provides us with the notion that – in this stegosaur at least – there was some degree of herding behavior, either seasonal or perennial. In other species, however, we have no such information – they may have been solitary creatures or gregarious. Perhaps the lack of evidence of complex, social behavior is reflected in the lack of large brains in these creatures. The fossil record is simply silent on this issue, so far.

Figure 6.8. The skeleton of *Kentrosaurus*, a spiny stegosaur from the Late Jurassic of Tanzania. (Photograph courtesy of the Institut und Museum für Geologie und Paläontologie, Universität Tübingen.)

Plates and spines

Whether or not stegosaurs engaged in herding, there are still some features of these animals that assuredly reflect their paleobiology: the spines and plates. As we have learned, the majority of stegosaurs have at least one row of osteoderms along the dorsal margin of each side of the body. And these osteoderms generally have the form of spines, either long and drawn-out spikes (as in *Kentrosaurus*; Figure 6.8) or as blunt cone-like affairs (as in *Tuojiangosaurus*). Only in *Stegosaurus* do the majority of osteoderms become shaped into often large, leaf-shaped plates. In all cases, at the end of the tail were pairs of long spines. All of these plates and/or spines, regardless of their position on the body, did not articulate directly with the underlying neural spines of the vertebrae, but instead were embedded in the skin (Figure 6.9).

Once thought to have solely protective and defense importance, the spines and plates of stegosaurs have begun to take on complex behavioral significance, relating not only to these aforementioned functions, but also to display and thermoregulation. On the one hand, enlarged spines and particularly plates would have provided these animals with a much larger, more formidable appearance. On the basis of studies of *Stegosaurus*, Bakker has argued strongly that the osteoderms were mobile at their bases and hence able to rotate from a folded down to an erect position, giving these animals a greater degree of protection from, and deterrence to, predators. Others, including V. de Buffrénil and colleagues at the University of Paris VII, disagreed. Working on the microscopic structure of the plates of *Stegosaurus*, they instead suggest that these

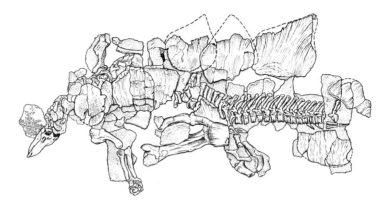

Figure 6.9. Diagram of one of the best skeletons of *Stegosaurus* as it was found in the field. Note that the plates do not articulate directly with the vertebrae.

plates were not particularly mobile, but rose nearly vertically on the back. In this position, the plates would have been quite useful in intraspecific display, as well as in thermoregulation.

As noted by Russian evolutionary biologist L. S. Davitashvili, by N.B. Spassov of the National Natural History Museum of Bulgaria, and most recently by Buffrénil and his colleagues, the shapes and patterns of plates and spines in stegosaurs are nearly always species specific. That is, more often than not we use the shape and size of the plates and spines as taxonomic identification markers (derived features, if you will) for stegosaur species. In all cases, osteoderms were arranged for maximal visual effect and thus have made their greatest impact during lateral display. If this interpretation of one of the functions of stegosaur osteo-derms is correct – and wouldn't it be great if we could identify sexual dimorphism in their size, shape, or placement pattern? – then it is likely that individual stegosaurs would have used these structures not only to tell each other apart, but also to gain dominance in territorial disputes and/or as libido-enhancers during the breeding season.

Yet our picture of the functional significance of these osteoderms, at least for plates, is incomplete. Here we turn to work by J. O. Farlow and colleagues. Again arguing that plates would have offered poor protection from predators, these researchers analyzed these structures as heat radiators and/or solar panels, for regulating body temperature. The plates of *Stegosaurus* are covered with an extensive pattern of grooves, while the insides are filled with a honeycomb of channels (Figure 6.10). These external grooves and internal channels most likely formed the bony walls for an elaborate network of blood vessels.

With such a rich supply of blood from adjacent regions of the body, Farlow and colleagues argued, these plates were well designed to cool the body by dissipating heat as air passed over them, or to warm the body by absorbing solar energy. In this model, the fine control for cooling and heating would be the regulation of blood flow to plates.

As a test of these ideas, a crude model of a stegosaur with plates along its back was placed in a wind tunnel and temperature changes were

Figure 6.10. Lateral view of one of the dermal plates of *Stegosaurus*. Note the great number of parallel grooves, presumably conveying blood vessels across the outer surface of the plate. (Photograph courtesy of the Royal Ontario Museum.)

monitored by a thermocouple placed inside the "animal." How these plates were arranged on the model becomes critical to their thermoregulatory performance. In the case of this experiment, two patterns were tested.

First, plates were positioned as symmetrical pairs. This arrangement proved to provide more than adequate heat dissipation. When the plates were placed in alternating positions, however, they functioned much better to dissipate internal heat loads. Thus we might expect that, on biophysical grounds, the plates of *Stegosaurus* were arranged in alternating pairs. Yet is there any evidence as to their actual arrangement, regardless of their relative ability to absorb or dump heat?

Suggestions of how stegosaur plates were positioned date back almost to the inception of studies of the group. For it was Marsh, in 1891, who first advocated not pairs of plates but a single row down the back. This hypothesis lasted only a decade, when F. A. Lucas replaced Marsh's idea with the suggestion that the back of *Stegosaurus* supported paired but staggered rows of plates. This hypothesis was strongly supported by C. W. Gilmore in his stegosaur monograph. Not to be outdone, R. S. Lull of Yale argued that Lucas and Gilmore were only half right: there were indeed two rows of plates, but they were symmetrically placed (he argued that the two Smithsonian workers had erred in their staggered interpretation, because the plates had slipped after death). Recently, K. Carpenter of the Denver Museum of Natural History described new *Stegosaurus* material that confirms this interpretation of paired plate arrangement.

One final point on the plates of *Stegosaurus*. It is very interesting that juveniles appear not to have developed osteoderms on their backs. If we are right in presuming that plates functioned in display and/or thermoregulation, then their absence in small, sexually immature individuals may reflect already adequate ability in dumping or absorbing heat and the relevance of looking big and sexy as maturity is reached.

We are left with the functional significance of the long, pointed parascapular spines and the pairs of terminal tail spikes. In both cases, these appear to be the main means of defense for stegosaurs. How the flanks of the animal were protected – even with the parascapular spines – is far from clear, and perhaps stegosaurs were particularly susceptible to attack on this account. Whatever the effectiveness of the parascapular spines, the tail spikes could surely have been swished from side to side on the powerful tail in order to injure or deter rear assaults by predators. Older reconstructructions show these spikes as pointing primarily upwards; however, more recently Carpenter has produced strong arguments that the spikes actually splayed out to the sides, producing a much more effective defensive weapon.

Plate and spine function can, and should, be integrated with stegosaur phylogeny (see below) to understand their evolutionary history within the group. For example, we might imagine that, basally in stegosaur history, portions of the neck and all of the back and tail of these animals were covered with backwardly projecting pairs of spines. This primitive condition may have been tied to a protective function, but it's equally likely that spines across the back and tail may have been part of the visual display complex that began with the small plates situated along the neck. Perhaps these two functions were not mutually exclusive; they may have been used to bluff and establish dominance both intra- and interspecifically.

Whatever the principal function(s) of this primitive stegosaur condition might have been, these large, spiny projections could not help collecting some degree of solar radiation and/or providing an avenue for dumping heat, even if they were initially covered with horn. As osteoderms became more plate-like down the back and at the same time show more evidence of vascularity, display and thermoregulatory functions become more important during stegosaur phylogeny. Finally, with the acquisition of a full complement of plates in *Stegosaurus*, there is a commitment of all osteoderms except the terminal spines to both display and thermoregulation.

Seen in this way, the story of stegosaur evolution is one of changing osteodermal patterns and functions. Moving from spiny osteoderms, whose primary functions are defensive and possibly display to a condition where display and thermoregulation are inextricably linked via plate-like osteoderms, stegosaurs appear to have mastered the business of looking and feeling hot.

The evolution of Stegosauria

Stegosauria is a monophyletic clade of ornithischian dinosaurs (Figure 6.11), defined as the common ancestor of *Huayangosaurus* and *Stegosaurus* and all of the descendants of this common ancestor. As shown by P. C. Sereno and Z.-M. Dong, the stegosaur clade can be diagnosed on the basis of a number of important features. These include back vertebrae with very tall neural arches and highly angled transverse processes (Figure 6.12), loss of ossified tendons down the back and tail, a broad and plate-like flange (the acromion process) on the forward surface of the shoulder blade, large and block-like wrist

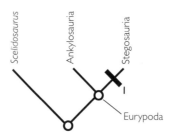

Figure 6.11. Cladogram of "higher" Thyreophora, emphasizing the monophyly of Stegosauria. Derived characters at **1** include: back vertebrae with very tall neural arches and highly angled transverse processes, loss of ossified tendons down the back and tail, a broad and plate-like acromion process, large and block-like wrist bones, elongation of the prepubic process, loss of the first pedal digit, and loss of one of the phalanges of the second pedal digit, and a great number of features relating to the development of osteoderms, and formation of long spines of plates from the shoulder toward the tip of the tail.

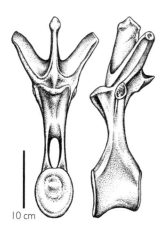

10 cm

Figure 6.12. Front and left lateral view of one of the back vertebrae of *Stegosaurus*. Note the great height of the neural arch.

bones, several changes in the pelvis including elongation of the prepubic process on the pubic bone, reduction of the toes on the feet (loss of the first digit, and loss of one of the toe bones of the second digit), and a great number of features relating to the development of osteoderms.

The sister group of Stegosauria consists of ankylosaurs, the armored dinosaurs best known from the Cretaceous Period of North America and central and eastern Asia, but also with a modest, but important, fossil record from the Jurassic of England (see Chapter 7). Together, stegosaurs and ankylosaurs make up a monophyletic group known as Eurypoda (*eury* – broad; *poda* – feet), a clade established through important work on the phylogeny of Ornithischia by Sereno. Eurypoda is united on the basis of many important shared, derived features, among them special bones that fuse to the margins of the eye socket, loss of a notch between the quadrate (the bone that buttresses the lower jaw against the skull roof) and the back of the skull, and enlargement of the forward part of the ilium (the upper bone of the pelvis). Together with a few other ornithischians – *Scelidosaurus*, *Emausaurus*, and *Scutellosaurus* – Stegosauria and Ankylosauria make up a diverse clade called Thyreophora (see Chapter 5).

Stegosaurs have been among the most resistent of dinosaur clades to cladistic analysis and, in the first edition of this book, we presented a very tentative evaluation of the shape of the stegosaur tree. Fortunately, newer research by P. M. Galton and P. A. Upchurch has brought our understanding of stegosaur interrelationships into line with that of other groups of dinosaurs.

Within Stegosauria, the basal split is between *Huayangosaurus* on the one hand and remaining species on the other (Figure 6.13). This divergence took place sometime before the latter half of the Middle Jurassic. *Huayangosaurus* itself has a number of uniquely derived features, among them the oval depression between the premaxilla and maxilla, the great number of cheek teeth, and the small horn on the top of the skull roof.

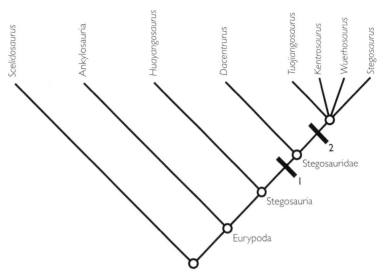

Figure 6.13. Cladogram of Stegosauria, with Ankylosauria and *Scelidosaurus* as successively more distant relatives. Derived characters include: at **1** large antitrochanter, long prepubic process, long femur, absence of lateral rows of osteoderms on the trunk; at **2** widening of the lower end of the humerus, an increase in femoral length, an increase in the height of the neural arch of the back and tail vertebrae.

Yet *Huayangosaurus* lacks a number of important derived features that are shared by a more inclusive group of stegosaurs. This group, called Stegosauridae, appears to be monophyletic and diagnosable on the basis of a large antitrochanter, a long prepubic process, a long femur (as compared with the length of the humerus), among other features.

Within Stegosauridae, *Dacentrurus* represents the most basal form. The remainder of Stegosauridae, termed Stegosaurinae, includes *Stegosaurus*, *Wuerhosaurus*, *Kentrosaurus*, and *Tuojiangosaurus*. Stegosaurines are united by a variety of modifications of the vertebral column, widening of the lower end of the humerus, and an increase in femoral length.

Unfortunately, it's unclear at present which taxon is more closely related to another within Stegosaurinae. This work, along with the inclusion of other taxa – *Monkonosaurus*, *Lexovisaurus*, *Chungkingosaurus*, and *Hesperosaurus*, among others – still remains to be done.

Stegosaurs meet history: a short account of their discovery

The earliest discoveries of stegosaurs were made in England in the early 1870s, but these went largely unrecognized because of the fragmentary nature of the bones. Some were studied by the great anatomist and vertebrate paleontologist, Sir Richard Owen, of the British Museum (Natural History), while others became the subject of papers by Harry Govier Seeley of Cambridge University and Franz Baron Nopcsa of Vienna, in the early twentieth century. Originally given names that have since been changed for a variety of reasons, these first European stegosaurs now bear the monikers *Dacentrurus* (named in 1902 by F. A. Lucas) and *Lexovisaurus* (named by R. Hoffstetter in 1957).

Dacentrurus, from Upper Jurassic beds of England, France, and Portugal, is known from only partial skeletons that include portions of the vertebral column, most of the forelimb, parts of the pelvis, a nearly complete hindlimb, and a few osteoderms. These last consist of small, plate-like elements from possibly over the front part of the trunk and two more spike-like elements near the sacrum and onto the tail. Unfortunately, no skull has yet been discovered. The vertebrae of *Dacentrurus* have relatively low neural arches and transverse processes. The forelimbs are very stocky, with massive distal elements (radius and ulna), yet relatively small, blocky feet. In the pelvis, there is a large process (the antitrochanter) on the lateral face of the ilium, which probably served as an muscle attachment site. In addition, the prepubic process of the pubis, a strip of the pubis that points toward the head of the animal, is long and the openings in the dorsal sacral shield, through which some of the spinal nerves reach the back, are relatively small. Compared with the rest of the hindlimb, the femur is proportionately very long.

The Middle Jurassic *Lexovisaurus* was a medium-sized stegosaur, ranging upward of 6 m in length. Originally named "*Omosaurus*" by Owen, it was found that the name was already occupied by another animal. Accordingly, it was renamed *Lexovisaurus*. As far as we can tell, it bore on its back a series of very large, thin, plate-like osteoderms that look very much like those of *Stegosaurus*. The position of these osteoderms is somewhat problematic, since they were not found in place with other skeletal remains of *Lexovisaurus*. Nevertheless, given their size and shape, they were probably positioned toward the posterior portion of the trunk and above the sacrum. At the end of the tail was at least one pair of terminal spines. In addition to these dorsally placed osteoderms, there were elongate parascapular spines over the shoulders; these spines are much larger than those of other stegosaur taxa within the clade.[1] From what is known of the vertebral column of *Lexovisaurus*, the neural arches of the back and tail vertebrae appear to be quite tall. The openings on the dorsal aspect of the sacral shield are small and the antitrochanter on the ilium is large. Finally, the forelimbs of *Lexovisaurus* are strongly built and the hindlimbs are long and massive.

Meanwhile, as these European stegosaurs were being discovered, across the Atlantic Ocean, things dinosaurian were beginning to brew – better yet, boil. For it was toward the end of the 1870s that a veritable wealth of dinosaur remains – including many virtually complete skeletons – were discovered at Como Bluff, Wyoming, and many other localities in the western USA. *Stegosaurus* was one of these new dinosaurs, a product of the Great Dinosaur Rush that was responsible for the early explorations of the North American Western Interior by universities and the federal government. These remains – along with trainloads of other kinds of newly discovered dinosaurs – made their way east, either to Othniel Charles Marsh of Yale University or to Edward Drinker Cope of

---

1 The length of these spines, however, seems to have varied. The specimen from France has quite large ones; the one from England does not. *Stegosaurus* has no parascapular spines at all.

Philadelphia. It was Marsh who bagged *Stegosaurus*. In Box 6.2, we recount the bizarre story of the Marsh–Cope feud, which fueled the greatest dinosaur rush in history.

Marsh published many papers on the remains of *Stegosaurus*, having at his disposal a number of specimens including a nearly complete skeleton. We know this North American stegosaur, the largest member of the clade (up to an enormous 9 m in length), from a wealth of material, all of it from the famous Morrison Formation of the North American Western Interior. The Upper Jurassic Morrison Formation has yielded not only *Stegosaurus*, the State Fossil of Colorado, but also a wealth of other dinosaurs, among them *Allosaurus*, *Ceratosaurus*, *Camptosaurus*, and *Diplodocus*. *Stegosaurus* has an elongate, narrow snout, which bears a roughened margin. In life, this was covered by a rhamphotheca. The external nares are relatively large, as is the eye socket. Around the upper rim of this socket are three roughened bones, the supraorbitals, which in life probably supported a horn-covered brow-ridge. Along the side of the face is a relatively small antorbital fenestra (remember, this is one of the shared, derived features of Archosauria). In the jaws are lots of teeth, as many as two dozen, in both upper and lower jaws. These teeth were relatively simple and triangular, with denticles on the fore and aft surfaces of the crown. Were there cheeks in *Stegosaurus*? We think so, since the tooth rows are deeply set in from the sides of the face.

Postcranially, the body slopes considerably upward from the head to the shoulders, in large part because of the very long hindlimbs (with proportionately long femora) when compared with the forelimbs. In *Stegosaurus*, as in many other stegosaurs, there are small openings on the dorsal aspect of the sacrum, a large antitrochanter on the ilium, and a long prepubic process. The upsloping of the back is especially accentuated by the ever-increasingly larger, plate-like osteoderms that adorn the dorsal margin of the body. This series decreases in size down the tail, culminating in two pairs of terminal tail spines.

From the 1870s and 80s, until the early twentieth century, little was heard about stegosaurs, either in terms of new discoveries or revisionary studies. In fact, it was not until 1914, more than 30 years after stegosaurs were recognized in Europe and the USA, that an attempt was made to study comprehensively any of these animals. This work, by C. W. Gilmore of the U.S. National Museum of Natural History emphasized the important finds from the western USA, but it also provided some details of the European stegosaurs. We do not know, but perhaps Gilmore had some regrets in publishing his study just when he did, for in the next year new and abundant stegosaur material, which rivalled anything yet found from the USA was discovered in Tanzania. Named *Kentrosaurus* by E. Hennig (University of Tübingen, Germany – no relation of Willi Hennig, the father of cladistic analysis) in 1915, this stegosaur represented just one of the many spectacular finds from the famous Tendaguru expeditions led by Werner Janensch of the Humboldt Museum of Natural History in Berlin in the early 1910s (Box 11.1). At 5 m long, *Kentrosaurus* is known from hundreds of bones, most of them found

BOX 6.2

Nineteenth century dinosaur wars: boxer versus puncher

One of the strangest episodes in the history of paleontology was the extraordinarily nasty and personal rivalry between late nineteenth century paleontologists Edward Drinker Cope and Othniel Charles Marsh (Figure B6.2.1). In many respects, it was a boxer versus puncher confrontation: the mercurial, brilliant, highly strung Cope versus the steady, plodding, bureaucratic Marsh. Their rivalry resulted in what has been called the "Golden Age of Paleontology," a time when the richness of the dinosaur faunas from western North America first became apparent; when the likes of *Allosaurus*, *Brontosaurus*, and *Stegosaurus* were first uncovered and brought to the world's attention. But the controversy had its down side, too. Who were these men, and why were they at each other's throat?

Cope was a prodigy, one of the very few in the history of paleontology. By the age of 18, he had published a paper on salamander classification. By 24, he became a Professor of Zoology at Haverford College, Philadelphia. Blessed with independent means, within four years he had moved into "retirement" (at the grand old age of 28) to be near Cretaceous fossil quarries in New Jersey. He quickly became closely associated with the Philadelphia Academy of Sciences, where he amassed a tremendous collection of fossil bones that he named and rushed into print at a phenomenal rate (during his life he published over 1,400 works). He

Figure B6.2.1. The two paleontologists responsibile for the Great North American Dinosaur Rush of the late nineteenth century. (a) Edward Drinker Cope of the Philadelphia Academy of Natural Sciences; and (b) Othniel Charles Marsh of the Yale Peabody Museum of Natural History. (Photographs courtesy of the American Museum of Natural History.)

was capable of tremendous insight, made his share of mistakes, and was girded with the kind of pride that did not admit to errors.

Marsh, 9 years older than Cope, was rather the opposite, with the exception that he, too, eventually rushed his discoveries into print almost as fast as he made them (some thought faster) and that he, too, did not dwell upon his mistakes. Marsh's own career started off inauspiciously; with no particular direction,

highly disarticulated. What we know of the skull is limited to portions of the braincase and skull roof. Hence we know nothing about the size or shape of the antorbital fenestra or external naris. For the rest of the skeleton, however, there is a wealth of information. The vertebrae all have high neural arches that supported as many as six pairs of plate-like osteoderms over the neck and anterior region of the back. Beyond, there were approximately five pairs of large spine-like osteoderms over the pelvis and down the tail. In addition, over each shoulder was a long, parascapular spine. Like *Stegosaurus*, the shoulder blade of *Kentrosaurus* is extremely large and is thought to have supported large muscles that helped to flex and extend the rest of the equally powerful forelimb. Muscular though they may have been, these forelimbs are relatively small when compared with the hindlimbs, as in all stegosaurs.

BOX 6.2 (cont.)

he reasoned that if he performed well at school he could obtain financial support from a rich uncle, George Peabody. This turned out to be perhaps the most significant insight in Marsh's life: Marsh persuaded Peabody to underwrite a natural history museum at Yale University (which to this day exists as the Yale Peabody Museum of Natural History), and, while he (Peabody) was at it, an endowed chair for Marsh at the Museum.

The careers of the two paleontologists moved in parallel: Marsh slowly publishing but acquiring prestige and rank, Cope frenetically publishing paper after paper. At first, there was no obvious acrimony, but this changed when Marsh apparently hired one of Cope's New Jersey collectors right out from under him. Suddenly, the fossils started going to Marsh instead of Cope. Then, in 1870, Cope showed Marsh a reconstruction of a plesiosaur, a long-necked, flippered marine reptile. The fossil was unusual to say the least, and Cope proclaimed his findings in the *Transactions of the American Philosophical Society*. Marsh detected at least part of the reason why the fossil was so unusual: the head was on the wrong end (the vertebrae were reversed). Moreover, he had the bad manners to point this out. Cope, while admitting no error, attempted to buy up all the copies of the journal. Marsh kept his.

Cope sought revenge in the form of correcting something that Marsh had done. The rivalry ignited, and the battle between the two spilled out into the great western fossil deposits of the Morrison

Formation. Both hired collectors to obtain fossils, the collectors ran armed camps (for protection against each other's poaching), and between about 1870 and 1890, eastbound trains continually ran plaster jackets back to New Haven and Philadelphia. There Marsh and Cope rushed their discoveries into print, usually with new names. The competition between the two was fierce, as each sought to "out-science" the other. Discoveries (and replies) were published in newspapers as well as scholarly journals, lending a carnival atmosphere to the debate. Because Philadelphia and New Haven were not that far apart by rail, it was possible for one of the men to hear the other lecture on a new discovery, and then rush home that night and describe it and claim it for himself. Because many of the fossils in their collections were similar, it was easy to do and each accused the other of it.

Both Cope and Marsh eventually aged, and, in Cope's case, his private finances dwindled. Moreover, a new generation of paleontologists arose that rejected the Cope–Marsh approach, believing, not unreasonably, that it had caused more harm than good. Both men ended their lives with somewhat tarnished reputations. History has viewed the thing a bit more dispassionately, and it is fair to state that the result ultimately was an extraordinary number of spectacular finds and a nomenclatorial nightmare that has taken much of the past 100 years to disentangle.[1]

_____
1 A superlative account of this rivalry can be found in A. J. Desmond's book, *The Hot Blooded Dinosaurs*.

The hindquarters must have been powerful: the antitrochanter is quite large and the prepubic process is long. On the dorsal surface of the sacrum, the openings that formerly accommodated spinal nerves to this region of the back have been obliterated.

Kentrosaurus is not the only stegosaur to have been discovered on the African continent. Originally from South African beds that date from the Late Jurassic–Early Cretaceous transition, there is an elongate snout fragment replete with very simple-looking upper teeth that has been called *Paranthodon*. This animal has had a rather checkered history. Originally thought to have been another sort of extinct reptile when it and a larger jumble of associated bones was described in 1876, Owen named it *Anthodon*. All of this material was restudied in 1912 by R. Broom of South Africa and yet again by Nopcsa in 1929. Both of these researchers noted the stegosaur affinities of the snout from the rest of the remains pertaining to *Anthodon*, and it was Nopcsa who provided the specimen with its final, proper name of *Paranthodon*.

Although China has proven to be a treasure trove of stegosaurs, it was not until 1959 that the first stegosaur, *Chialingosaurus*, was discovered and described by China's premier vertebrate paleontologist, C.-C. Young. Known from a partial skull and skeleton, *Chialingosaurus* is a medium-sized form from the Late Jurassic of Sichuan Province. Its skull is high and narrow, and there are fewer teeth in the jaws than in other stegosaurs. The limbs of *Chialingosaurus* are relatively slender and on the back are small, plate-like osteoderms, but their extent and position are not yet known.

At the time of the discovery of *Chialingosaurus*, Young did not realize what riches China – particularly central Sichuan – possessed. As it turns out, China has yielded five different species, with probably more on the way. These stegosaurs exploded onto the dinosaur scene in the 1970s and 80s, mostly through the considerable work of Dong Zhiming and co-workers. In close succession, these paleontologists discovered and named *Wuerhosaurus* from Xinjiang and *Tuojiangosaurus*, *Huayangosaurus*, and *Chungkingosaurus* from Sichuan.

Described first in 1973, the 7 m long *Wuerhosaurus* is the youngest of the stegosaurs from China (from the Tugulu Group of Xinjiang Autonomous Region; Early Cretaceous in age). We know very little about this stegosaur because of the fragmentary nature of its remains. What we do know, though, is that the forelimb was small, the hips relatively flaring, and the antitrochanter large. The openings on the upper surface of the sacrum are completely closed. A few osteoderms have also been recovered; these plate-like elements are large, long, and low, but their position on the back is unknown.

The late 1970s saw the description of *Tuojiangosaurus*, another Late Jurassic form from Sichuan Province. Along the neck, back, and forward half of the tail of this stegosaur are 17 pairs of osteoderms. Over the neck and forward region of the back, these elements are circular and plate-like. From there back, the osteoderms become quite large (up to 75 cm in height), conical, and even spine-like. Down the tail, these spines become smaller, but at the end are two or four pairs of quite elongate spines. The skull is extremely long and low, with very small external nares. Like *Stegosaurus*, three supraorbitals rim the upper margin of the eye socket. The front of the upper jaws is edentulous (without teeth) and the pre-maxillary margin is well ridged for attachment of a rhamphotheca. There is no evidence of an antorbital fenestra on the side of the face; it apparently closed over during the evolution of the group. Both upper and lower jaws house as many as 27 cheek teeth, the record for stegosaurs. Further back in the skeleton, the holes in the dorsal sacral shield are relatively small. Both fore- and hindlimbs are massively built, but there is still the great disparity in size between the two. On the ilium is a large antitrochanter and the prepubic process of the pubis is elongate.

The early 1980s saw the discovery and description of *Huayangosaurus*, a 4 m long, Middle Jurassic stegosaur from Sichuan. This stegosaur has a relatively short, deep skull. Forming the front of the tooth row are seven conical premaxillary teeth, a condition that appears to be absent in other stegosaurs. There is a small, oval depression between the premaxilla and maxilla, behind which are many cheek teeth; more, apparently, than in any other thyreophoran. Above these teeth is a triangular antorbital fenestra. Surrounding the upper margin of the eye socket are three supraorbitals. Unique to *Huayangosaurus* among stegosaurs, there is a modest horn on the top of the skull roof just above the eye. A small external mandibular fenestra occupies the lateral face of the lower jaw.

Unlike other stegosaurs, the vertebral column of *Huayangosaurus* does not appear so strongly flexed upwardly behind the head nor are the neural arches as high. There are large openings on the dorsal surface of the sacrum, the ilium has a very small antitrochanter, the prepubic process is short, all limbs are robustly constructed, and the femur is relatively short (as compared with the remainder of the hindlimb). In addition, the toes of the hindfoot appear to have been slender and somewhat more splayed apart than in other stegosaurs. Finally, running down the back of the neck and along the dorsal aspect of trunk are two rows of osteoderms, each shaped like a blunt, conical spine. There is also a lateral row of oval, keeled osteoderms across the trunk, absent in other stegosaurs but much like those in more primitive thyreophorans. At the tip of the tail, there are two pairs of tall spines, while over each shoulder is a sharp parascapular spine.

Within a year of the description of *Huayangosaurus*, Dong and colleagues unleashed *Chungkingosaurus*, one of the smallest stegosaurs yet discovered (3–4 m long). Hailing from the Late Jurassic of Sichuan, the skull of *Chungkingosaurus* is relatively high, with large external nares. Unfortunately, we have no idea about the nature of the antorbital fenestra because only the front of the snout is preserved. Behind the head, the vertebral column is reasonably complete. The neural arches of these vertebrae are relatively short and the holes in the sacral shield are large, in both cases again more as in *Huayangosaurus* than in other stegosaurs. The osteoderms over the neck and most of the back are large, thick, and plate-like. These are gradually replaced by spine-like osteoderms on the tail. At the far end of the tail are four pairs of terminal tail spines. The pelvis is somewhat reminiscent of *Huayangosaurus* in having large openings on the dorsal side of the sacrum. In contrast, however, the lateral face of the ilium bears a medium-sized antitrochanter and the prepubic process is long.

In the midst of this flurry of new stegosaur discoveries in China, P. Yadagiri and K. Ayyasami described a very important early Late Cretaceous stegosaur, *Dravidosaurus*, in 1979. This, the sole stegosaur from India and the only stegosaur of Late Cretaceous age, is known from extremely fragmentary material. Some scientists have argued that *Dravidosaurus* is not a stegosaur at all; instead they believe it to be a fully sea-going plesiosaur. Nevertheless, it is still regarded as a stegosaur by P. M. Galton, of the University of Bridgeport, Connecticut, one of the world's foremost specialists on this group. For example, the eye socket is rimmed by two supraorbital bones and there are a few large, triangular, thick-based osteoderms found over the sacrum (which is otherwise very poorly preserved). *Dravidosaurus* possessed unusually shaped tail spines, each with uniquely expanded mid-shafts.

The most recent stegosaur discovery was reported in 2001 by Carpenter and colleagues. Named *Hesperosaurus* (*hesper* – western), this form hails from the Morrison Formation (Late Jurassic in age) from Wyoming and appears to be a basal member of Stegosaurinae.

Although *Hesperosaurus* marks the last stegosaurian discovery anywhere in the world, its description has not been the last word on stegosaurs. Seventy-five years after Gilmore provided the first comprehensive study of these animals, Galton presented a more modern synthesis of available information on these plated dinosaurs in 1990. Within two years of this work, another important study appeared that analyzed the phylogenetic relationships of *Huayangosaurus* to other stegosaurs. This research, by Sereno and Dong, is the first to present *Huayangosaurus* within a cladistic framework of Stegosauria. The same treatment was provided by Carpenter and co-workers for their new stegosaur *Hesperosaurus*. Finally, comprehensive work by Galton and Upchurch, reviewing the stegosaur fossil record, including anatomy, phylogeny, and paleobiology, constitutes the most recent work on this group. From these sorts of studies, we are beginning to have a much clearer sense of stegosaurs as living, evolving denizens of the Mesozoic.

Important readings

Alexander, R. McN. 1989. *Dynamics of Dinosaurs and Other Extinct Giants*. Columbia University Press, New York, 167pp.

Bakker, R. T. 1986. *Dinosaur Heresies*. William Morrow, New York, 481pp.

Bakker, R. T. 1987. Return of the dancing dinosaur, In Czerkas, S. J. and Olsen, E. C. (eds.), *Dinosaurs Past and Present*. Natural History Museum of Los Angeles County, pp. 38–69.

Buffrenil, V. de, Farlow, J. O. and de Riqles, A. 1986. Growth and function of *Stegosaurus* plates: evidence from bone histology. *Paleobiology*, **12**, 459–473.

Carpenter, K. (ed.) 2001. *The Armored Dinosaurs*. Indiana University Press, Bloomington, IN, 512pp.

Davitashvili, L. S. 1961. [*The Theory of Sexual Selection*]. Izdatel'stov Akademia Nauk SSSR, Moscow (in Russian), 537pp.

Desmond, A. J. 1975. *The Hot-Blooded Dinosaurs*. The Dial Press, New York, 238pp.

Galton, P. M. and Upchurch, P. A. 2004. Stegosauria. In Weishampel, D. B., Dodson, P. and Osmólska, H. (eds.), *The Dinosauria*, 2nd edn. University of California Press, Berkeley, pp. 343–362.

Giffin, E.B. 1991. Endosacral enlargements in dinosaurs. *Modern Geology*, **16**, 101–112.

Gilmore, C. W. 1914. Osteology of the armored Dinosauria in the United States National Museum, with special reference to the genus *Stegosaurus. Bulletin of the U.S. National Museum*, **89**, 1–136.

Hopson, J. A. 1977. Relative brain size and behavior in archosaurian reptiles. *Annual Reviews of Ecology and Systematics*, **8**, 429–448.

Hopson, J. A. 1980. Relative brain size in dinosaurs – implications for dinosaurian endothermy. In Thomas, R. D. K. and Olson, E. C. (eds.) 1980. *A Cold Look at the Warm-Blooded Dinosaurs*. AAAS Selected Symposium no. 28, pp. 278–310.

Jerison, H. J. 1973. *Evolution of the Brain and Intelligence*. Academic Press, New York, 482pp.

Sereno, P. C. 1986. Phylogeny of the bird-hipped dinosaurs (Order Ornithischia). *National Geographic Society Research*, **2**, 234–256.

Sereno, P. C. and Dong, Z.-M. 1992. The skull of the basal stegosaur *Huayangosaurus taibaii* and a cladistic analysis of Stegosauria. *Journal of Vertebrate Paleontology*, **12**, 318–343.

Spassov, N. B. 1982. The "bizarre" dorsal plates of *Stegosaurus*: ethological approach. *Comptes rendus de l'Académie Bulgare Sciences*, **35**, 367–370.

Thulborn, R. A. 1982. Speeds and gaits of dinosaurs. *Palaeogeography, Palaeoclimatology, and Palaeoecology*, **38**, 227–256.

CHAPTER 7

Ankylosauria: mass and gas

Nature has a penchant for inventing armor. Today, we have armadillos, pangolins, turtles, and pill-bugs. And from the past, there were glyptodonts (large, squat, relatives of sloths and armadillos that died out about 10,000 years ago) and glyptosaurs (lizards now extinct for 40 million years), and the long-gone trilobites. But nobody did it like ankylosaurs, the Mesozoic armor-plated, spiked-shouldered experts who elevated hunkering self-defense to an art.

As their name implies, ankylosaurs (*ankylo* – fused; *saurus* – lizard) were encased in shell-like dermal armor (Figure 7.1), a pavement of bony plates and spines – each embedded in skin and interlocked with adjacent plates – that formed a continuous shield across the neck, throat, back, and tail. In many cases, it covered the top of the head and cheeks. The armament varied from species to species, and is used to identify particular ankylosaurs, even from scrappy material.

Under the armor, the ankylosaur body is round and broad, and clearly designed for housing a large gut to digest food. Likewise, the head was low and broad, and equipped with simple, leaf-shaped teeth for pulverizing whichever plants an ankylosaur chose to feed on (Figure 7.2).

These stocky quadrupeds were rarely over 5 m in length, although some (such as *Ankylosaurus*) ranged upward of 9 m. Like stegosaurs, the limbs were short, with the hindlimb exceeding the length of the forelimb by 50%. These proportions are in keeping with the general construction of the body and the presumed slow pace of the animals.

Ankylosauria was at its acme toward the end of the Cretaceous, but somewhat like the proverbial trail of bread crumbs, there are evolutionary links – often based on pretty fragmentary remains – back to the Middle Jurassic. Fortunately, there have been concerted efforts to understand the anatomy, diversity, and phylogeny of all of these forms. From a flurry of efforts in the 1980s, individually by W. P. Coombs of Amherst College in Massachusetts, T. Maryańska of the Muzeum Ziemi (Warsaw, Poland), and P. C. Sereno of the University of Chicago, it is clear that the ankylosaur clade consists principally of two great clades: Nodosauridae and Ankylosauridae.

Members of Nodosauridae had relatively long snouts, well-muscled shoulders, flaring hips, and pillar-like limbs. All of course were armor covered and many also had tall spikes and spines on their backs and shoulders. Nodosaurids are known principally from the Northern Hemisphere (North America and Europe), although new discoveries in Australia and Antarctica have extended the geographical range of these animals deep into the Southern Hemisphere.

Members of Ankylosauridae also give the impression of impregnability. All are well armored, but there are fewer tall spines along the body. The tail ends in a massive bony club, in some instances with

Figure 7.1. *Euoplocephalus*, the armored, club-tailed ankylosaur.

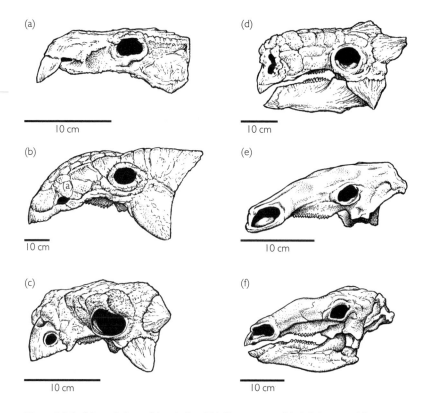

Figure 7.2. Left lateral view of the skulls of (a) *Shamosaurus*, (b) *Ankylosaurus*, (c) *Pinacosaurus*, (d) *Tarchia*, (e) *Silvisaurus*, and (f) *Panoplosaurus*.

several paired knobs along its length. The head is shorter and broader in ankylosaurids than in nodosaurids and there are large triangular plates attached to the rear corners of the skull (what are called squamosal "horns").

Ankylosaur lives and lifestyles

Going their way

Ankylosaurs have a worldwide distribution, predominantly from North America and Asia, but also from Europe, Australia, South America, and Antarctica (Figure 7.3). The best-preserved of these finds come from Asia, particularly Mongolia and China, where spectacular specimens are nearly complete and articulated, in most cases preserved in an upright pose or on their sides. In contrast, in North America, only partial skeletons have yet been found, and these are often upside-down, sometimes in rocks deposited along the sea shore or even in rocks representing open marine environments. Nobody ever claimed that these armored dinosaurs were ocean-going animals, but the North American forms at least must have lived in terrestrial habitats sufficiently close to the sea that their bloated or partially dismembered carcasses might have been carried out with the tide. The flipping upside-down presumably comes from the heavy nature of their armored backs: a floating ankylosaur carcass should turn over prior to settling out on the bottom of a stream channel, lake, or ocean floor.

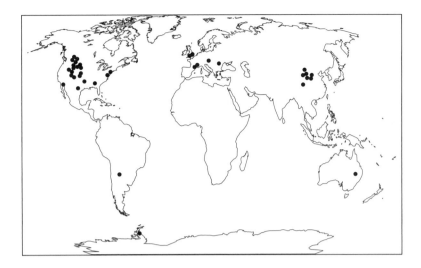

Figure 7.3. Global distribution of Ankylosauria.

In addition to these general preservational considerations, ankylosaur finds most commonly consist of individual skeletons or isolated partial remains; there is only one ankylosaur bonebed known. Perhaps this indicates that these animals had solitary habits or lived in very small groups. Even from our incomplete window on the past, it appears reasonably certain that ankylosaurs did not enjoy the company of huge herds.

Food and mouths to feed

In whatever size groups, it is clear that ankylosaurs had a very low browsing range, foraging no more than a meter or so above the ground. But what were they foraging for? Bugs? F. Nopcsa once suggested that ankylosaurs were insectivorous, but this idea is no longer given much credence. (How many beetles would it take to keep a 3500 kg ankylosaur going? And for how long?) Instead, it is now thought that these animals fed exclusively on plants. Why this is so comes from a look at the digestive system, from the beak, mouth, and teeth at the front end, as well as the far extreme of the gut.

Up front, it is the cutting edge of the beak that made a first impression on plant food. As noted earlier for ornithischians in general, the beak was covered by horn, making this rhamphotheca both strong and sharp. Primitively in ankylosaurs, the beak is scoop shaped but relatively narrow (although slightly broader than in stegosaurs) and remains so in the nodosaurid clade and in *Shamosaurus*. In all other ankylosaurids, by contrast, the beak becomes very broad (Figure 7.4a) and, in a couple of these forms, the premaxillary teeth are lost. As noted by K. Carpenter of the Denver Museum of Natural History, the difference in beak shape may indicate a degree of feeding differentiation among ankylosaurs. A narrow beak in nodosaurids suggests that these ankylosaurs fed in a somewhat selective manner, plucking or biting at particular kinds of foliage and fruits and transferring them into the mouth with the sharp edge of the rhamphotheca. In contrast, the very broad beak of ankylosaurids may

(a)

(b)

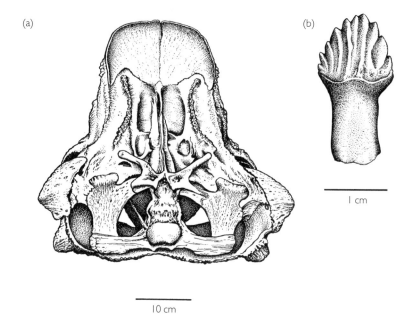

1 cm

10 cm

Figure 7.4. Palatal view of the skull of (a) *Euoplocephalus* and (b) a tooth of *Edmontonia.*

imply less selective feeding, in which plant parts were indiscriminately bitten off from the bush or pulled from the ground.

How the food was then prepared for swallowing is a bit of a mystery. Like stegosaurs (and pachycephalosaurs; see Chapters 6 and 8), the triangular teeth of both nodosaurids and ankylosaurids are small, not particularly elaborate, and less tightly packed than those of other ornithischian dinosaurs (Figure 7.4b). However, tooth wear indicates that chewing involved not only puncturing of foods, but also grinding action as well. In addition, it is likely that ankylosaurs had a long, flexible tongue (they have large hyoid bones – which support the root of the tongue – in their throats) and an extensive secondary palate that allowed them to chew and breath at the same time. Moreover, deeply inset tooth rows suggest well-developed deep cheek pouches to keep whatever food was being chewed from falling out of the mouth. Likewise, the jaw bones themselves were relatively large and strong (although lacking enlarged areas for muscle attachment). Many ankylosaur jaw features – except for tooth design and placement – suggest that ankylosaurs were reasonably adept chewers.

Perhaps the paradox of simple teeth placed in strong, cheek-bound jaws can be understood by looking not at how much chewing was done prior to swallowing (there obviously was some), but at the rear end of the animal, where the majority of plant digestion must have been accomplished by gut fermentation. Here, bacteria lived symbiotically within the stomach (or stomachs, because ankylosaurs may well have had a series of them). The stomach (or stomachs) would have served as a great fermentation vat, decomposing even the toughest, woody plant material. Among modern mammals, this method of breaking down tough plant material is well known in ruminants such as cows.

The amount of food needed to sustain these large animals must have been great, and consequently the gut must have been huge. Huge guts mean equally huge abdomens. In ankylosaurs, the very deep rib cage circumscribed an enormously expanded abdominal region and this can only mean that digestion took place in a very large, perhaps highly differentiated, fermentation compartment in these armored dinosaurs. R. T. Bakker called this hind-gut fermentation system "after-burner" digestion, analogizing, we fear, jet exhaust with that of fermenting ankylosaurs.

The combination of browsing at low levels and having anatomy indicative of chewing and fermenting places limits on what kinds of plant ankylosaurs may have fed on. At the levels where ankylosaurs concentrated, the undergrowth consisted of a mixture of low-stature ferns, gymnospermous plants such as cycads and bennettitaleans, and, during the Late Cretaceous, shrubby angiosperms (see Figure 16.11). What a rich array of plants to choose from, even within the first few meters above the ground!

Defensive moves

Ankylosaurs may not have been particularly adept at making such choices, however, because their brain power was close to the bottom of the dinosaur range (only sauropods had smaller brains for their size; see Chapters 6 and 11, and Box 15.4). According to J. A. Hopson's calculations, ankylosaurs had slightly less brain mass per kilogram of body weight than stegosaurs and much less than all other dinosaurs except sauropods. Hand in hand with slow thinking, ankylosaurs were among the slowest moving of all dinosaurs for their body weight. According to R. A. Thulborn's calculations from limb size and estimated body mass, these animals were able to run no faster than 10 km/h and walked at a considerably more leisurely pace (about 3 km/h).

Although slow on their feet, other aspects of the limb skeleton suggest that these Mesozoic plodders could actively defend themselves against predators. The ankylosaur shoulder region – particularly in nodosaurids – was exceedingly strongly muscled, much more so than in many other four-legged dinosaur groups. And relatively longer and wider hindlimbs may have had the effect of dropping the more forward center

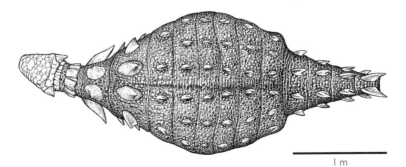

1 m

Figure 7.5. Dorsal view of the body armor of *Euoplocephalus*.

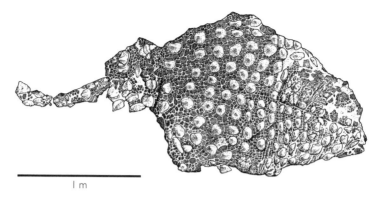

1 m

Figure 7.6. Dorsal view of the body armor of *Sauropelta*.

of gravity of the animal. Could these features mean that the fore- and hindlimbs were used in conjunction with body armor and, in ankylosaurids, a stiffened tail and terminal club? In the context of defense against predators, the answer is a resounding "Yes."

Consider other aspects of defense. First, in all ankylosaurs the entire upper surface of the body – the head, neck, torso, and tail – is covered by a pavement of bony plates (Figures 7.5 and 7.6), a condition quite unlike that in stegosaurs or more basal thyreophorans. Nodosaurids, but not ankylosaurids, also retain the formidably sharp parascapular spines that were also seen in stegosaurs. Ankylosaurids were even more modified, producing armor literally everywhere, even over the eyes. And of course there is the formidable ankylosaurid tail-club (Figure 7.7), paired masses of bone set at the end of a tail (itself about half the length of the body), in some cases augmented by spikes along the length of the tail.

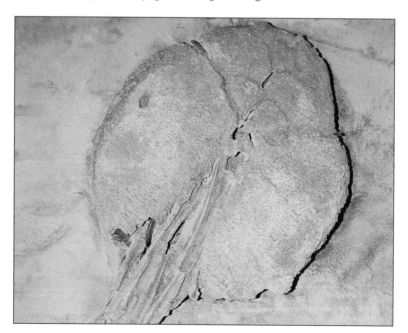

Figure 7.7. The bony tail-club of *Euoplocephalus*. (Photograph courtesy of M. Brett-Surman.)

How this tail and its terminal club worked is surely a masterful mechanical feat, fully as skillful as any mace-wielding knight. While the front part of the tail was relatively free to swing side to side, the rear half was stiffened by modified vertebrae, as well as by a series of longitudinally running tendons. These latter provided firm attachment for the powerful muscles, but more especially they added considerable stiffening of the end of the tail. Using the forward flexibility of the tail vertebrae and the stiffened distal end, the club could have been forcefully swung side to side without the whiplash effect of over oscillation of the club and the rest of the body.

It is now easy to imagine how an ankylosaur might have behaved while under attack from one of its contemporary predators, among whom were the likes of *Deinonychus*, *Tarbosaurus*, and *Tyrannosaurus*. It was either offense or defense: tail clubbing or hunkering down. Who knows what neural pathways or synapses might determine one or the other of these options for an ankylosaur, but ultimately the game must have been to persuade the predator to GO AWAY. The only active defense available to ankylosaurids was a club bash to the shins. In order to effectively wield their weaponry, these animals must have first planted their hindlimbs and then rotated their forequarters with their strong forelimb muscles, ever keeping watch on the threatening predator. For nodosaurids, this was a head-first (or shoulder-first) affair, but in ankylosaurids there was a bit of crocodile defense about them: "Look me in the eyes, but watch out for my tail."

If such an offensive strategy fails to dissuade a predator, then the ankylosaur fail-safe was assuredly to hunker down and wait out the attack. With its legs folded under its body, a 3,500 kg ankylosaur would have been very difficult to flip over. Safe under protective armor, both nodosaurids and ankylosaurids were virtually impregnable to predators. They were, in short, the best-defended fortresses of the Mesozoic.

The evolution of Ankylosauria

Ankylosauria is defined as that clade of dinosaurs consisting of the common ancestor of *Panoplosaurus* (a nodosaurid) and *Euoplocephalus* (an ankylosaurid) and all of the descendants of this common ancestor (Figure 7.8). As we learned in Chapter 6, Stegosauria is the closest relative to Ankylosauria; together they form the Eurypoda. More distant relationships are with *Scelidosaurus*, *Emausaurus*, and *Scutellosaurus* within the context of Thyreophora (see introductory text to Part II: Ornithischia).

Reflecting the importance of heavy armoring to ankylosaurs, it is not surprising that the accoutrements of armor and/or its support comprise the majority of derived features uniting the clade Ankylosauria. These animals all share, among other features, closure of the antorbital and upper temporal openings, ossification and fusion of a keeled plate onto the side of the lower jaw, fusion of some of the first couple of tail vertebrae to the sacral vertebrae and ilium, rotation of the ilium to form flaring blades, closure of the hip joint, and of course the development of a dorsal shield of symmetrically placed bony plates (both large and small) and spines.

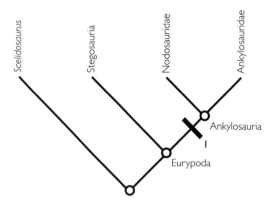

Figure 7.8. Cladogram of "higher" Thyreophora, emphasizing the monophyly of Ankylosauria. Derived characters include: at 1 closure of antorbital and upper temporal openings, ossification and fusion of keeled plate onto side of lower jaw, fusion of first tail vertebrae to sacral vertebrae and ilium, rotation of ilium to form flaring blades, closure of hip joint, development of dorsal shield of symmetrically placed bony plates and spines.

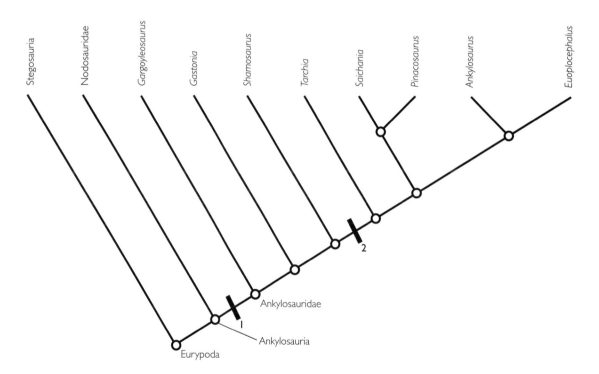

Figure 7.9. Cladogram of Ankylosauridae, with its two closest relatives, Nodosauridae and Stegosauria. Derived characters include: at 1 pyramidal squamosal boss, shortening of premaxillary palate, presence of premaxillary notch, rostrolaterally directed mandibular ramus of pterygoid, tail-club; at 2 anterodorsal and posteroventral arching of palate, vertical nasal septum, rugose and crested basal tubera; at 3 caudal end of postaxial cervical vertebrae dorsal to cranial end, fusion of sternals.

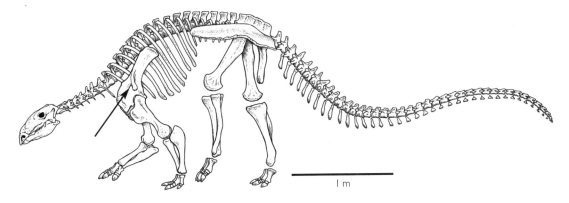

1 m

Figure 7.10. Left lateral view of the skeleton of *Sauropelta* without its armor shield. Note the projection on the scapula (arrowed), known as the acromial process.

Evolution within Ankylosauria has followed two principal pathways since their origin sometime in the Jurassic: Ankylosauridae and Nodosauridae. All ankylosaurs belong to one or the other of these two clades.

Ankylosauridae

Having already worked through what it takes to be a member of Ankylosauria, who indeed are the constituent nodosaurids and ankylosaurids? To begin with, ankylosaurids have a pyramidal squamosal boss ("horn"), shortening of the front of the palate, a notch in the premaxilla, modification of the palate, and a tail club (Figure 7.9). Recent work by Matt Vicaryous and co-workers indicates that *Gargoyleosaurus*, *Minmi*, and *Gastonia* are successively more closely related to remaining ankylosaurids. Forming an unnamed taxon, these remaining ankylosaurids are united by an arched palate, vertical nasal septum, and modifications of the base of the braincase. They begin with a small clade of *Shamosaurus* and *Gobisaurus*, followed in successively closer relationships by *Tsagantegia*, *Tarchia*, a small grouping of *Ankylosaurus* and *Euoplocephalus*, *Talarurus*, *Pinacosaurus*, and *Saichania*. These last three ankylosaurids are united by fusion of the sternal bones and changes in the shape of the neck vertebrae.

Nodosauridae

Turning to the other great clade of ankylosaurs, nodosaurids share a number of derived features, most obviously the knob-like process (the acromial process; Figure 7.10), that extends laterally from the shoulder blade and indicates that this region was heavily muscled. Basally, Nodosauridae consists of *Cedarpelta* and the clade comprising all remaining forms (Figure 7.11). Early on in the history of this clade, we encounter a small grouping of *Sauropelta* and *Silvisaurus*, followed thereafter by *Pawpawsaurus*, *Panoplosaurus*, and *Edmontonia*. The two last-mentioned nodosaurids possess one or two flat polygonal ornamentation across front of the skull, oral margin of premaxilla continuous

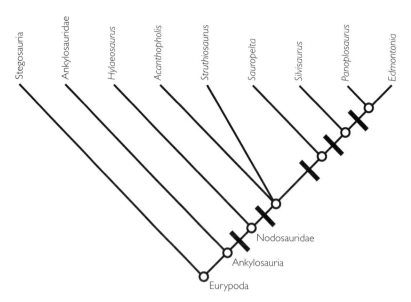

Figure 7.11. Cladogram of Nodosauridae, with its two closest relatives, Ankylosauridae and Stegosauria. Derived characters include: at **1** knob-like acromion on scapula; at **2** distinct pattern of one or two flat polygonal ornamentation across anterior region of snout, oral margin of premaxilla continuous with maxillary tooth row, anterodorsal palatal arch, rugose, crest-like basal tubera.

with maxillary tooth row, and modification of the palate and braincase.

Anklyosaurs meet history: a short account of their discovery

Ankylosaurs were there at the very beginning of the history of dinosaur studies. Along with *Iguanodon* and *Megalosaurus* (an ornithopod and a theropod, respectively), *Hylaeosaurus* (*hylaios* – forest-dwelling) was also a charter member of Sir Richard Owen's Dinosauria. This first-named ankylosaur hails from the Tilgate Forest area of southern England, from Lower Cretaceous rocks (approximately 130 million years old). Very little detail is known of the anatomy of *Hylaeosaurus* despite efforts by its namer, Gideon Mantell, and a number of later workers. After the initial discovery, no further finds helped to enlarge our understanding of this animal. As a result, it stands today almost as completely known as it did in the mid-1800s: mostly the front half of the skeleton and parts of the armor shield.

It took another 34 years to find a rear end: *Polacanthus* (*polys* – many; *akantha* – spine) was described by T. H. Huxley in 1867. What is known of *Polacanthus* consists of the rear end of the animal, replete with many spines. At the same time that *Polacanthus* was discovered, two other new ankylosaurs were also recognized. Again in 1867, Huxley – most famous for debating in the service of Darwinian evolution – named *Acanthopholis* (*pholis* – scale) from mid-Cretaceous beds of England. At roughly the same time (1871) but some 1,200 km away, E. Bunzel of the University of Vienna was beginning to sort out a new collection of bones from the Gosau Formation (Upper Cretaceous; approximately 75 Ma) from eastern

Austria. In this collection were some interesting skeletal elements that he thought belonged either to a bird or a crocodile. These Bunzel called *Struthiosaurus* (*strouthion* – ostrich). It was not until this material was restudied by English paleontologist H. G. Seeley, and especially when new material of *Struthiosaurus* was discovered in the Transylvanian region of what was then Hungary (now Romania) by Nopcsa, that the true ankylosaur identity of this dinosaur was finally revealed. Finally, to close out the century, Richard Lydekker of the British Museum (Natural History) in London named *Sarcolestes* (*sarkos* – flesh; for flesh-robber) for an unusual lower jaw from the Middle Jurassic (approximately 160 Ma) of central England. Little did Lydekker realize at that time that this piece of jaw had nothing to do with a flesh-eating dinosaur; thanks to the much later work of Peter Galton of the University of Bridgeport, Connecticut, it was revealed in the 1980s that *Sarcolestes* was an ankylosaur.

Discoveries of North American ankylosaurs began in the late 1880s. It was then (1889) that Yale paleontologist O. C. Marsh announced a new armored dinosaur, *Nodosaurus* (*nodus* – knob). Discovered in Wyoming in Lower Cretaceous rocks, *Nodosaurus* is represented by only an incompletely preserved postcranial skeleton and armor.

North of the USA–Canada border, what was eventually to be named *Euoplocephalus* (*eu* – well; *hoplon* – shield; *kephale* – head), one of the all-time best known of these armored dinosaurs, was discovered in the great badlands of the Red Deer River of Alberta and given its first name "*Stereocephalus*" in 1902 by L. M. Lambe of the Geological Survey of Canada. Unfortunately, this name had already been given to an insect,[1] so in 1910 Lambe gave it its presently recognized name, *Euoplocephalus*.

Thereafter, on a nearly regular basis (on the order of a name every 10 years), new ankylosaurs began to be discovered and named. *Ankylosaurus* – the namesake of the group – was discovered in Upper Cretaceous rocks in Montana and named by Barnum Brown of the American Museum of Natural History in 1908. *Panoplosaurus* (*pan* – all), like *Euoplocephalus*, was found in Alberta and also named by Lambe. C. M. Sternberg of the National Museum of Canada continued the Lambe tradition, naming *Edmontonia* (for the Edmonton Formation of Alberta) in 1928.

The history of ankylosaur studies turned eastward in the 1930s. Part of the booty of the great Central Asiatic Expedition – that great paleontological field effort conducted by Roy Chapman Andrews of the American Museum of Natural History in the 1920s (see Box 7.1) – were some splendid ankylosaur remains. C. W. Gilmore named these in 1933: *Pinacosaurus* (*pinak* – plank) for an armored dinosaur from the Djadochta Formation (Late Cretaceous in age) from Mongolia. Next came *Talarurus* (*talaros* – basket; *oura* – tail), another ankylosaur from Upper Cretaceous beds of Mongolia (this time collected by the Soviet–Mongolian Paleontological

1 Insects continue to plague dinosaur nomenclature. As recently as 1993, the Mongolian Cretaceous bird "*Mononychus*" was described. Unfortunately, the name was already occupied by a beetle. Consequently, "*Mononychus*" had to be renamed *Mononykus*, the name this remarkable bird carries to this day (see Chapter 14).

BOX 7.1

Indiana Jones and the Central Asiatic Expeditions of the American Museum of Natural History

He stands in the middle of the remote, rugged Mongolian desert: high leather riding boots, riding pants, broad-brimmed felt hat, leather-holstered sidearm hanging from a glittering ammunition belt. He carries a rifle and knows how to use it. Nobody else dresses like him, but then nobody else is the leader of the American Museum's Central Asiatic Expeditions to Mongolia (a place which, at the time of the expeditions, the 1920s, could have been the moon). He is Roy Chapman Andrews, who 50 years later will be the inspiration, it is most plausibly rumored, for Indiana Jones (Figure B7.1.1).

Andrews always knew that he was a man with a destiny. Although he began his career at the American Museum of Natural History (AMNH) modestly (he scrubbed floors), training in mammalogy (an M.A.), sheer will, charisma, and a very good idea carried him the distance. He had traveled extensively, spoke several Asian languages more or less fluently (at a time when very few Westerners did), and had fabulous contacts in Beijing (then called Peking).

His idea was simple: to run an expedition to what was then known as Outer Mongolia and to see what he could see. Andrews' timing was superb: the Director of the AMNH, the powerful H. F. Osborn, had concluded that the cradle of humanity was located in Outer Mongolia, and so Andrews was effectively offering Osborn the opportunity to prove his thesis right (the possibility that Osborn could be wrong did not seem to be of concern). The logistics of the expedition were extravagant: Dodge cars, resupplied by a caravan of camels would bear the brunt of the expedition. The expedition itself would consist of a range of earth scientists – paleontologists, geologists, and geographers – to explore the Gobi Desert, the huge desert that forms the vast southern section of Mongolia (then called "Outer" Mongolia, as

Figure B7.1.1. Roy Chapman Andrews, explorer, adventurer, and leader of what he called the "New Conquest of Central Asia." (Photograph courtesy of the American Museum of Natural History.)

if to emphasize its remoteness) and northern China.

The journey was not without its risks. The Gobi Desert is a place of temperature extremes, beset by relentless strong winds. Politically, at the time, the region was an uproar. China, the base of operations, was torn by civil strife. And in 1922, the year of the first of three expeditions, a revolution shook Mongolia. Moreover, only one fossil, a rhinoceros tooth, had ever been found in Mongolia.

As it turned out, the Central Asiatic Expeditions were an unqualified success. Although Osborn's theory was not supported, Andrews brought back a wealth of fossils, including dinosaur material, that made Osborn's error easy to forget. Among the most famous dinosaur finds of his expedition, for example, were *Protoceratops* (the species name of this famous dinosaur is *andrewsi*) and eggs; the first time that dinosaur egg nests were ever found. Other incredible finds included *Velociraptor* and a group of tiny Mesozoic mammals (still the rarest of the rare). Andrews and his field parties also found the largest land mammal and the largest carnivorous land mammal of all time (both Cenozoic in age). Other fossils were obtained whose significance was not completely understood. For example, it was only in 1992 that a specimen of *Mononykus*, collected by Andrews' scientists in the 1920s, was finally correctly identified. All in all, it was quite a haul.

Andrews and his parties survived the Mongolian revolution of 1922, but eventually the expeditions came to an end when the political situation in China became too unstable and travel too dangerous. Andrews, himself, eventually went on to get the job held much earlier by Osborn, Director of the AMNH. He assured his place in history, however, by leading the Central Asiatic Expeditions.

Expeditions) named by E. A. Maleev of the Paleontological Institute in Moscow. As we shift back to North America, T. H. Eaton of the University of Kansas discovered and named *Silvisaurus* (*silva* – forest) from Lower Cretaceous rocks of Kansas in 1960, and in 1970 J. H. Ostrom of the Yale Peabody Museum of Natural History named *Sauropelta* (*pelte* – shield) from the Early Cretaceous Cloverly Formation of Wyoming and Montana.

Despite this pace, it was the end of the 1970s and 80s that were the bonanza years for ankylosaur discovery. In 1977, two ankylosaurs – *Tarchia* (Mongolian: *tarchi* – brain) and *Saichania* (Mongolian: *saichan* – beautiful) – were collected by the Polish–Mongolian Expeditions of the late 1960s and 1970s, and more particularly the important studies of Teresa Maryańska of the Muzeum Ziemi (Museum of the Earth) in Warsaw, Poland. *Minmi* (named for Minmi Crossing, Queensland) surprised the world when R. E. Molnar of the Queensland Museum described this first ankylosaur from Australia, indeed from the Southern Hemisphere! The 1980s closed out with two new ankylosaurs from the Late Cretaceous of Mongolia: *Shamosaurus* (*shamo* – Chinese for desert, a reference to the Gobi Desert) and *Maleevus* (named in honor of E. A. Maleev), both named by T. A. Tumanova of the Paleontological Institute in Moscow (in 1983 and 1987, respectively).

From the 1990s to the present, a wide array of new ankylosaurs has been discovered. Several have come from the Cretaceous of Asia. *Tsagantegia* (collected from Tsagan Teg in the Gobi Desert of Mongolia) named by Tumanova, hails from Mongolia, while four new genera are known from China: *Shanxia* (named for Shanxi Province, where it was found) discovered by P. M. Barrett and co-workers, *Tianzhenosaurus* (Tianzhen lizard) discovered by Q. Pang and Z. Cheng, *Gobisaurus* (named for the Gobi Desert) described by M. Vickaryous and co-authors from Canada and China, and *Liaoningosaurus* (named for Liaoning Province) described by X. Xu, X. Wang, and X. You.

Interesting though these Asian forms are, new discoveries in the USA made since 1990 have helped fill in many of the blanks in the history of ankylosaurs during the Late Jurassic through the mid-Cretaceous. Of these, *Mymoorapelta* (named for Mygatt-Moore Quarry), described by J. I. Kirkland and Carpenter, and *Gargoyleosaurus* (gargoyle lizard) described by Carpenter and collaborators, represent the best of the Jurassic record of ankylosaurs. Important new discoveries in Texas (*Texasetes* (Texas resident) named by Coombs, and *Pawpawsaurus* (named for the Paw Paw Formation) described by Y. Lee) provide information on Early Cretaceous ankylosaurs from the Gulf Coast region. Continued work by Carpenter, Kirkland, and their several co-investigators in the mid-Cretaceous of Utah has yielded *Gastonia* (named for Robert Gaston, who discovered it), *Animantarx* (living fortress), and *Cedarpelta* (named for the Cedar Mountain Formation). Finally, three new ankylosaurs have been described from the Late Cretaceous of Kansas, New Mexico, and California: *Niobrarasaurus* (named for the Niobrara Chalk Formation by Carpenter and co-authors), *Nodocephalosaurus* (node head lizard; named by R. Sullivan) and *Aletopelta* (wandering shield; named by T. Ford and Kirkland).

Important readings

Bakker, R. T. 1986. *Dinosaur Heresies*. William Morrow, New York, 481pp.

Carpenter, K. 1982. Skeletal and dermal armor reconstruction of *Euoplocephalus tutus* (Ornithischia: Ankylosauridae) from the Late Cretaceous Oldman Formation of Alberta. *Canadian Journal of Earth Sciences*, **19**, 689–697.

Carpenter, K. 1984. Skeletal reconstruction and life restoration of *Sauropelta* (Ankylosauria: Nodosauridae) from the Cretaceous of North America. *Canadian Journal of Earth Sciences*, **21**, 1491–1498.

Carpenter, K., Dilkes, D. and Weishampel, D. B. 1995. The dinosaur fauna of the Niobrara Chalk Formation. *Journal of Vertebrate Paleontology*, **15**, 275–297.

Coombs, W. P., Jr 1978. Theoretical aspects of cursorial adaptations in dinosaurs. *Quarterly Review of Biology*, **53**, 393–418.

Farlow, J. O. 1978. Speculations about the diet and digestive physiology of herbivorous dinosaurs. *Paleobiology*, **13**, 60–72.

Hopson, J. A. 1977. Relative brain size and behavior in archosaurian reptiles. *Annual Review of Ecology and Systematics*, **8**, 429–448.

Hopson, J. A. 1980. Relative brain size in dinosaurs – implications for dinosaurian endothermy. In Thomas, R. D. K. and Olson, E. C. (eds.), *A Cold Look at the Warm-Blooded Dinosaurs*. Westview Press, Boulder, CO, pp. 287–310.

Maryańska, T. 1977. Ankylosauridae (Dinosauria) from Mongolia. *Palaeontologia Polonica*, **37**, 85–151.

Sereno, P. C. 1986. Phylogeny of the bird-hipped dinosaurs (Order Ornithischia). *National Geographic Society Research*, **2**, 234–256.

Vicaryous, M., Maryańska, T. and Weishampel, D. B. 2004. Ankylosauria. In Weishampel, D. B., Dodson, P. and Osmólska, H. (eds.), *The Dinosauria*, 2nd edn. University of California Press, Berkeley, pp. 363–392.

For information on the Central Asiatic Expeditions:

Andrews, R. C. 1929. *Ends of the Earth*. G. P. Putnam's Sons, London, 355pp.

Andrews, R. C. 1933. Explorations in the Gobi Desert. *National Geographic Magazine*, **63**, 653–716.

Andrews, R. C. 1953. *All About Dinosaurs*. Random House, New York, 146pp.

Preston, D. J. 1986. *Dinosaurs in the Attic*. St Martin's Press, New York, 244pp.

CHAPTER 8

Pachycephalosauria: ramrods of the Cretaceous

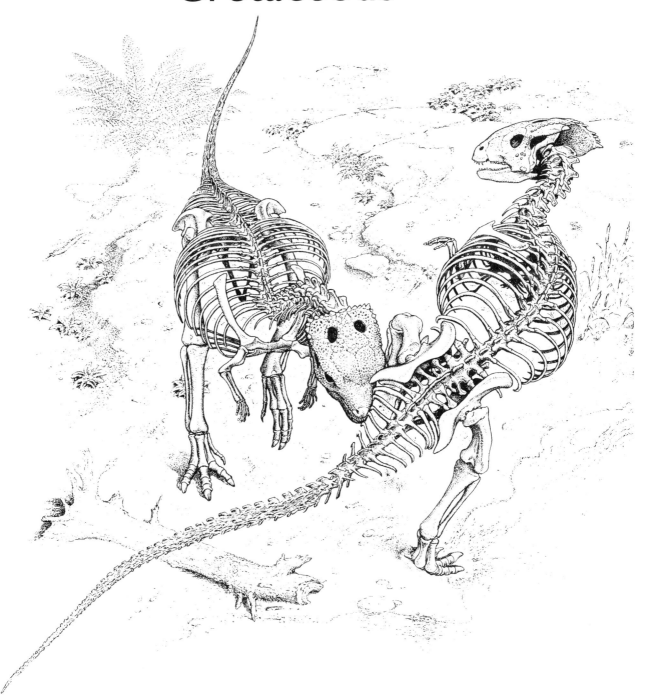

Through a dense thicket come the deep sounds of thuds, slaps, and scuffling. Beyond these shrubs in a large clearing are a dozen or more pachycephalosaurs – *Homalocephale* – their broad thickened heads fringed with an array of small knobs and horns. Many are slowly foraging for succulent leaves and fruits in the undergrowth, while two of the largest individuals, some 2.5 m long, are kicking up a storm of dust in the center of the group. At regular intervals, they turn toward each other, then rapidly lunge forward, colliding head-to-head, head-to-sides, head-to-thighs (Figure 8.1). After more than an hour, one tires and is forced from the group by the victor.

Fact or fiction? How can we possibly tell? What sort of data would we look to in order to assess these ideas about pachycephalosaur behavior? What *was* the lifestyle of these unusual animals, many of which had large domes on their heads?

Pachycephalosaurs, bipedal ornithischians so far known from only the continents in the Northern Hemisphere and from Cretaceous rocks, all had thickened skulls, but not all were domed. Some like the Asian *Homalocephale* had flattened skulls. These flat-headed pachycephalosaurs were thought to represent a clade different from the round-headed forms, but, as we will see later in this chapter, they actually represent an evolutionary trend in pachycephalosaurs to thicken and then dome the top of the head (Figure 8.2).

Figure 8.1 (see p. 147). The flat-headed, thick-headed *Homalocephale*, best-known of all pachycephalosaurs.

Pachycephalosaur lives and lifestyles

We begin with the global distribution of pachycephalosaurs (Figure 8.3). All are known from northern hemisphere continents, suggesting that their origin is related to the northern hemisphere supercontinent of Laurasia. In fact, all of the primitive members of Pachycephalosauria (e.g., *Wannanosaurus*, *Goyocephale*, and *Homalocephale*) are known from central or eastern Asia, which implies that they arose in this part of the world.

Figure 8.2. The left side of the skull of *Stegoceras*. (Photograph courtesy of H.-D. Sues.)

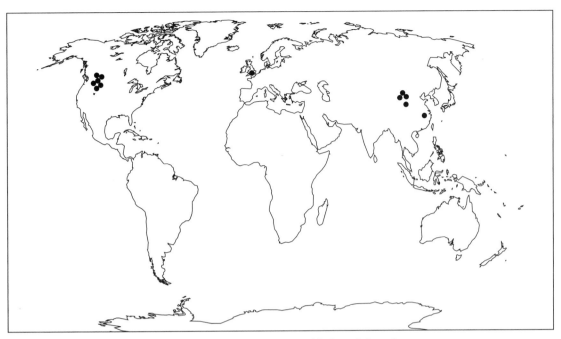

Figure 8.3. Global distribution of Pachycephalosauria.

Concentrating for a moment on the material from North America, most of what we know about pachycephalosaurs comes from isolated skull caps, many of which are highly water-worn. In fact, only single specimens of *Stegoceras* and *Pachycephalosaurus* are represented by more than just these skull caps, the former by a well-preserved skull and partial postcranial skeleton and the latter by a nearly complete skull. Still, these skull caps are very abundant in Upper Cretaceous rocks of North America, suggesting that some pachycephalosaurs – particularly *Stegoceras* – may have constituted up to 10% of the dinosaur fauna of the time. What is controlling the bias against the preservation of better specimens? C. M. Sternberg and, later, P. Dodson have suggested that North American pachycephalosaurs may have lived some distance away from those rivers and lakes whose sediments now preserve their remains. Alternatively, the habitats of the North American forms might have been far away from the kinds of environment that are preserved as sedimentary rocks; for example, if they lived in more upland or mountainous settings, only the most robust of their bones – their skull caps – would survive the long journey in the rivers that drained these regions to the flatlands where burial (and deposition) occurred.

In contrast, when we turn to Asia, the situation is quite different. Here, the combination of desert conditions and fluvial environments of Mongolia and China must have created favorable conditions for pachycephalosaur preservation. The nearly complete skulls and associated skeletons clearly lack traces of the kinds of long-distance transport seen in North American forms, suggesting that these animals must have lived very close to where they were eventually buried and preserved.

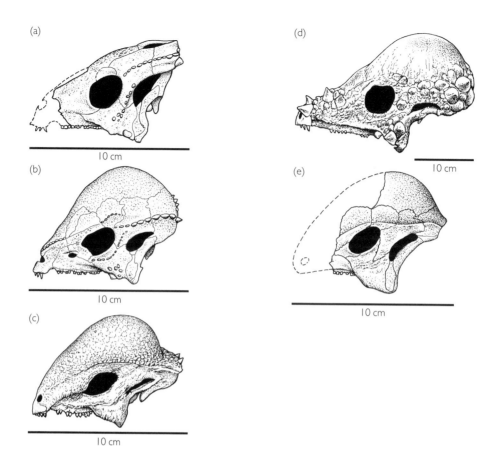

Figure 8.4. Left lateral view of (a) *Homalocephale*, (b) *Prenocephale*, (c) *Stegoceras*, (d) *Pachycephalosaurus*, and (e) *Tylocephale*.

Feeding

Like many herbivorous dinosaurs, pachycephalosaurs must have spent a great deal of their time foraging on many of the low-growing plants of the time (because of their stature, pachycephalosaurs probably browsed no more than a meter or so above the ground). The herbivorous habits of these animals are evident not only from their teeth but also the impressive volume of their abdominal region. At the front of the jaws were simple, peg-like gripping teeth, the last of which were sometimes enlarged in a canine-like fashion (Figure 8.4). It is likely that these teeth were surrounded by a small, horny rhamphotheca as in all ornithischians. Further back, the cheek teeth of pachycephalosaurs were uniformly-shaped, with small, triangular crowns (Figures 8.4, 8.5). The front and back margins of these crowns bear coarse serrations, the better for cutting or puncturing plant leaves or fruits. This kind of chewing produced a variety of kinds of tooth wear, implying that different pachycephalosaurs fed on different kinds of vegetation.

At the other end of the feeding apparatus, pachycephalosaurs must have had a large internal vat for fermenting their food (Figure 8.6). The rib cage is very broad, a great girth that extends backward to the base of the tail. These anatomical modifications of the more primitive condition seen

Figure 8.5. An upper cheek tooth of *Pachycephalosaurus*.

5 mm

in ceratopsians (see Chapter 9), ornithopods (see Chapter 10), and other ornithischians suggest a backward migration and enlargement of the digestive tract to occupy a position between the legs and under the tail. Much like the case for thyreophorans (see introductory text to Part II: Ornithischia, and Chapter 9), simple styles of chewing may have combined with more extensive chemical digestion via the development of a huge gut to solve the problem of making a living as a plant-eating dinosaur.

Thoughts of a pachycephalosaur

Pachycephalosaur neuroanatomy suggests that, despite having only an average-sized brain for their body size, these animals may have had a quite acute sense of smell; the olfactory lobes of the brain were enlarged. The front half of the brain (the cerebrum) is highly flexed relative to the rest of the brain (to give it a horizontal orientation) as it was nestled inside the skull, while the back half (the pontine region) was less flexed than in other dinosaurs. With a smaller degree of pontine flexure, this region of the brain is downwardly inclined. As reported by E. B. Buchholz (formerly Giffin), the reduction in pontine flexure appears to reflect the rotation of the back of the skull (the occiput) to face not only backward, but also slightly downward. She has also shown that that the most extreme degree of occipital inclination is associated with more prominent doming of the frontal–parietal skull cap. And prominent doming, of course, is the feature that has led paleontologists to suggest that pachycephalosaurs head-butted.

Social behavior

The idea that pachycephalosaurs used their thickened skull roofs as battering rams first came from suggestions made in 1955 by E. H. Colbert and independently by the Russian evolutionary biologist L. S. Davitashvili in 1961. In the years since, head-butting among pachycephalosaurs has been analyzed in considerable detail particularly by P. M. Galton and H.-D. Sues in the 1970s.[1]

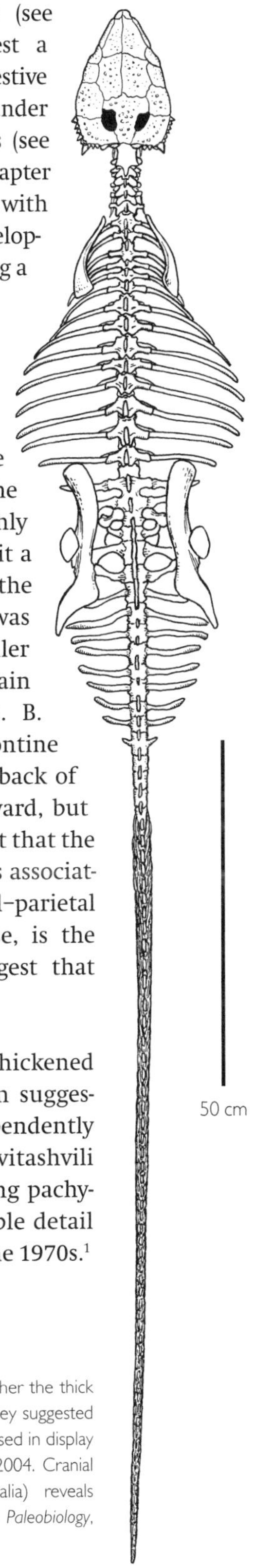

50 cm

Figure 8.6. Dorsal view of the skeleton of *Homalocephale*.

1 Very recently M. Goodwin and J. Horner have challenged whether the thick skulls were able to sustain heavy blows; using bone histology, they suggested that the domes grew very fast and therefore were more likely used in display rather than combat. (See Goodwin, M. B. and Horner, J. R. 2004. Cranial histology of pachycephalosaurs (Ornithischia; Marginocephalia) reveals transitory structures inconsistent with head-butting behavior. *Paleobiology*, **30**, 253–267.

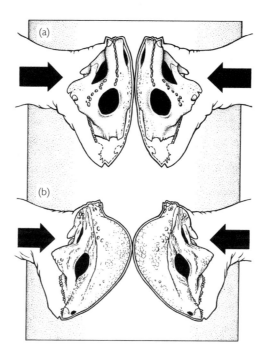

Figure 8.7. Head-on pushing and butting in (a) *Homalocephale* and (b) *Stegoceras*.

Let's look at what Galton and Sues had to say about pachy-cephalosaurs as living battering rams (Figure 8.7). Each noted that the outside of the very thick dome is often smooth. Some specimens bear what look like scars on this external surface. Internally, the structure of the dome is very dense, consisting of fine bony columns that radiate so as to be approximately perpendicular to the external surface of the dome. Such an arrangement is ideal for resisting forces that come from strong and regular thumps to the top of the head and transmitting such forces around the brain, much like an American football player's helmet channels forces around the head. Adding strength to this functional interpretation of bone structure is Sues's simulation of how such forces would pass through the dome. Using special clear plastic cut to resemble a cross-section of the dome of *Stegoceras* (Figure 8.8), Sues stressed this plastic model in way that simulated head-butting and, *voilà*, the stress lines (which can be seen under ultraviolet light) had the same orientation as the columnar bone. The match between stress lines and bony columns strongly suggests that the latter have optimal orientation to resist the former.

Even though there is a match between stress-line and bone-column directions in Sues's functional model, the microscopic arrangement of the bone within the dome suggests that something else besides shock absorption was going on. In a paper on the histological structure of the dome of *Stygimoloch*, a pachycephalosaur from the Late Cretaceous of Wyoming, M. Goodwin and co-workers discovered that the enlarged frontal bone contains abundant microscopic openings for blood vessels.

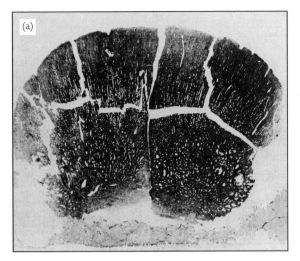

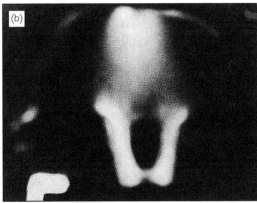

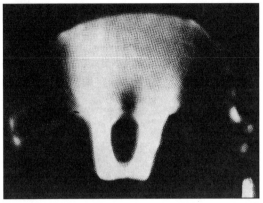

Figure 8.8. (a) Vertical section through the dome of *Stegoceras*. Note the radiating organization of internal bone. (b) Plastic model of the dome of *Stegoceras*, in which forces were applied to several points along its outer edge and seen through polarized light. Note the close correspondence of the stress patterns produced in this model and the organization of bone indicated in (a). (Photographs courtesy of H.-D. Sues.)

Because of this pattern of vascularity, these authors inferred that the skull of pachycephalosaurs were not well designed to sustain either front or side impacts and that domes functioned principally as display structures. We don't yet know how many other pachycephalosaurs had this kind of bone histology, although it is unlike that seen in the skulls of other dinosaurs that have been investigated.[2]

If domes were used in some sort of intraspecific behavior, whether display or as a battering ram, we might expect that these structures should show some degree of sexual dimorphism. The first suggestion of such dimorphism in domes was made by Galton and W. P. Wall, but by far the best treatment is by R. E. Chapman and several of his colleagues. Their work assessed variation in the size and shape of the frontal–parietal dome using a large sample of a single pachycephalosaur species, *Stegoceras validum*. Many of these domes were from juveniles, while others – the largest specimens – obviously came from old individuals. On the whole, variation within the *Stegoceras validum* sample was fairly homogeneous – none of the specimens stood out from the main group, which is what one would like when assessing a single species known only from the fossil record.

---

2 Goodwin and Horner have recently done just that, which promises to give us a much more complex story about pachycephalosaur behavior.

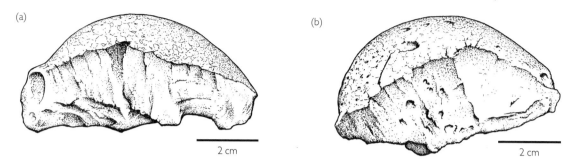

(a)

(b)

2 cm

2 cm

Figure 8.9. Two forms of the dome of *Stegoceras*. The shallower dome is thought to pertain to a female (a), while the deeper dome may pertain to a male (b).

Some heterogeneity in size and shape occurs in the domes of *Stegoceras*. In fact, the sample can be segregated into two groups on the basis of relative sizes of domes and braincases (Figure 8.9). One group, which shows slight acceleration in the rate of growth of the dome relative to the braincase, naturally formed a larger dome. At the same time, these larger-domed individuals had a slightly greater growth rate for dome thickness as well. Those individuals with larger, more convex domes were arbitrarily designated by Chapman and collaborators as males. The other group, distinguished by having thinner, flatter domes (as a consequence of lower growth rates than in the other, more prominently domed group), was presumed to be female.[3] It is additionally interesting that these assignments to male or female come out to be a one-to-one ratio. Since these *Stegoceras* individuals are thought to be sympatric (living in the same place at the same time), such a sex ratio is to be expected. Finally, in terms of the subject at hand – the functional significance of differences between sexes – it is the males with their large and more convex domes that are better designed to initiate and receiving head-butting blows, while the females have domes less well developed and in fact very similar in the general profile to juvenile or young adult males, exactly those individuals not yet likely to engage in head-butting. If Chapman and his colleagues are correct, then we have a much more subtle perspective on the relationship of dome size and shape in males and females to the use of the domes in head-butting. But what else can be said about pachycephalosaur battering rams in terms of other regions of the body? We turn next to the orientation of the head on the neck.

As we have already noted, the back of the pachycephalosaur skull is progressively rotated forward beneath the skull roof. In this way, the line of action of any impact force to the top of the head passes close to the

3 Why is maleness assigned to the larger-domed forms, and femaleness assigned to the smaller-domed forms? Is there any biological validity to this, or are vertebrate paleontologists simply atavistic chauvinists? While we might not exonerate all of our colleagues from the charge, the fact is that, in birds, the closest living relatives of pachycephalosaurs, it is predominantly the males who are larger and who are disposed toward sexual display, mostly through coloration and behavior. For this reason, those pachycephalosaurs with large domes, for whom head-butting in sexual display may be indicated, are reasonably – if arbitrarily – inferred to have been males.

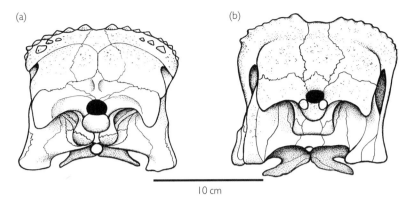

Figure 8.10. Rear view of the skull of (a) *Homalocephale* and (b) *Stegoceras*.

occipital condyle (the bony joint between the skull and neck vertebrae). With the head in a downward position – the only position that makes sense for head-butting – rotation of the back of the skull minimizes the chance of violent rotation or even dislocation of the head on the neck. These sorts of injury represent the Cretaceous equivalent of something like automobile whiplash: immediately debilitating and often life-threatening. Therefore, we might expect some corrective measures to be seen in the neck region as well. Unfortunately, the neck itself is not known in any pachycephalosaur. It is nevertheless clear from the back of the skull of these animals (Figure 8.10) that the extremely well-developed and very strong neck musculature was used to position the head at the right attitude for head-butting and to resist some potentially injurious rotations of the skull.

What about further down the animal? Are there skeletal measures to ensure that the body did not torque around itself, injuring the spinal cord and the nerves that extended from it? The answer is a resounding "Yes." Unique to pachycephalosaurs, the tongue-and-groove articulations between vertebrae along the back are very much like the carpentry joints of the same name that give strength to, and make rigid, adjoining pieces of wood. In a similar fashion, the pachycephalosaur tongue-and-groove condition would have provided a great deal of rigidity to the back, preventing the kinds of violent lateral rotations of the body that would otherwise have been suffered at the time of impact.

The skull and the back vertebrae all speak strongly for head-butting in pachycephalosaurs. With the phylogeny of the group in mind (see below), we can now develop some of the history of this kind of behavior in the clade. Most primitively, pachycephalosaurs all bear thickened, yet flat heads. This feature alone strongly suggests that these animals appear to have been head-pushers. Such encounters were probably made with the vertebral column held horizontally, an obvious advantage when competing males approached one another, and in any case it was the primitive condition for these bipedal ornithischians. It is heartening to note that analogous kinds of pushing match are also found in modern marine iguanas, which use these types of confrontation to establish social hierarchies. Perhaps the social structuring described here for primitive pachycephalosaurs had an even wider phylogenetic distribution, as ceratopsians are also thought to have evolved a similar social order. If so, then perhaps social hierarchies and head-on confrontations constitute the ancestral behavioral pattern for at least Marginocephalia and retained in all members of Pachycephalosauria.

It was not until the evolution of fully domed forms that battering-ram behavior appears to have developed. Again, approaches of competing males were probably done with the vertebral column held horizontally. Perhaps these males faced each other and made dome-to-dome blows or else charged at each other and lowered their heads just before impact. But these head-to-head collisions did not come cost-free: very severe injuries to the head and neck were to be expected simply because these animals lacked self-correcting mechanisms of the kind seen in modern head-butters such as goats and big-horn sheep that would have kept the tremendous forces aligned with the rest of the skeleton. Glancing blows would have been the worst for these head-on battering rams. Debilitating, or even lethal, injuries to the brain or spinal cord were to be avoided at all cost. So without precision head-butting, these animals instead may have butted each other along their flanks. It was much safer that way and the results were presumably the same – winners and losers.

Whatever region of the body that pachycephalosaurs butted, a myriad of anatomical modifications took place to reduce the possibility of injury. As we have already outlined, the tremendous forces applied to the top of the dome would have been transmitted by the reoriented occiput through the long axis of the neck, thereby reducing the chances that the skull was dislocated by the jolt of head collision. Any additional tendency toward head dislocation would have been prevented by the large and powerful muscles of the neck. Finally, whiplash to the back would have been prevented by the rigid construction of the vertebral column. All in all, these animals were very well designed for the rough and tumble lifestyle of the ram-rods of the Late Cretaceous.

Integrated with head-butting is a suite of features related to visual display. First, there are the canine-like teeth. These were probably used in threat display or biting combat between rival individuals, much as pigs and primitive deer do today. If we only knew more about them, we might discover that these teeth may be sexually dimorphic, large in males and less prominent, more normal-sized in females. Alas, all this remains speculation. More informative are the knobby and spiny osteoderms that covered the snout, the side of the face, and most extensively on the back of the parietal–squamosal shelf. Assuredly these distinctive knobs and spines – especially in *Stygimoloch* – were used to show off, males alternately displaying to females and threatening rival males.

Sexual selection The "glue" holding all of these evolutionary modifications together is sex. The establishment of dominance hierarchy gives some males – those that send the right signals about their qualifications for breeding –

preferred access to females. And it is the females who choose not only their mates but by their very actions what those signals are. Males that show off in ways that are acceptable to females are those that gain reproductive access. These same males must also fend off competitors for access to females. To do so requires some kind of threat display and/or offensive weaponry that is used to establish dominance between greater and lesser males. In modern bighorn sheep it is the curl of horns, while in modern anole lizards it is the colorful dewlap that unfolds under the chin. In general, this practice of female choice and its effect on establishing dominance hierarchies constitute what is called sexual selection, selection not between all individuals within a species but between males alone. Sexual selection emphasizes features related to display and combat, principally of males to females and males to males. In pachycephalosaurs, this emphasis is on the prominence of domes, knobs, and spikes, structures that acted in ritual display and, should the need arise, in actual violent clashes. The winner, presumably the male with the best-fashioned head-thumper, either in terms of showiness or safest design, perpetuates its family line, but nevertheless must be ever vigilant for other males that want to knock his block off – or at least knock him off his block.

The evolution of Pachycephalosauria

Today we recognize over a dozen pachycephalosaur species, principally from the Northern Hemisphere and all but two (*Yaverlandia* and *Stenopelix*) from the Late Cretaceous. With all of these different forms, their taxonomy and more especially their phylogenetic relationships – both with other ornithischians and among themselves – have been the focus of much attention.

The most obvious thing about pachycephalosaurs is their thick-headedness. With clear modifications of the skull, it is not surprising that Pachycephalosauria has long been considered a monophyletic group. Where the controversy lies is "Which other ornithischian group represents the next closest relatives of pachycephalosaurs?" As we have seen, the interpretation of derived versus primitive characters used to support a monophyletic group depends on the group's next closest relatives. You can't have one without the other, as they say. L. M. Lambe originally thought pachycephalosaurs had affinities with stegosaurs (which in those days also included ankylosaurs), while more modern studies have placed these animals within, or as close relatives to, ornithopods, ankylosaurs, or ceratopsians. It is this latter relationship that now appears to be the best supported (see introductory text to Part II: Ornithischia, and discussion later in this chapter).

With this information in hand, it is possible to define and diagnose Pachycephalosauria (Figure 8.11). This clade consists of the common ancestor of *Stenopelix* and *Pachycephalosaurus*, and all of the descendants of this common ancestor. Ceratopsians appear to be most closely related to this pachycephalosaur clade and together Pachycephalosauria and Ceratopsia form the clade called Marginocephalia. More distant relationships within Ornithischia are with Ornithopoda, Thyreophora,

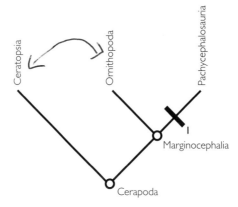

Figure 8.11. Cladogram of Cerapoda, emphasizing the monophyly of Pachycephalosauria. Derived characters include: at **1** thickened skull roof, frontal excluded from orbital margin, tubercles on posterolateral margin of squamosal, thin, plate-like basal tubera, double ridge-and-groove articulations on dorsal vertebrae, elongate sacral ribs, caudal basket of fusiform ossified tendons, ilium with sigmoidal border, medial process on ilium, pubis nearly excluded from acetabulum, tubercles on squamosal, broad expansion of squamosal onto occiput, free ventral margin of the quadratojugal eliminated by contact between jugal and quadrate.

and *Lesothosaurus* (see introductory text to Part II: Ornithischia, and Chapter 10).

Members of the pachycephalosaur clade share a host of derived features, most of them cranial. Most important are the thickened skull roof (either table-like or domed), modified cheek region that shrouds the jaw joint, extensive ossification of the orbit (including additional bony elements fused to its upper margin), shortening of the floor of the braincase, expansion of the back of the skull, abundant and strongly developed ornamentation of the external surfaces of the skull, special ridge-and-groove articulations between articular processes on the back and tail vertebrae, elongate sacral ribs, a "basket-work" of ossified tendons that cover the end of the tail, and reduction of the pubic bone to where it does not contribute to the formation of the hip joint. Within Pachycephalosauria, there are a few poorly known taxa at the base of the clade (Figure 8.12). Although it lacks skull material and has been referred to numerous other ornithischian groups, *Stenopelix*, from the Early Cretaceous of Germany, is probably best placed as the most basal pachycephalosaur. It possesses elongate sacral ribs and a pubis that is nearly excluded from the acetabulum but lacks a medial process on the ilium and a sigmoidal dorsal iliac margin, features otherwise found in all remaining pachycephalosaurs for which they are known.

Yaverlandia – the only other Early Cretaceous pachycephalosaur (this time from England) – was a small animal (length of the skull cap is 45 mm). The top of its head is thickened into two small domes, one on each side of the frontal bones. The upper surface of each dome is pitted. Because *Yaverlandia* is known only from these frontals, it is difficult to position this form within the cladogram of Pachycephalosauria.

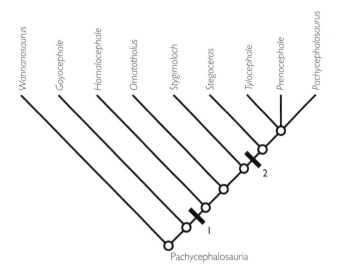

Figure 8.12. Cladogram of Homalocephaloidea. Derived characters include: at **1** broad parietal bone, broad medial process on ilium; at **2** nasal and postorbitals incorporated into dome.

However, it seems to be basal within the clade, having either an unresolved relationship with the *Wannanosaurus* + *Goyocephale* clade or constituting an unresolved pachycephalosaur at its basal node with *Stenopelix*.

The remainder of pachycephalosaur taxa has been subject to several cladistic analyses, beginning with P. C. Sereno's 1986 study, followed by research published by T. Maryańska, Goodwin and co-workers, Williamson and Carr, and Maryańska, Chapman, and Weishampel, each of which has produced roughly consistent cladograms. Pachycephalosaurs more derived than *Stenopelix* (and *Yaverlandia*?) emphasize the growth of the frontal dome, yet share a number of derived features involved with the development of the nerves of the sense of smell that are not seen in *Yaverlandia*. The basal taxon within this less-inclusive clade is *Wannanosaurus*. Excluding this flat-headed Chinese form, remaining pachycephalosaurs constitute *Goyocephale*, a clade diagnosed by reduction of the upper temporal fenestra and a rectangular postacetabular process on the ilium. In the same way, the next less inclusive clade – the most recent common ancestor of *Homalocephale* and *Pachycephalosaurus* and all of its descendants – is called Homalocephaloidea, diagnosed by having a broad parietal bone and a broad medial process on the ilium.

Successively more restricted clades are found within Homalocephaloidea. The first, more inclusive, includes *Stygimoloch*, *Stegoceras*, *Tylocephale*, *Prenocephale*, and *Pachycephalosaurus* as terminal taxa and is as yet unnamed. This clade is diagnosed by doming of the thickened frontal–parietal part of the skull roof, obliteration of the sutures between these bones, and closure of the upper temporal fenestrae.

Thereafter is an unresolved relationship among the two species of *Stegoceras* and a clade formed by *Tylocephale*, *Prenocephale*, and *Pachycephalosaurus*, together forming Pachycephalosauridae. In these forms, the nasal and postorbitals are incorporated into the domed skull roof. Finally, *Tylocephale*, *Prenocephale*, and *Pachycephalosaurus* form an unresolved clade within Pachycephalosauridae, recognized by having the dome formed by the frontal, parietal, postorbital, and squamosal bones, loss of the squamosal and postorbital platform, and the presence of tubercles on the nasals. According to a new study by Williamson and Carr, the newly discovered *Sphaerotholus* is also a member of this latter clade.

Pachycephalosaurs meet history: a short account of their discovery

Pachycephalosaurs are now so thoroughly associated with the moniker "dome-headed dinosaurs," it's hard to imagine that human acquaintance with these animals started out in a sorry state of confusion. The story begins in 1856, with Joseph Leidy's studies of some of the first fossil vertebrate material to come back to the Academy of Natural Sciences in Philadelphia from the earliest government explorations of the wilds of the Western Interior of the USA. One of these specimens, discovered in the Judith River Formation of Montana (Late Cretaceous in age), was a curious-looking, triangular, "cuspy" tooth no larger than half a centimeter in height. Leidy thought that this tooth came from a fossil lizard, calling it *Troodon* (*troo* – wound; *odon* – tooth).

Similar kinds of teeth began turning up at the turn of the century through the paleontological explorations of Lambe in Alberta and J. B. Hatcher in Wyoming. Despite the ever-increasing North American sample of these Late Cretaceous teeth, it took Franz Baron Nopcsa, Hungary's best-known dinosaur paleontologist, to suggest that these teeth belonged to a theropod dinosaur instead of a lizard. Nevertheless, that didn't stop a number of dinosaur paleontologists from thinking otherwise. The crux of this difference of opinion was Lambe's 1902 discovery of some rather peculiar, yet fragmentary skull fragments in the same Judith River beds where he had been finding *Troodon*-like teeth. These specimens consisted of domed skull caps that were exceedingly thick. Moreover, the caps were distinctively ornamented with prominent tubercles. The animal from which these skull fragments came he called *Stegoceras* (*stego* – cover; *keras* – horn).

Nearly a quarter of a century later, C. W. Gilmore described the first complete skull and partial skeleton of a dome-headed dinosaur, which he referred to as *Troodon validus*. No longer than 2 m, this animal had a 20 cm long skull, with an exceedingly thick, knobby-looking frontal–parietal region. On the back of this thickened skull cap was a short platform (the parietal–squamosal shelf) ornamented with knobs and bumps. On beyond, the skeleton was much like many other bipedal ornithischians, having relatively short forelimbs (the hand remains unknown), long four-toed hindlimbs, stiff vertebral column, and long tail. Since he gave his new dome-headed dinosaur the same generic name as Leidy's dinosaur tooth, Gilmore apparently thought the teeth in the jaws of his new specimen were quite similar to the one Leidy had described 75 years previously, the result of which was that dome-headed dinosaurs began being called troodontids. The troodontids, Gilmore felt, had close relationships with ornithopods, perhaps even being members of the clade.

Despite the importance of Gilmore's study, it took another 20 years for a new dome-headed dinosaur to be described. In 1943, as part of their larger study of the existing material from these dinosaurs, the legendary dinosaur hunters Barnum Brown and Eric Schlaikjer described a new form – *Pachycephalosaurus* (*pachys* – thick; *kephale* – head; *saurus* – lizard) – which consisted of only a skull. But what a skull: nearly complete, it approached 65 cm in length, with a dome of solid bone 20 cm thick. A skull this size would ride at the end of an estimated 8 m long body. Hailing from the Upper Cretaceous strata of Wyoming, South Dakota, and Montana, *Pachycephalosaurus* was one of the centerpieces of this detailed study, which also included an important review of what was known of these animals up to the 1940s and the first suggestion that these enigmatic dome-heads had evolutionary affinities with Ceratopsia.

Still, Brown and Schlaikjer continued to think that *Troodon* was a dome-headed dinosaur. It took further studies to sort out this identity problem. In 1945, Sternberg argued that Leidy's original *Troodon* tooth came from a small theropod dinosaur (see Chapter 12). More importantly for the dinosaurs at hand, the animal described by Gilmore was correctly renamed *Stegoceras*, and *Pachycephalosaurus* became the namesake of the entire group of dome-heads, which was christened Pachycephalosauridae. This view was seconded by L. S. Russell, who studied the original material of *Troodon* and additional jaw specimens from Upper Cretaceous rocks of Alberta. Pachycephalosaurs finally had come into their own.

Although North American pachycephalosaur specimens continued to be discovered over the ensuing decades, it was not until the 1970s that new kinds of these dome-headed dinosaurs were recognized elsewhere. The first and only pachycephalosaur from England was named by Galton in 1971. Called *Yaverlandia* (from Yaverland Point), this animal is known only from a small skull fragment, collected from the famous Wealden beds of the Isle of Wight, off the southern coast of England.

Thereafter, beginning in the mid-1970s, came a virtual flood of new pachycephalosaurs, this time from Mongolia and China. The first of these was Maryańska and Osmólska's 1974 study of Asian pachycephalosaurs, which unleashed in one fell swoop three new forms, including the first of the flat-headed pachycephalosaurs. Two of these came from the Late Cretaceous Nemegt Formation of Mongolia: *Prenocephale* (*prenes* – sloping) and *Homalocephale* (*homalos* – even). The former – consisting of a complete skull, but partial postcranial skeleton – is the more spectacular in terms of the beautiful preservation of the skull. Apparently all that needed to be done to clean it up after it had been found was to blow a few sand grains out of its nasal cavity.

In contrast to the dome-headed *Prenocephale*, *Homalocephale* is a flat-headed pachycephalosaur. Known from a nearly complete skull, and more importantly a virtually complete, articulated skeleton, *Homalocephale* is perhaps the best known among all pachycephalosaurs. Still lacking a hand (to this day, none has been found for a pachycephalosaur), the rest of the forelimb is quite small, much shorter than the hindlimb. The backbone appears to have been quite rigid, given that each vertebra is locked with the next via tongue-and-groove articulations. Elsewhere, the rib cage is exceedingly rotund, the hips are broad, and the front section of the tail is wide and lacks chevrons along the undersurface of the vertebrae. The tail ends in a criss-cross of ossified tendons, suggesting that this region was very stiff in life.

The last of Maryańska and Osmólska's new pachycephalosaurs is *Tylocephale* (*tyle* – swelling). Collected from the Barun Goyot Formation of Mongolia, this full-domed pachycephalosaur is known so far from only a partial frontal–parietal specimen that appears to be somewhat similar to *Stegoceras*. We have no knowledge yet what the rest of the skeleton of this animal is like.

On the other side of the Mongolian border from where Maryańska and Osmólska had been working, Chinese pachycephalosaurs began showing up in the late 1970s. *Wannanosaurus* (Chinese name for a southern part of Anhui Province), so far known only from a partial skull and associated skeletal fragments, was described in 1977 by L. Hou. It comes from the Xiaoyan Formation (Upper Cretaceous) of Anhui Province, China. This was followed shortly thereafter with the discovery of *Micropachycephalosaurus* (*micro* – small). Named by the Dean of Chinese dinosaur paleontology, Z.-M. Dong, in 1978, *Micropachycephalosaurus* is presently known only from a lower jaw and an associated, but fragmentary postcranial skeleton from the Wang Formation (Upper Cretaceous) of Shandong, China. In this same paper, Dong coined the name Homalocephalidae for those pachycephalosaurs with a thick, yet flattened, skull roof such as *Homalocephale*, *Wannanosaurus*, and possibly *Micropachycephalosaurus*. The full-domed pachycephalosaurs, of course, require a taxonomic name as well, which Dong suggested is Gilmore's Pachycephalosauridae. A year later, another new North American pachycephalosaur reared its unimpeachably ugly head. *Gravitholus* (*gravis* – heavy; *tholos* – dome) was named by Wall and Galton in 1979. Consisting of only part of a dome, this nevertheless distinctive pachycephalosaur was collected from the Judith River Formation of Alberta.

The last four pachycephalosaurs so far to be discovered include *Goyocephale* (*goyo* – decorated), *Stygimoloch* (*Stig* (*Styx*) – river of Hades (Hell Creek); *moloch* – demon), *Ornatotholus* (*ornatus* – adorned), and *Sphaerotholus* (*sphaira* – ball). *Goyocephale* – a flat-headed pachycephalosaur – was described in 1982 by A. Perle, Maryańska and Osmólska. This animal is known from a fragmentary skull and nearly complete postcranial skeleton that came from Upper Cretaceous strata in Mongolia. *Stygimoloch* was originally described by Sues and Galton in 1983 and consists of a domed skull cap and postcranial fragments collected by a variety of different individuals from Upper Cretaceous layers in central and eastern Montana and adjacent parts of Wyoming. *Ornatotholus* was also described by Sues and Galton in 1983; it consists of dome fragments found in the Judith River Formation of both Alberta and Montana. Finally, Williamson and Carr described *Sphaerotholus* from the Upper Cretaceous Kirtland Formation of northwestern New Mexico – it is presently known from several partial skulls.

Important readings

Chapman, R. E., Galton, P. M., Sepkoski, J. J. and Wall, W. P. 1981. A morphometric study of the cranium of the pachycephalosaurid dinosaur *Stegoceras*. *Journal of Paleontology*, **55**, 608–616.

Colbert, E. H. 1955. *Evolution of the Vertebrates: A History of the Backboned Animals Through Time*. Wiley, New York, 479pp.

Davitashvili, L. S. 1961. [*The Theory of Sexual Selection*]. Izdatel'stov Akademia Nauk SSSR, Moscow (in Russian), 537pp.

Galton, P. M. 1971. A primitive dome-headed dinosaur (Ornithischia: Pachycephalosauridae) from the Lower Cretaceous of England, and the function of the dome in pachycephalosaurids. *Journal of Paleontology*, **45**, 40–47.

Giffin, E. B. 1989. Pachycephalosaur paleoneurology (Archosauria: Ornithischia). *Journal of Vertebrate Paleontology*, **9**, 67–77.

Maryańska, T., Chapman, R. E. and Weishampel, D. B. 2004. Pachycephalosauria. In Weishampel, D. B., Dodson, P. and Osmólska, H. (eds.), *The Dinosauria*, 2nd edn, University of California Press, Berkeley, pp. 464–477.

Maryańska, T. and Osmólska, H. 1974. Pachycephalosauria, a new suborder of ornithischian dinosaurs. *Paleontologica Polonica*, **30**, 45–102.

Sereno, P. C. 1986. Phylogeny of the bird-hipped dinosaurs (Order Ornithischia). *National Geographic Society Research*, **2**. 234–256.

Sues, H.-D. 1978. Functional morphology of the dome in pachycephalosaurid dinosaurs. *Neues Jahrbuch für Geologie und Paläontologie, Monatshefte*, pp. 459–472.

Sues, H.-D. and Galton, P. M. 1987. Anatomy and classification of the North American Pachycephalosauria (Dinosauria: Ornithischia). *Palaeontographica A*, **198**, 1–40.

Wall, W. P. and Galton, P. M. 1979. Notes on pachycephalosaurid dinosaurs (Reptilia: Ornithischia) from North America, with comments on their status as ornithopods. *Canadian Journal of Earth Sciences*, **16**, 1176–1186.

CHAPTER 9

Ceratopsia: horns and all the frills

From the time of their discovery in the second half of the 1800s to the present day, there has hardly been a group of dinosaurs that has evoked more fascination than ceratopsians. The quadrupedal, horned, frilled, rhinoceros-like ceratopsians of the Late Cretaceous (Figure 9.1) have one of the most outstanding fossil records of any dinosaur group. Diverse and abundant, these horned-faced dinosaurs must have dominated the landscape, foraging in the shrubs and lumbering or even dashing across the open terrain in places now known as Alberta, Montana, and Wyoming.

Most of our knowledge about these animals comes from the latest Cretaceous of North America, a time when ceratopsian evolution reached its height. It was also a time when these animals were at their largest, in many cases boasting large, aggressive horns. While these late survivors ranged upward of 6 or 7 tonnes, a host of non-horned ceratopsians from slightly earlier in the Cretaceous, hailing from Asia, weighed in at between 25 and 200 kg. Even without horns, it is still easy to recognize the ceratopsian familial stamp. That recognition factor – perhaps the most significant derived feature shared by all ceratopsians – is the uniquely-evolved bone at the tip of the upper jaw, the rostral bone. This new bone gives the snout its parrot-like beak appearance (Figure 9.2).

In all ceratopsians, the hooked beak of the snout together with the predentary (which is also pointed), strongly suggests that these herbivores were capable of a great deal of selective feeding – in fact, much like that we see in living parrots. However, unlike parrots, the rest of the jaw apparatus indicates that these animals were masters of the business of chewing – often through slicing and dicing – the plants they chose to ingest.

Along with a distinctive snout, the horns and frills make ceratopsians instantly recognizable. The horns are probably the most impressive. Some ceratopsians had two prominent horns, one over each eye, and a third – in some cases large and in others small – over the nose. One even had a large mass of roughened bone over the nose and eyes.

Equally important is the ceratopsian frill. Extending from the back of the skull and made up of the parietal and squamosal bones, the frill varied considerably in size. In addition, the margin of the frill was ornamented in some genera by long spikes or by extra bones, known as epoccipitals, which formed a sort of scalloping to the otherwise smooth rim. Regardless of trim, the ceratopsian frill could be quite large: in *Torosaurus*, the 2.7 m long skull is nearly two-thirds frill.

What went on behind the head in ceratopsians is no less interesting. Primitively bipedal, these animals evolved a quadrupedal stance early in their phylogenetic history. Nevertheless, they retained relatively long and powerful hindlimbs, while at the same time developing hooves on the toes of both fore- and hind feet. Ceratopsians, like other ornithischians, had a meshwork of ossified tendons across the back, pelvis, and tail.

Figure 9.1. *Chasmosaurus*, a ceratopsian from the Late Cretaceous of the Western Interior of North America.

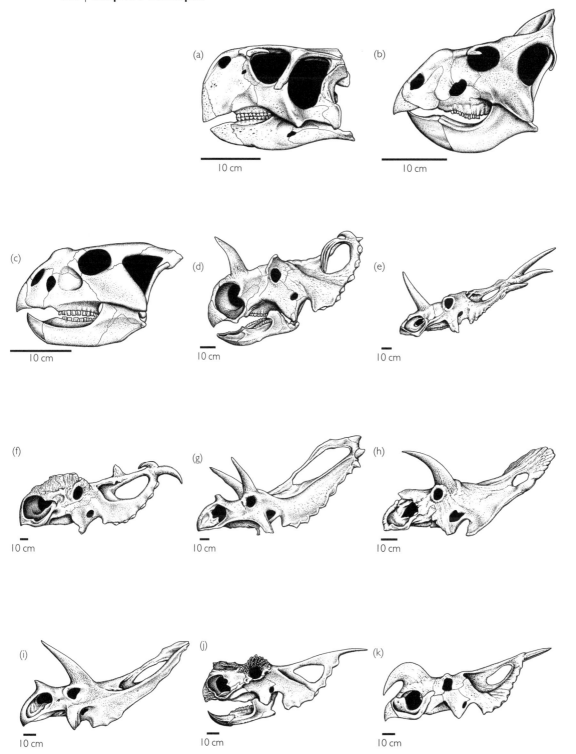

Figure 9.2. Left lateral view of the skull of (a) *Psittacosaurus*, (b) *Leptoceratops*, (c) *Bagaceratops*, (d) *Centrosaurus*, (e) *Styracosaurus*, (f) *Pachyrhinosaurus*, (g) *Pentaceratops*, (h) *Arrhinoceratops*, (i) *Torosaurus*, (j) *Achelousaurus*, and (k) *Einiosaurus*.

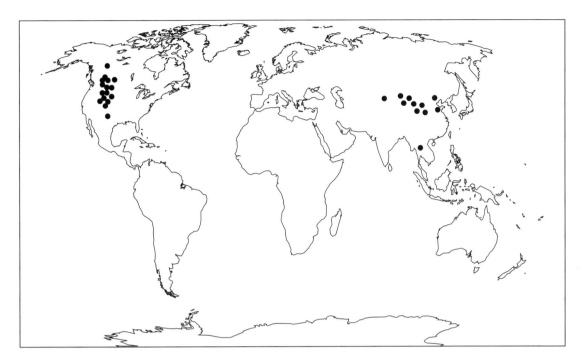

Figure 9.3. Global distribution of Ceratopsia.

The meshwork, a holdover from the primitive ornithischian condition, presumably lessened the stress and strain of weight-bearing during both walking and running.

Ceratopsians are a monophyletic group because of the unique rostral bone at the front of their upper beak. Ceratopsia is presently divided into the small, bipedal psittacosaurids from Asia and neoceratopsians, a huge clade consisting of basal forms from Asia and North America such as *Protoceratops*, *Leptoceratops*, and *Montanoceratops* on the one hand, and the large ceratopsids from North America such as *Triceratops* and *Centrosaurus* on the other hand.

Ceratopsian lives and lifestyles

Feeding

Ceratopsians, now known from great collections made from eastern Asia to western North America (Figure 9.3), come dressed and ready to eat: by chewing. A hooked rhamphotheca, dental batteries (a dense cluster of functional and replacement teeth) in both upper and lower jaws, a sturdy coronoid process, and evidence for the existence of fleshy cheeks all reflect sophisticated adaptations for the manipulation of food in the mouth.[1]

Ceratopsians chewed in ways no other vertebrates appear to have tried. The business end of the ceratopsian mouth was the narrow, parrot-

---

1 This anatomical arrangement – a cropping beak followed by a diastema, dental batteries towards the rear of the mouth, strong muscles to manipulate these, and muscular cheeks to keep things in the mouth – is similar to that seen in the herbivorous hadrosaurids (Chapter 10), as well as many groups of herbivorous mammals (such as cows, sheep, and horses, to name three of the most familiar).

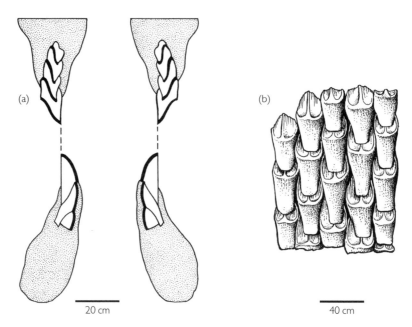

20 cm

40 cm

Figure 9.4. Cross-section through the upper and lower jaws of *Triceratops*, illustrating, (a) high-angle slicing-and-dicing motion of the teeth and (b) internal view of the dental battery in the lower jaw of *Triceratops*.

like beak. The hooked rim of the rostral bone (covered by a sharper cornified rhamphotheca) suggests the capability for careful selection of the plants for food. Because the earliest members of the clade all had the pointed hooked beak, some degree of selectivity must have been present at the very beginning of the history of ceratopsians and was maintained by members of the group until the end. Likewise, all ceratopsians clearly were chewers, grinding leaves and stems between the occlusal surfaces of their sturdy dentition. Worn teeth were constantly replaced, so that the active surface of the dental battery was continually refurbished. The grinding action was bequeathed from the common ancestor of all ceratopsians, and through the phylogenetic history of the group, the angle of occlusion became more and more vertical, until, in the latest forms, a near-vertical slicing and dicing along the side of the tooth, characteristic of the later large ceratopsids, was achieved (Figure 9.4).

The force behind this high-angle mastication derived from a great mass of jaw-closing musculature, which in the frilled forms crept through the upper temporal opening and onto the base of the frill. The other end of this muscle attached to a great, hulking coronoid process on the mandible. All in all, the chewing apparatus in ceratopsians was among the most highly evolved in all vertebrates.[2]

---

2 Interestingly, if the kinds of chewing we've described here were not enough for the likes of *Psittacosaurus*, then a packet of gastroliths lodged in the gizzard doubly pulverized its meal. Only in *Psittacosaurus* among ceratopsians (in fact among all ornithischians) are gastroliths known. Perhaps such a unique occurrence represents a special (that is, derived) feeding strategy in this most primitive of ceratopsians.

On beyond the mouth, the remainder of the digestive tract does not appear to have been disproportionately large in ceratopsians, a likely consequence of their masticatory prowess. Nevertheless, it must have been sufficiently voluminous to accommodate the continual passage of the great quantities of foliage that formed the diet of these animals.

Both small and bipedal ceratopsians, as well as even the largest quadrupedal forms never browsed particularly high above the ground. At their largest (*Triceratops* and *Torosaurus*), browse height was probably no more than 2 m and no one has seriously entertained the possibility that ceratopsids were able to rear up on their hind legs to forage at higher levels (as has been suggested for some sauropods and stegosaurs). Nevertheless, they may have been able to knock over trees of modest size in order to gain access to choice leaves and fruits.

Which plants were preferred by ceratopsians remains a mystery. Once thought to be feeders on the fibrous fronds of cycads and palms, the majority of ceratopsians are rarely found in the same areas as these kinds of plants. The principal plants whose statures match browsing heights of ceratopsians were a variety of shrubby angiosperms, ferns, and perhaps small conifers. In fact, it has been argued by two paleo-botanists, S. Wing and B. Tiffney, that ceratopsians, along with their cohort of other large yet low-browsing, generalist-feeding herbivorous dinosaurs were doing a reciprocally advantageous evolutionary waltz with early flowering plants (see Chapter 16). Suffice it to say for the present that herbivorous dinosaur feeding habits may have contributed to the extraordinary rise of flowering plants during the Late Cretaceous. This hypothesis and a related one are discussed in Chapters 15 and 16.

Locomotion

The legs of horned dinosaurs are unlike those of any mammal – living or extinct – and thus exactly how ceratopsians cruised over their Cretaceous landscapes remains a matter for some conjecture (Figure 9.5). Did they thunder along like enraged rhinos, or were their legs (and perhaps metabolism, as well) built in such a way that this kind of locomotion was impossible? Although the matter remains unresolved, we are closer to developing a picture of the locomotor skills of these horned dinosaurs.

As befits its primitive position within Ceratopsia, *Psittacosaurus* appears to have retained the fully bipedal limb posture found among other ornithischians. The same condition is found in *Leptoceratops* and *Microceratops*, but thereafter ceratopsians assumed a quadrupedal posture. How fast these animals may have traveled is not known with much precision, especially as there are no known trackways for any of these animals (why this should be so remains a mystery). However, those estimates that have been made – based principally on biomechanical analyses of the limbs – range from 20 to 50 km/h. A problem with this kind of analysis, however, is that the orientations of the front limbs in quadrupedal ceratopsians are not fully understood. R. T. Bakker has argued strongly for what is essentially a mammal-like posture; that is, fully erect front limbs that mimic the well-known posture of the back

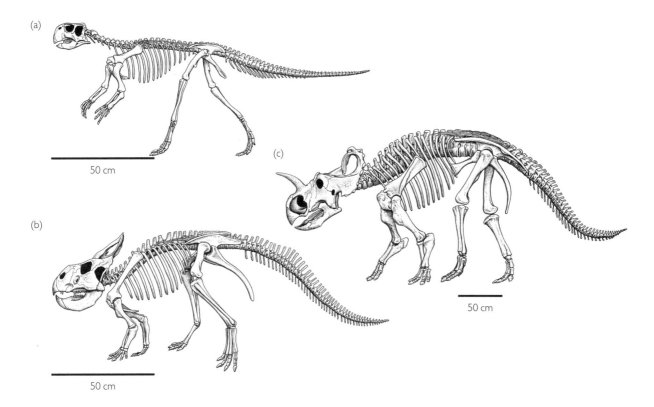

(a)

50 cm

(b)

50 cm

(c)

50 cm

Figure 9.5. Lateral view of the skull and skeleton of (a) *Psittacosaurus,* (b) *Protoceratops,* and (c) *Centrosaurus.*

limbs. Others have argued that this cannot be so, and that a more sprawling posture of the front legs is indicated (see Chapter 15).

For the present, R. A. Thulborn has presented the most detailed speed estimates. His work indicates that average walking speed for both large and small ceratopsians was somewhere between 2 and 4 km/h, while maximum running speeds ranged from 30 to 35 km/h (see also Box 15.3).

Horns, frills, and ceratopsian behavior

At virtually *any* speed, a *Triceratops* juggernauting across the Late Cretaceous countryside was apt to pack one serious wallop if something got in the way, especially when fully equipped with horns. Long presumed to function for individual defense against predators, horns and their relationship to frills and defense were examined by E. H. Colbert in 1951. He began his investigation by observing that, in living African bovids (e.g., impalas, antelopes, and gnus), it is not so much the shape of the horns of these animals that is important but rather their presence that is critical to their survival. In this way, he interpreted the various cranial ornaments as being independently evolved, but more or less equally satisfactory solutions to the same problem: resisting predators during interspecific (*inter* – between; between different species) combat. Thus we have bovid horns as weapons in defense against predators. By analogy, ceratopsian horns were thought to have functioned to ward off predators at close quarters and in the classic Late Cretaceous All-Star Game between *Tyrannosaurus* and *Triceratops*, *Triceratops* could come up the winner by virtue of its prominent brow and nose horns. The frill presumably added additional insurance against lethal bites to the neck region.

Subsequent interpretations have also drawn on analogy with horned mammals such as bovids, but have instead centered on the intraspecific (*intra* – within; among members of the same species) functioning of horns; that is, their behavioral context in display, ritualized combat, defense of territories, and establishment of social ordering. Thus variation in horns and antlers do count, first at the level of making sure that species recognize each other and hence prevent matings between different species and second by aiding in the establishment of within-species dominance hierarchies (that is, who within a population gains in reproductive and/or resource rights) or, possibly, in defending territories.

The link between dominance, defense, and horns comes from important studies of mammals in their natural habitats, research that has gained increasing prominence ever since humans began taking stock of how much they have disturbed virtually all terrestrial (and aquatic) ecosystems. For example, we know that in the case of almost all horned mammals, larger males tend to have a reproductive advantage over smaller males. Simply put, they tend to breed more often than do small individuals, most likely because females choose them more often than not. Dominance in these mammals (and in other tetrapods) is accentuated by the development of structures that "advertise" the size of the animal; these obviously include horns and antlers, as well as the horn-like ossicones of giraffe and the nasal horns of rhinoceroses. Indeed, these structures increase visibility and hence the probability that the owners of the horns and antlers fend off competitors and impress females (in most cases). On the basis of sexual selection (selection that favors display-related structures in one or both sexes within a given species), these features should come to be highly linked with reproductive success. Put on a good show and your genes have a greater chance of making their way into subsequent generations.

In short, the variety of horn and antler shapes are now known to reflect (1) species-recognition mechanisms that aid in preventing interspecies matings and (2) intraspecific differences in displays and ritualized fighting behavior. Can such interpretations of intraspecific behavior and sexual selection in mammals shed light on the development of horns and frills in ceratopsians? Several dinosaur paleontologists, among them L. S. Davitashvili, J. O. Farlow, P. Dodson, R. E. Molnar, N. B. Spassov, J. H. Ostrom, and P. Wellnhofer, think so. As does Scott Sampson of the University of Utah, who has provided the most comprehensive and detailed discussion of the behavioral significance of cranial ornamentation in ceratopsid ceratopsians.

Let's begin by examining the common thread of these investigations into the evolution of ceratopsian horns and frills from the perspective of their intraspecific behavioral significance. No one has ever doubted that ceratopsian horns were used for combat; the question has been at whom they were aimed. Using modern horned mammals as analogues, the large nasal and brow horns of ceratopsians are thought to have functioned primarily during within-species combat (as in territorial defense and establishing dominance hierarchies). Similarly, the development of elaborate scallops and spikes along the frill margin in many of the more highly derived ceratopsians separates one species from another. However, what we dinosaur paleontologists use to recognize species aren't necessarily the same criteria that ceratopsians used to size each other up: the size and shape of horns and frills. Can we provide any more evidence that the behavioral hypothesis is likely to be true?

Before answering this question, we must first look to another hypothesis that attempts to explain the evolution of ceratopsian frills (but not horns). In 1966, Ostrom suggested that the frill provided a platform for the attachment of the major mass of jaw-closing muscles; these were thought to have extended through the upper temporal fenestra and onto the upper surface of the frill. Thus, in long-frilled ceratopsians, the jaw musculature must have been very extensive and consequently exceedingly powerful. The phylogenetic changes in the size and shape of the frill among ceratopsians, Ostrom suggested, reflected changes in the attachment and action of muscles running between the frill and the lower jaw, ultimately increasing the power of the bite.

Dodson and others have remained suspicious of such an explanation, because it focuses only on the significance of frills in aiding chewing. For example, as the frill of *Protoceratops* grows during the lifetime of the animal, from hatchling to old adults, the jaw muscles show a decrease in their mechanical ability to produce high bite forces. Likewise, there is a marked sexual dimorphism in the size and shape of adult frills, suggesting that jaw mechanics is unlikely to be the sole factor governing frill morphology. Instead, it is probable that ceratopsian frills answered to other important functions to account for such aspects of ceratopsian biology as sexual dimorphism, ontogenetic patterns, and taphonomic occurrences.

As has been mentioned, the earliest dinosaur egg nests to be found were then ascribed to *Protoceratops*, but recent discoveries in the Gobi Desert clearly indicate that these belong to oviraptorid theropods. However, *Protoceratops* hatchlings are known from complete skeletons that occur in what can only be interpreted as nests (Figure 9.6); these hatchlings are about 25 cm long. Virtually all of the remaining growth stages have been documented, making the ontogenetic development of *Protoceratops* one of the best understood. This is particularly relevant to the question at hand ("What is the function of frills?"), because it allows us to ask "When during development does the frill begin to grow?" and "When does it become most expansive?" Thanks to statistical studies of growth in *Protoceratops*, it is now clear that frill development takes off when individuals, apparently both males and females, are approaching fully adult body sizes, reaching their maximum shortly thereafter. In the

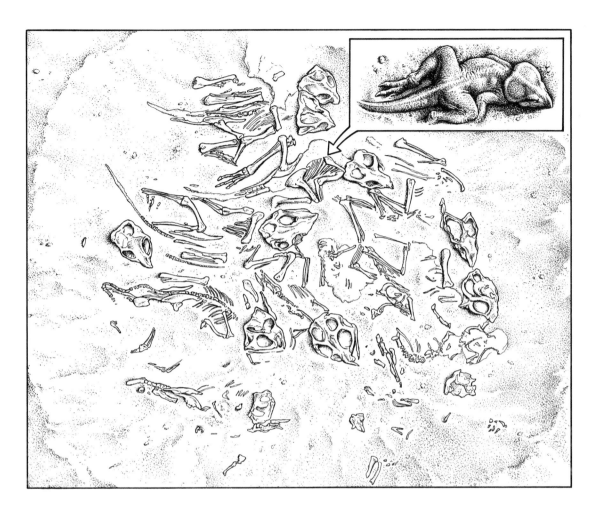

Figure 9.6. A nest of hatchling *Protoceratops* from the Late Cretaceous of Mongolia.

group arbitrarily designated "males," the frill becomes inordinately larger and showier than in the "females" (Figure 9.7). This pattern strongly suggests that the onset of frill growth occurs with sexual maturity and therefore that there is a reproductive connection to frill size and shape. Sounds like sexual selection?

This pattern, in which frills take on their greatest prominence once sexual maturity was reached, is now thought to occur also in other ceratopsians, among them *Centrosaurus* and *Chasmosaurus*. In many of these forms, the development of scallops and spikes on the frill margin would enhance the dimorphic nature of the frill.

Taken as individual features, frills apparently lend themselves to explanations involving within-species display. They even seem to develop in synchrony with the attainment of sexual maturity. But what can be said about the social context of ceratopsians, the arena in which the frill functioned? At the very least, it can be claimed with much justification that many if not all ceratopsians lived in large herds, at least during part

(a)

(b)

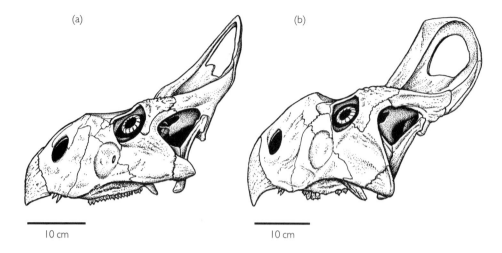

10 cm

10 cm

Figure 9.7. Sexual dimorphism in *Protoceratops*. Note in (a), a presumed female, the frill is less showy and the nasal ridge is less prominent, quite the opposite of (b), a presumed male.

if not all of the year. This justification comes from our ever-increasing catalog of ceratopsian bonebeds. These mass accumulations of single species of ceratopsians are known for at least nine separate species, including several in which the minimum number of individuals may exceed 100. Herding ceratopsians is consistent with these bonebeds. Moreover, such gregariousness also makes sense when one is putting frills and horns into their behavioral context. Territoriality, ritualized combat and display, and the establishment of dominance hierarchies are to be expected in animals that are thrown together in highly social circumstances such as herds. Perhaps in ceratopsians, we are seeing an example of the most complex of dinosaur intraspecific behavior.

Still, we are left with a series of ruminations about frills, horns, population density, and behavior. Is there nothing like a "smoking gun" from the fossils themselves that might give us a clue that we're on the right track? What might be expected of threatening, displaying, and combating ceratopsians that could be recognized in the fossil record are injuries, such as puncture wounds inflicted on faces, frills, and bodies of competing "males"? In fact, puncture wounds are preserved in at least five forms. These pathologies, not only on the cheek region, but also in the frill provide strong evidence of the blood-letting that comes from head-on engagements between competing members of the same species.

Given the possibility that ceratopsians may have had complex social behaviors involving display, ritualized combat, the establishment of dominance hierarchies, and defense of territories, it comes as a bit of a surprise that their brain size is not at all large (see Box 15.4). For example, despite being near opposites in terms of body size and display-related anatomy, both *Protoceratops* and *Triceratops* had brains less than the size expected of a similarly sized crocodilian or lizard. Cerebrally, they were above average as compared with sauropods, ankylosaurs, and stegosaurs, but commanded proportionally less gray matter than either ornithopods

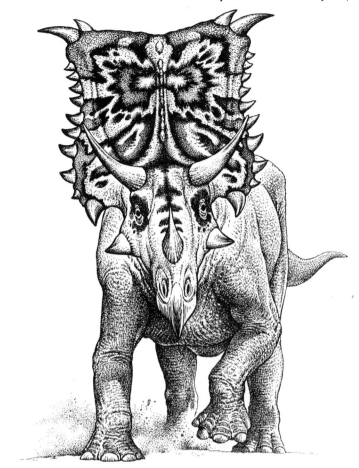

Figure 9.8. "Back off": frill display in *Chasmosaurus*.

or theropods. As J. A Hopson has suggested, perhaps these brain size measures indicate that ceratopsians had relatively unhurried and uncomplicated lifestyles.

Even without high levels of brainpower, there is strong evidence that the peaceful life of ceratopsians was interrupted – at least intermittently – by resounding clashes. Farlow and Dodson outlined the evolution of intraspecific behavior among ceratopsians. To begin with, display- and combat-related behavior seems to have been present primitively among neoceratopsians and perhaps even among all ceratopsians. With their frills serving as a visual dominance rank symbol and their small, yet sharp nasal horns acting as weaponry, display in *Leptoceratops*, *Protoceratops*, and *Montanoceratops* perhaps involved swinging the head from side to side. Should this ritual have failed to impress, these animals may have rammed their horns full tilt into the flanks of their opponent.

The more derived ceratopsids share more elaborate frills and either nasal or brow horns. Among the long-frilled chasmosaurines (e.g., *Chasmosaurus*, *Pentaceratops*, and *Torosaurus*), the display function of the frill may have been exaggerated. The very long frills of these dinosaurs could have provided a very prominent frontal threat display exhibited not only by inclining the head forward (Figure 9.8) but also by nodding or

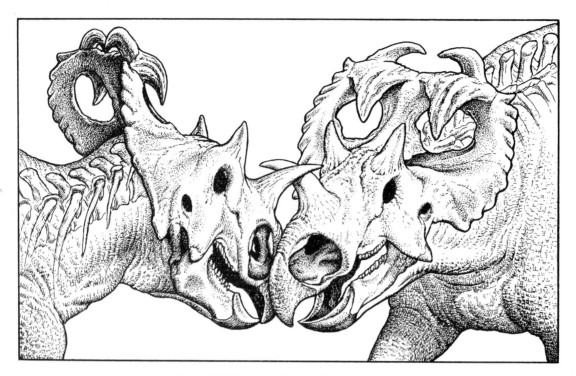

Figure 9.9. "Crossing of the horns": combat between male *Centrosaurus*.

shaking the head from side to side. Should such a display have failed to send the message to one or the other of the opponents, combat may have involved frontal engagement of the nasal and brow horns, with shoving and wrestling determining the winner and loser of the contest. In contrast, most of the short-frilled centrosaurines (such as *Centrosaurus*, *Avaceratops*, and possibly *Pachyrhinosaurus*) were rather rhinoceros-like in their appearance and probably in their behavior as well (Figure 9.9). We presume that opponents tried to catch each other on their nasal horns, thus reducing to a degree the amount of damage inflicted to the eyes, ears, and snout. Nevertheless, the possibility of injury assuredly would have been very much greater in short-frilled ceratopsians than in other taxa.

In the context of frills, horns, and behavior discussed above, two anomalous forms stand out in the herd. First, *Styracosaurus* was a short-frilled centrosaurine whose frill appears inordinately large because of long spikes along its margin. Given its otherwise rhinoceros-like profile, *Styracosaurus* also was provided with the exceptionally distinctive display qualities of the frill. Should frill-wagging have proved to be an insufficient threat or deterrent, *Styracosaurus* could then have relied on the kinds of horn-locking and head-pushing that may have characterized other centrosaurines.

In contrast, *Triceratops* was a chasmosaurine ceratopsid whose frill is secondarily shortened, suggesting that its threat display may have been less utilized than in other chasmosaurines. Instead, combat may have been similar to what is primitively found in chasmosaurines (engagement of brow and nasal horns, followed by shoving and wrestling), but in which the solid frill served as a shield against the parry and thrust of the opponents' horns.

The cephalization of weaponry and display seems almost to drive the evolution of ceratopsian dinosaurs. In this diverse group, we witness a world where display and competition were all important, where – when push came to shove – it may have been better to nod vigorously than to cross horns.

The evolution of Ceratopsia

As was indicated earlier, Ceratopsia is a monophyletic taxon that consists of the common ancestor of members of Psittacosauridae and Neoceratopsia and all the descendants of this common ancestor (Figure 9.10). Among the rich array of unambiguous derived features shared by this ceratopsian clade, the most important include a rostral bone, a skull that is narrow at the beak end and flaring and deep in the cheek region (Figure 9.11), a frill composed principally of the paired parietal bones, and a strongly vaulted palate beneath the beak.

Ceratopsians and pachycephalosaurs (Chapter 8) appear to share a unique common ancestor, thus forming a larger clade that P. C. Sereno has called Marginocephalia (see introductory text to Part II: Ornithischia). More distant relationships are with Ornithopoda (Chapter 10) and Thyreophora (Chapters 6 and 7).

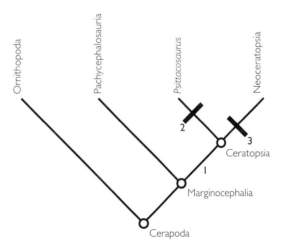

Figure 9.10. Cladogram of Cerapoda, emphasizing the monophyly of Ceratopsia, *Psittacosaurus*, and Neoceratopsia. Derived characters include: at **1** rostral bone, a high external naris separated from the ventral border of the premaxilla by a flat area, enlarged premaxilla, well-developed lateral flaring of the jugal; at **2** short preorbital region of the skull, very elevated naris, loss of antorbital fossa and fenestra, unossified gap in the wall of the lacrimal canal, elongate jugal and squamosal processes of postorbital, dentary crown with bulbous primary ridge, manual digit IV with only one phalanx, manual digit V absent; at **3** enlarged head, keeled front end of the rostral bone, much reduced quadratojugal, primary ridge on the maxillary teeth, development of humeral head, gently decurved ischium.

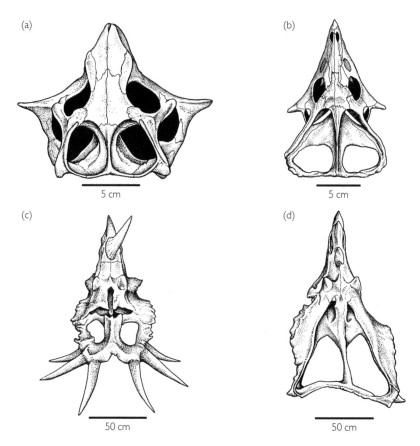

Figure 9.11. Dorsal view of the skull of (a) *Psittacosaurus*, (b) *Protoceratops*, (c) *Styracosaurus*, and (d) *Chasmosaurus*.

Ceratopsia primitively consists of Psittacosauridae and the much more diverse monophyletic Neoceratopsia (Figure 9.12). Psittacosauridae is presently the most species-rich, genus-poor clade yet known among dinosaurs: it consists of a single genus (*Psittacosaurus*) and 10 species. Thus far, only the bare rudiments of the evolutionary relationships of these species are known. Still, members of this clade share as many as 12 derived features, among them a short snout, highly positioned external nares (the opening in the skull accommodating the nostrils), loss of the antorbital opening, and loss of the fifth digit on the hand. Also characteristic of all species of *Psittacosaurus* are their adult size (about 2 m), making them one of the smallest of all ceratopsians. Juveniles of the *Psittacosaurus* clade are known; hatchlings were no more than 23 cm long, about the size of an adult pigeon. In addition, a recently discovered *Psittacosaurus* skeleton provides evidence of its epidermal covering. In addition to the irregular pavement of large and small scales surrounding most of the body, more than 100 long, bristle-like structures extend vertically from the top of the base of the tail. Some regard these filaments as homologous with the integumentary filaments seen in theropods such as *Sinosauropteryx* and true feathers present in *Caudipteryx*, *Microraptor*,

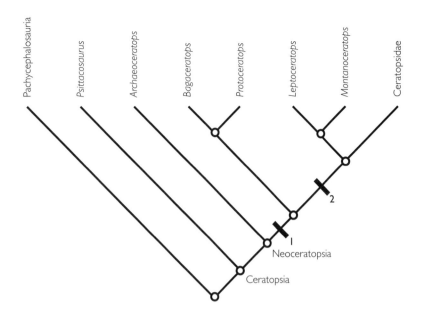

Figure 9.12. Cladogram of basal Neoceratopsia, with the more distantly related *Psittacosaurus* and Pachycephalosauria. *Derived characters include: at* **1** *elongated preorbital region of the skull, an oval antorbital fossa, triangular supratemporal fenestra, development of the syncervical (fusion of cervical vertebrae); at* **2** *greatly enlarged external nares, reduced antorbital fenestra, nasal horn core, frontal eliminated from the orbital margin, supraoccipital excluded from foramen magnum, marginal undulations on frill augmented by epoccipitals, more than two replacement teeth, loss of subsidiary ridges on teeth, teeth with two roots, ten or more sacral vertebrae, laterally everted shelf on dorsal rim of ilium, femur longer than tibia, hoof-like pedal unguals.*

and true birds. Others have argued that they are not structurally similar to the filaments in theropods, so that they are probably not homologous. In either case, however, they were probably used in some sort of display behavior, particularly if they were colored, as has been speculated.

Another basal ceratopsian from Asia – *Chaoyangosaurus* from the Late Jurassic or Early Cretaceous of Liaoning, China – may be positioned directly above or below *Psittacosaurus*. Although the upper part of the skull is not yet known, it does preserve a rostral bone and a widely projecting cheek region, attributes of Ceratopsia. At the same time, numerous features ally it either directly with *Psittacosaurus*, as the closest known relative of Neoceratopsia, or as a taxon more basal than *Psittacosaurus* within Ceratopsia. At present, we place it in an unresolved, basal position within Ceratopsia.

Neoceratopsia, the remaining clade of ceratopsians, is clearly monophyletic, based on at least 10 important shared, derived characters. These include a sharply keeled rostral bone and a predentary that both end in a point, reduction or loss of premaxillary teeth, loss of the external opening in the lower jaw (the external mandibular fenestra), and a very short projection of the lower jaw beyond the jaw joint (the retroarticular process).

Thanks to recent studies by B. Chinnery and D. Weishampel, P. J. Makovicky, and H.-L. You and P. Dodson, we have a good appreciation for the structure of taxa among basal neoceratopsians (Figure 9.12). At its base, this clade consists of *Archaeoceratops* and an unnamed clade that can be broken down into two further clades. One of these, formed of *Asiaceratops*, *Montanoceratops*, *Udanoceratops*, and *Leptoceratops* has been called Leptoceratopsidae. The other clade, *Microceratops*, *Protoceratops*, *Bagaceratops*, and *Zuniceratops*, represents the closest relatives of Ceratopsidae, and within this group *Protoceratops* and *Bagaceratops* are each other's closest relatives. In this context only is there a monophyletic Protoceratopsidae formed of these two taxa.

Ceratopsidae   Now that we have dealt with the question of primitive neoceratopsians, we can finally turn to Ceratopsidae (Figure 9.13). This monophyletic clade, which includes Centrosaurinae (those ceratopsids with short squamosals) and Chasmosaurinae (those with long squamosals), is supported by upward of 50 important diagnostic features, among

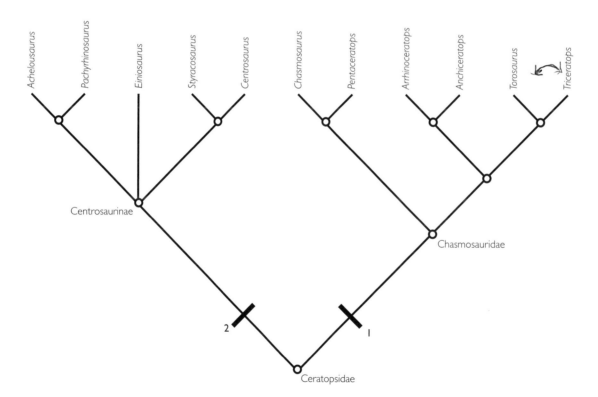

Figure 9.13. Cladogram of Ceratopsidae. Derived characters include: at **1** enlarged rostral, presence of an interpremaxillary fossa, triangular squamosal epoccipitals, rounded ventral sacrum, ischial shaft broadly and continuously decurved; at **2** premaxillary oral margin that extends below alveolar margin, postorbital horns less than 15% skull length, jugal infratemporal flange, squamosal much shorter than parietal, six to eight parietal epoccipitals, predentary biting surface inclined steeply laterally.

them enlarged external nares set into well-developed excavations on the snout, folding of bones on the top of the head to form a secondary skull roof, reduced upper temporal opening, dental batteries, and an everted (sticking out) dorsal border on the ilium.

As a monophyletic clade, chasmosaurines all uniquely share a suite of modifications of the snout, nasal horn, and external nares. In addition, the frill becomes enlarged and there is a distinctive change in the pattern of ornamentation of the margin of the frill. Within this clade, the most primitive members are *Chasmosaurus* and *Pentaceratops*, which together appear to be each other's closest relatives. They share a number of derived features, among them huge openings in the parietal part of the frill and large, flat epoccipital bones along the frill margin.

All remaining, more derived members of Chasmosaurinae consist of the clade containing *Anchiceratops, Arrhinoceratops, Torosaurus, Diceratops,* and *Triceratops.* Sharing as many as seven derived features (among them, further modification of the snout, the development of additional sinuses in the head, and expansion of the ischium), this group has as its primitive members the small clade of *Chasmosaurus* and *Pentaceratops*, followed by another small clade that comprises *Anchiceratops* and *Arrhinoceratops* (themselves united by having a square frill that has very numerous traces of blood vessels on its undersurface).

That leaves us with a triumvirate of chasmosaurines: *Torosaurus, Diceratops,* and *Triceratops.* This clade, which shares a number of new modifications of the snout and nasal horn, has *Triceratops* as its primitive member and *Diceratops* and *Torosaurus* as the "top of the tree."

The other great ceratopsian clade is Centrosaurinae. This group ancestrally acquired a number of important features, among them a large nasal horn, long and narrow opening in the top of the skull roof, short brow horns, short squamosals, and broad, rounded epoccipitals decorating the rim of the frill. Centrosaurinae has at least five members: *Achelousaurus, Einiosaurus, Centrosaurus, Pachyrhinosaurus,* and *Styracosaurus.* The evolutionary interconnectedness of these taxa is as yet not well resolved. On the one hand is a small clade of *Achelousaurus* and *Pachyrhinosaurus*, with *Styracosaurus* and *Centrosaurus* on the other; the fifth taxon, *Einiosaurus*, remains unresolved at the base of Centrosaurinae.

Ceratopsians meet history: a short account of their discovery

As the Western Interior of the USA began yielding its fossil riches to the paleontologists of the day (or rather, to their hard-working collecting parties; see Box 6.2), the first of what came to be known as ceratopsians was discovered. The initial finds, however, were not auspicious. The first was a partial pelvis discovered near Green River, Wyoming, from Upper Cretaceous rocks. Named *Agathaumas* (*aga* – very; *thauma* – wonder) by E. D. Cope in 1872, this material would prove to be of little value in understanding the affinities and biology of these great dinosaurs. *Monoclonius* (*mono* – single; *klon* – sprout, referring to the root of the tooth), recovered from Montana and named in 1876 by Cope on the basis of a fragmentary pelvis, some vertebrae, a few teeth, and a parietal bone, also provided little sense of Ceratopsia-ness. Eleven years later, O. C. Marsh announced the discovery of a pair of very large bony ceratopsian horn cores near Denver, Colorado, that he referred to as *Bison alticornis*!

It wasn't until the late 1880s that the true identity of the horn cores, pelvis, and collection of *Monoclonius* bones and teeth was finally revealed. The enlightenment was due to none other than *Triceratops* (*treis* – three; *kerat* – horn; *ops* – face), originally uncovered from Upper Cretaceous rocks in Wyoming. Named by Marsh in 1889 on the basis of an imperfect skull from a very huge animal, *Triceratops* is now known from innumerable complete and partial skulls, as well as a number of well-preserved skeletons that have been collected from Alberta and Saskatchewan in the north to Colorado in the south.

Two years later, Marsh named another Late Cretaceous ceratopsian – *Torosaurus* (*toreo* – perforate, referring to the holes in the frill) – again from Wyoming. Today it is the most widespread of dinosaurs, having been documented from additional sites in Saskatchewan, Montana, South Dakota, Colorado, Utah, New Mexico, and Texas.

The first quarter of this century established that ceratopsians were dinosaurs to be reckoned with. Discovered with a vengeance, particularly in Alberta during the years of the great Canadian dinosaur rush, most of the spoils went to the National Museum of Canada (now the Canadian Museum of Nature) in Ottawa, the Royal Ontario Museum in Toronto, and the American Museum of Natural History in New York. In 1904, *Centrosaurus* (*kentron* – sharp-point) gave us – and L. M. Lambe – a glimmer of the wealth of ceratopsians when it was first discovered. This was followed shortly thereafter by the discovery of *Diceratops*, collected from Upper Cretaceous strata of Wyoming and named by J. B. Hatcher (see below). Over the years, this ceratopsian came to be thought of as *Triceratops*, but C. Forster suggests that it is a valid taxon, distinct among ceratopsian genera.

These early discoveries as well as aspects of ceratopsian anatomy and paleoecology were first presented in comprehensive fashion in a large study begun by Marsh, and later taken over by Hatcher, Marsh's principal collector and an accomplished researcher in his own right, after Marsh died. The project eventually fell to Richard Swan Lull (also at Yale) to complete after Hatcher's tragic death from typhoid fever at the age of 43. Despite its morbid history, this monograph – finally published in 1907 – stands today as one of the most important references on those ceratopsians that were known at the turn of the century.

Yet it was not until the 1910s and 1920s that ceratopsian riches of Canada were to be fully realized. In 1913 and 1914 alone, *Styracosaurus* (*styrak* – spike), *Chasmosaurus* (*chasma* – wide opening), *Anchiceratops* (*anchi* – close), and *Leptoceratops* (*leptos* – slender) were discovered, described, and named by Lambe and by Barnum Brown. All the while but across the U.S. border, C. W. Gilmore was providing Montana with another ceratopsian, this time *Brachyceratops* (*brachys* – short).

The world of ceratopsians opened widely in the 1920s. Hailing from New Mexico, *Pentaceratops* (*penta* – five) was described by H. F. Osborn in 1923, while on the other side of the globe the Gobi Desert was beginning to yield its dinosaurian treasures through the toil of the American Museum of Natural History's Central Asiatic Expeditions. The most famous of all was *Protoceratops* (*protos* – first), announced to the world in 1923 by W. Granger and W. K. Gregory. Also from the Gobi Desert and no less important were the remains of *Psittacosaurus* (*psittakos* – parrot), described by Osborn in 1923. Osborn, a formidable figure in the history of vertebrate paleontology and head of Vertebrate Paleontology (and later Director) at the American Museum of Natural History, provided the first evidence of the early history of ceratopsian dinosaurs. *Psittacosaurus* is now also known from a variety of localities in China and even from Gorno-Altayask Autonomous Region in eastern Russia. A return to North America completes the research cycle of the 1920s. In 1925, W. A. Parks described yet another long-frilled ceratopsian, *Arrhinoceratops* (*a* – without; *rhin* – nose), from the great fossil beds of Alberta.

In the years that intervened between the 1920s and the present, new studies of ceratopsians came out in what might best be described as fits and starts. In 1933, Lull published his revision of ceratopsian dinosaurs, emphasizing the anatomy of *Monoclonius* (now known to be *Centrosaurus*), as well as the burgeoning record of horned dinosaurs from the great Canadian dinosaur rush. Yet it was not until 1942 that another new ceratopsian, *Montanoceratops* (from Montana) was announced to the world. Originally described by B. Brown and E. Schlaikjer as a new species of *Leptoceratops* from the Late Cretaceous of Montana, it took nine years until C. M. Sternberg, paleontologist for the National Museum of Canada, recognized it as a new kind of ceratopsian. He then baptized it with the name of its host state. Meanwhile (1950), he was also was busy with the announcement of another ceratopsian from Upper Cretaceous rocks of Alberta. *Pachyrhinosaurus* (*pachy* – thick) was one of the ugliest (or most magnificent, depending upon one's perspective) of ceratopsians. *Pachyrhinosaurus* came replete with masses of exceedingly roughened bone extending from the top of the snout to over both eyes. Even without the adornment of prominent nose or brow horns, here was a ceratopsian to be reckoned with.

Three years after C. M. Sternberg's description of *Pachyrhinosaurus*, B. Bohlin – paleontologist with the joint Sino-Swedish Expedition to the Gobi Desert – published his account of *Microceratops* (*micro* – small). Much like *Leptoceratops* and *Protoceratops*, this small ceratopsian was without horns, but sported a modest frill behind the head. *Microceratops* is now known from several localities in northern China and southern Mongolia.

It was another 22 years before these Asian ceratopsians were again in the limelight, this time as the result of a comprehensive study of new (and some old) material from the Gobi Desert. In 1975, T. Maryańska and H. Osmólska, the two principal dinosaur specialists in the Polish–Mongolian Palaeontological Expedition (1963–1971) provided the first comprehensive description of the anatomy, paleobiology, and evolution of *Protoceratops*, *Microceratops*, and their brethren, including a new form that they called *Bagaceratops* (Mongolian: *baga* – small).

0

Recently, a variety of new kinds of ceratopsians have been described. The first, *Avaceratops* (to honor Ava Cole, the wife of the collector of these fossils), comes from a 1986 study by Dodson and represents the first advanced horned ceratopsid to have been discovered since Sternberg's discovery of *Pachyrhinosaurus*. Nearly ubiquitous *Protoceratops*-like ceratopsians continued to tumble out of the rocks. One, discovered in Kazakhstan by L. Nessov and co-workers, was named *Turanoceratops* (from Turan, Kazakhstan) in 1989. Another – *Breviceratops* (*brevis* – short) – was originally thought by Maryańska and Osmólska in 1975 to be a new species of *Protoceratops*. It was S. Kurzanov who rechristened it *Breviceratops* in 1990.

Ceratopsian discoveries continued to be made throughout the rest of the 1990s and into the new millennium in both Asia and North America. In Mongolia, *Udanoceratops* (from Udan Sair) was the product of on-going field research conducted by Kurzanov, and *Graciliceratops* (*gracilis* – slender) represents a renaming of material from the Gobi Desert by P. C. Sereno in 2000 that was originally thought to belong to *Microceratops*. On the other hand, *Archaeoceratops* ("ancient ceratops") and *Chaoyangsaurus* (named for the founder of vertebrate paleontology in China, C.-C. Young) are newly discovered Chinese forms, the former named by Z. Dong and Y. Azuma in 1997 and the latter by X. Zhao, Z. Cheng, and X. Xu in 1999; both have proved pivotal in our understanding of ceratopsian relationships.

In 1995, S. D. Sampson described the latest ceratopsids to be discovered – *Achelousaurus* and *Einiosaurus* – both from the upper beds of the Two Medicine Formation (Upper Cretaceous) of Montana. These unusual-looking animals bear some of the strangest horns ever seen on a ceratopsian. While these same strata have produced the spectacular hadrosaurid nesting sites, North American ceratopsian eggs, babies, and nests remain virtually unknown.

Important readings

Dodson, P. 1976. Quantitative aspects of relative growth and sexual dimorphism in in *Protoceratops*. *Journal of Paleontology*, **50**, 929–940.

Dodson, P. 1992. Comparative craniology of the Ceratopsia. *American Journal of Science*, **293-A**, 200–234.

Dodson, P. 1996. The *Horned Dinosaurs*. Princeton University Press, Princeton, NJ, 346pp.

Dodson, P., Forster, C. A. and Sampson, S. D. 2004. Ceratopsidae. In Weishampel, D. B., Dodson, P. and Osmólska, H. (eds.), *The Dinosauria*, 2nd edn. University of California Press, Berkeley, pp. 494–513.

Farlow, J. O. and Dodson, P. 1975. The behavioral significance of frill and horn morphology in ceratopsian dinosaurs. *Evolution*, **29**, 353–361.

Hatcher, J. B., Marsh, O. C. and Lull, R. S. 1907. The Ceratopsia. *U.S. Geological Survey Monograph*, **49**, 1–300.

Lull, R. S. 1933. A revision of the Ceratopsia or horned dinosaurs. *Peabody Museum of Natural History Memoires*, **3**, 1–175.

Maryańska, T. and Osmólska, H. 1975. Protoceratopsidae (Dinosauria) from Mongolia. *Palaeontologia Polonica*, **33**, 133–182.

Ostrom, J. H. 1964. A functional analysis of the jaw mechanics in the dinosaur *Triceratops*. *Postilla*, **88**, 1–35.

Ostrom, J. H. 1966. Functional morphology and evolution of ceratopsian dinosaurs. *Evolution*, **20**, 290–308.

Ostrom, J. H. and Wellnhofer, P. 1986. The Munich specimen of *Triceratops* with a revision of the genus. *Zitteliana*, **14**, 111–158.

Sereno, P. C. 1986. Phylogeny of the bird-hipped dinosaurs (Order Ornithischia). *National Geographic Society Research*, **2**, 234–256.

You, H.-L. and Dodson, P. 2004. Basal Ceratopsia. In Weishampel, D. B., Dodson, P. and Osmólska, H. (eds.), *The Dinosauria*, 2nd edn. University of California Press, Berkeley, pp. 478–493.

CHAPTER 10

Ornithopoda: the tuskers, antelopes, and "mighty ducks" of the Mesozoic

Ornithopods had it all. Some had tusks projecting from the corners of their mouths, some had spikes on their thumbs, some had more teeth than just about any other kind of animal, some sported hollow crests atop their heads, and all had long tails that projected straight back from their hips. Many must have had the grace of a running antelope.

Ornithopods had one of the longest reigns of all dinosaur groups, lasting for most of the Mesozoic. From the Early Jurassic, when they first show up in the fossil record, until the end of the Cretaceous, when all went extinct, ornithopods evolved as one of the most diverse clades among dinosaurs, boasting nearly 100 species at present count. In so doing, they also managed to reach all of the regions of the globe, from Asia, Europe, and North America in the north to South America, Australia, and Antarctica in the south.

Some ornithopods have become so well known that they border on the famous. *Iguanodon*, for example, a founder member of Sir Richard Owen's original 1842 Dinosauria, is an Early Cretaceous ornithopod known from isolated skeletal parts to huge quantities of skeletons, throughout Europe as well as from North America and Asia.

But even better known than *Iguanodon* are the hadrosaurids, a group of ornithopods whose fossil record extends back some 90 million years prior to the close of the Cretaceous. Commonly called "duck-billed" dinosaurs (Figure 10.1), hadrosaurids are known from extremely abundant fossil remains, ranging from piles of bones in single-species bonebeds, to fully articulated skeletons, to isolated elements. The preservation of their remains can be spectacular, including skin impressions and ossified tendons, and delicate bones such as sclerotic rings, stapes, and hyoid bones. Paleontologists also have at their hands the remains of hadrosaurid eggs and growth series that range from hatchlings through juveniles and "teenagers" to adults. Finally, hadrosaurid footprints and trackways abound in some parts of the world. Thus these dinosaurs, as well as many other kinds of ornithopods, offer a smorgasbord of information about their anatomy, biology, and evolution.

Ornithopods presently fall into four groups that are nested within each other. Most primitive are heterodontosaurids, best known from *Heterodontosaurus* itself. The remaining ornithopods are called euornithopods ("true ornithopods"), which primitively include such forms as *Hypsilophodon*, *Orodromeus*, and *Thescelosaurus*. Nested still higher within euornithopods are iguanodontians, formed of *Tenontosaurus*, *Dryosaurus*, *Iguanodon* (naturally), and hadrosaurids, among other taxa. Nearly all hadrosaurids can be grouped into those with hollow crests atop their heads (lambeosaurines) and those that have solid crests or are flat-headed (hadrosaurines).

Figure 10.1. *Edmontosaurus*, a Late Cretaceous hadrosaurid ornithopod from the Western Interior of North America, looking at you, over its left shoulder, and to the right.

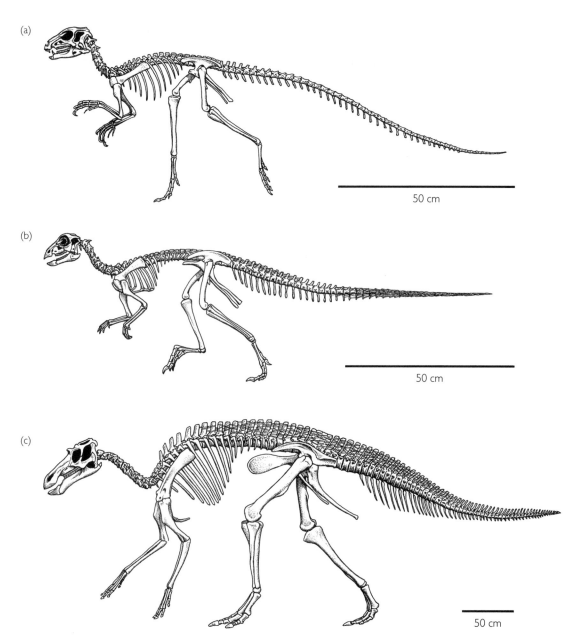

Figure 10.2. Right lateral view of the skull and skeleton of (a) *Heterodontosaurus*, (b) *Hypsilophodon*, and (c) *Maiasaura*.

In body size, ornithopods run the gamut. Early in their history, they were generally small (ranging from 1 to 2 m in length) and only iguanodontians attained great body size; one of the species of *Iguanodon* and the vast majority of hadrosaurids ranged upward of 10–12 m. Not all iguanodontians were towering; *Dryosaurus* was a comparatively tiny 3–4 m in length. Youngsters of course were smaller still: hatchling material *Orodromeus*, *Dryosaurus*, and a few hadrosaurids suggest individuals of approximately 30 cm length (Figure 10.2). An ever-increasing abundance of hatchling and juvenile specimens of other taxa are also providing considerable information on the ways and rates that dinosaurs grow – information that a generation ago was only dreamed about, but now is at our fingertips.

Because of their abundance and often-spectacular preservation, ornithopods have been the subject of considerable research. They have figured widely in studies of Mesozoic paleoecology, the evolution of vertebrate herbivory, and archosaurian phylogeny. They have even begun to play a role in our understanding of dinosaurian reproduction and life history strategies. All of these subjects will be touched on in this chapter.

Ornithopod lives and lifestyles

Going their way

Ornithopods ranged from near the paleoequator to such high-paleo-latitude occurrences as the north slopes of Alaska, the Yukon, and Spitsbergen in the Northern Hemisphere, and Seymour Island, Antarctica, and the southern coast of Victoria, Australia in the Southern Hemisphere (Figures 10.3 and 10.4). Local conditions in these regions varied widely, such that we can safely assume that the many ornithopods lived in quite diverse habitats, with variable climates. For example, the Lower Jurassic sediments of southern Africa from which *Heterodontosaurus* and its relatives have been recovered are indicative of a semi-arid, probably seasonal (wet-dry) climate, altogether quite inhospitable.

Ornithopods elsewhere in the world and at other times are known from a vast array of terrestrial depositional environments, ranging from

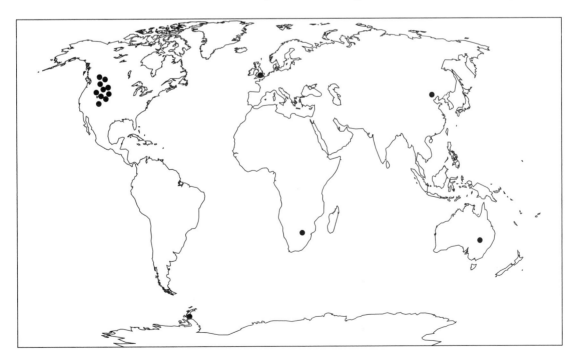

Figure 10.3. Global distribution of Heterodontosauridae and basal Euornithopoda.

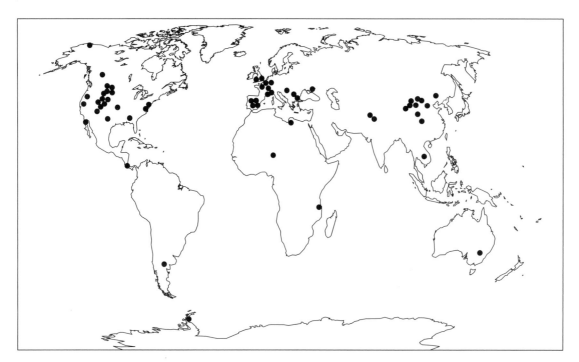

Figure 10.4. Global distribution of Iguanodontia.

upper coastal-plain deposits, to lower coastal-plain channels and overbank deposits, to delta-plain sediments. Several ornithopods, hadrosaurids mostly, are known from islands and even rare marine occurrences, where they are thought to represent the remains of bloated carcasses swept out to sea.

Some degree of habitat partitioning among ornithopods has been reported; that is, several genera or species appear to have divided the available ecospace into domains that do not overlap with each other. For example, J. R. Horner (Museum of the Rockies, Bozeman, Montana) has noted that, in western North America, there are many hadrosaurine taxa that tend to be found in near-marine deltaic sediments. In contrast, other hadrosaurines, but more especially the vast majority of lambeosaurine taxa, are restricted to lower coastal-plain sediments that were deposited inland from these near-marine environments. Finally, other hadrosaurines – specifically *Maiasaura* – have been found nowhere else but in upper coastal-plain sediments and are therefore thought to be endemic to these environments.

Ornithopods varied widely in their anatomy and geographical distribution, and hence in the way they presumably carried out their daily lives (see Box 10.1). These animals are now thought to be predominantly bipedal terrestrial animals, although with different agility and speed. As might be expected, many of the smaller forms – heterodontosaurids, basal euornithopods, and a few iguanodontians – must have been fast running, although they also may have adopted a quadrupedal stance when foraging or standing still. Some of the larger ornithopods, such as

BOX 10.1

Hypotheses that didn't go the distance

In the history of the study of ornithopods, habitats and anatomy conspired to put some of these animals in exotic places and give them unusual locomotor skills. For example, hadrosaurids were once regarded as amphibious, in part because the tail was long and deep (great for sculling in the water), the hand appeared to be webbed, and jaws were deemed too weak to handle anything but soft aquatic vegetation. Not true in all three cases. In a similar fashion, for over 100 years, *Hypsilophodon* was regarded as a tree-dweller. Upon close scrutiny by P. M. Galton, however, this animal was found to have no specializations for this particularly demanding mode of life.

The combination of a strongly seasonal African habitat and some basic heterodontosaurid anatomy created a dilemma – and ultimately a solution – for R. A. Thulborn in 1978. Heterodontosaurids, he believed, chewed by moving the lower jaw forwards and backwards relative to the upper jaw. Yet, evidence of tooth replacement that he expected (given that heterodontosaurids fed on very abrasive food) simply did not exist. To replace the teeth gradually would have impaired their ability to feed,

he reasoned, so the teeth could only have been replaced *en masse*. How could this be accomplished? Thulborn argued that heterodontosaurids must have estivated (lain dormant), most likely during the dry season. While dormant, the formerly functional teeth fell out and were replaced, to be worn down while the animal was active and feeding during each wet season.

Several years after Thulborn's estivation hypothesis appeared, J. A. Hopson re-examined heterodontosaurid jaw mechanics and tooth replacement patterns. As it turns out, heterodontosaurids chewed transversely, not forward and backward, so that tooth replacement was reduced, but not lost, in these animals. The combination of these two aspects of heterodontosaurid feeding are mutually compatible and certainly do not call for periods of dormancy to accommodate rapid tooth replacement. Thus anatomical support for Thulborn's hypothesis disappeared. There is no compelling reason to believe that heterodontosaurids engaged in estivation during the harshness of the southern African climate of the Early Jurassic.

Iguanodon, may have engaged in extensive quadrupedal locomotion. These forms have a solidly built wrist and hand that clearly was capable of considerable weight support. Interestingly, for these quadrupedal ornithopods, juvenile individuals of the same species may have been more bipedal than their adult counterparts. The same appears to have been true of *Tenontosaurus*. Here we have an indication of shifts in locomotion with age. By and large, the larger iguanodontians were also primarily terrestrial bipeds, although not as fast running as the smaller ornithopods.

In all cases, the tail was long, muscular, strengthened by ossified tendons, and held at or near horizontal, making an excellent counterbalance for the front of the animal. In general, the powerful hindlimbs tend to be at least and sometimes more than twice the length of the forelimbs.

How fast could these dinosaurs have traveled? Larger iguanodontians such as hadrosaurids may have been able to reach 15 to 20 km/h during a sustained run, but upward of 50 km/h over a short distance. Quadrupedal galloping appears to have been unlikely, given the rigidity of the vertebral column and the limited of movement of the shoulder against the ribcage and sternum. For smaller ornithopods, running speeds were higher. Maximum speeds were probably on the order of 60 km/h.

Fast running, maybe, but were they smart? According to J. A. Hopson, yes, they were. In fact, Hopson suggested that they were as smart or smarter than might be expected of living crocodilians if they were scaled up to dinosaur size. For example, *Leaellynasaura*, a basal euornithopod from Victoria, Australia, was apparently quite brainy and had acute vision, as suggested by prominent optic lobes.[1] In general, ornithopod braininess may relate to greater reliance on acute senses for protection that, in the absence of extensive defensive structures, may have been their only recourse. Moreover, brain size in these dinosaurs may relate to their complex behavioral repertoire, which we will discuss below.

What did the dominantly bipedal ornithopods do with their hands? To the degree that we can tell (based on proportions of limb elements, development of muscle scars, presumed habitat, etc.), *Heterodontosaurus* may have used its powerful forelimbs and clawed hands to grab at vegetation or to dig up roots and tubers. The forelimbs and hands of many basal euornithopods appear to have been less powerful than those of either heterodontosaurids or iguanodontians. Nevertheless, because they were not usually used in weight support, the hands were free to grasp at leaves and branches, bringing foliage closer to the mouth so that it could be nipped off by the toothed beak. By contrast, *Iguanodon*, *Altirhinus*, *Ouranosaurus*, and to a degree *Camptosaurus* have a very specialized hand as compared with those of other ornithopods (Figure 10.5). The first digit is conical and sharply pointed, likely to have been used as a stiletto-like, close-range weapon or for breaking into seeds and fruits. In contrast, the outer finger, digit V, was capable of some degree of opposition against the middle three digits, all of which were hoofed. By curling around to face the palm, the fifth digit was opposable (very much as the thumb is in humans[2]). The same cannot be said for hadrosaurids. Their reduced hands, with three hoofed fingers joined together in a thickened pad, hardly had any way to function other than as a support while the animal was standing on all fours. Manual dexterity was not a hadrosaurid specialty.

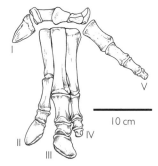

Figure 10.5. The hand of *Iguanodon*. Note the spiked thumb.

Feeding and food

No other group of dinosaurs has been the subject of as much research on feeding as ornithopods. Not only are their skulls marvels of intricate mechanics, but also in some cases (the so-called hadrosaurid "mummies") stomach contents have been fossilized within the abdominal region. These spectacular specimens apparently dried before burial and replacement (see Chapter 1). Preserved are beautiful skin impressions, dried, stretched tendons and muscles, and fossilized remnants of the last supper in the gut. Another reason why the feeding

---

1 The animal had an estimated encephalization quotient (EQ; see Box 15.4) of 1.8; Hopson estimated that the average EQ of other ornithopods is about 1.5.

2 Human success is sometimes ascribed to an opposable thumb, but, as you can see in this book, dinosaurs invented opposable digits at least twice: once in ornithopods and once in theropods.

mechanisms of ornithopods have been so extensively studied is that the group obviously had some unique and remarkably sophisticated ways of processing food. Best of all, these have left a tangible, easily preserved imprint in the design of the teeth and jaws. Suggestions about how ornithopods chewed their food were made as early as 1895 and several detailed studies of ornithopod jaw mechanics have been published since 1900. In addition, it was ornithopods that gave us our first clues about the presence of fleshy cheeks, of great utility to animals that chew their food and, as we know by now, a condition that is found in all genasaurian ornithischians.

So what did ornithopods eat? The hadrosaurid "mummies," at least in life, ate twigs, berries, and coarse plant matter. This correlates nicely with their size: ornithopods are thought to have been active foragers on ground cover and low-level foliage from conifers and in some cases from deciduous shrubs and trees of the newly evolved angiosperms (Box 16.3).

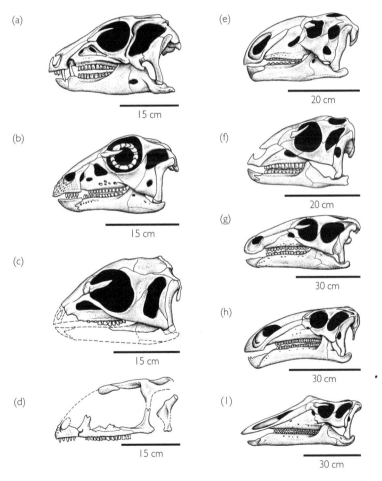

Figure 10.6. Left lateral view of the skull of (a) *Heterodontosaurus*, (b) *Hypsilophodon*, (c) *Yandusaurus*, (d) *Zephyrosaurus*, (e) *Tenontosaurus*, (f) *Dryosaurus*, (g) *Camptosaurus*, (h) *Iguanodon*, and (i) *Ouranosaurus*.

Browsing on such vegetation appears to have been concentrated within the first meter or two above the ground, but the larger animals must have been capable of reaching vegetation as high as 4 m above the ground.

Eating coarse, fibrous food requires some no-nonsense equipment in the jaw to extract enough nutrition for survival, and ornithopods had what it took (Figures 10.6, 10.7, and 10.8). In general, the group came equipped with a beak in the front for cropping vegetation, a well-developed block of teeth (the dental battery) for shearing coarse plant matter (Figure 10.9), a large, robust coronoid process for serious chewing muscles, and, as we have seen, a tooth row that was deeply set in, indicating that large fleshy cheeks were present. But beyond these basics, different ornithopods had different modifications of the jaw, and different kinds of jaw motion are believed to have been used for the processing of food.

The first modern treatment of ornithopod jaw mechanics in ornithopods was an extensive study of the cranial anatomy – including

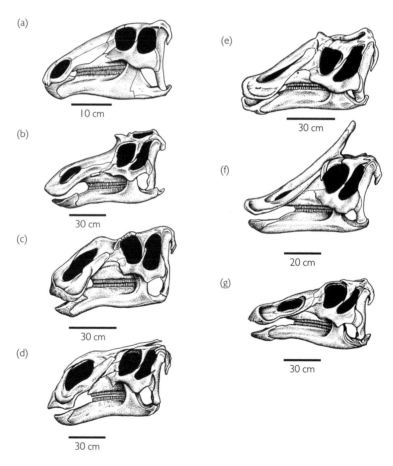

(a)

10 cm

(b)

30 cm

(c)

30 cm

(d)

30 cm

(e)

30 cm

(f)

20 cm

(g)

30 cm

Figure 10.7. Left lateral view of the skull of (a) *Telmatosaurus*, (b) *Maiasaura*, (c) *Gryposaurus*, (d) *Brachylophosaurus*, (e) *Prosaurolophus*, (f) *Saurolophus*, and (g) *Edmontosaurus*.

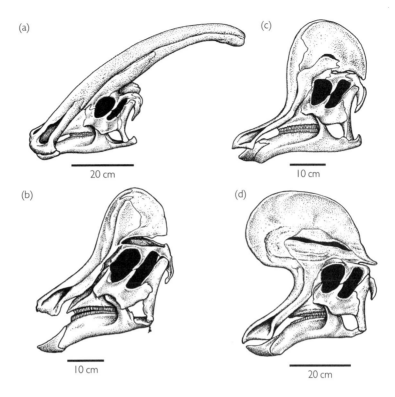

Figure 10.8. Left lateral view of the skull of (a) *Parasaurolophus*, (b) *Hypacrosaurus*, (c) *Corythosaurus*, and (d) *Lambeosaurus*.

skeletal, as well as muscular, vascular, and nervous – of North American hadrosaurids published in 1961 by J. H. Ostrom. Using these four anatomical perspectives, which provided the basis for reconstructing the pattern of chewing in these Late Cretaceous ornithopods, Ostrom suggested that hadrosaurids chewed back to front – in what is called propalinal jaw movement – on both sides of the mouth at the same time.

Other dinosaur paleontologists have suggested otherwise, at least for different ornithopods. P. M. Galton noted that *Hypsilophodon* may have chewed in much the same way as many mammals do today – side-to-side on one side of the mouth at a time. R. A. Thulborn regarded chewing in heterodontosaurids as similar to what Ostrom suggested for hadrosaurids: bilateral propalinal jaw movement.

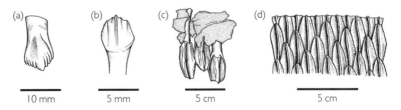

Figure 10.9. Upper tooth of (a) *Lycorhinus*, (b) upper tooth of *Hypsilophodon*, (c) three upper teeth of *Iguanodon*, and (d) lower dental battery of *Lambeosaurus*.

(a)

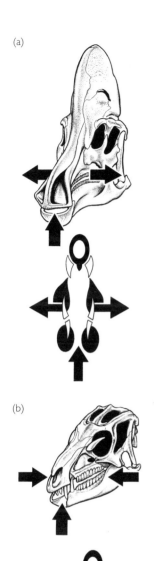

(b)

Figure 10.10. (a) Jaw mechanics in Euornithopoda, showing lateral mobility of the upper jaws (pleurokinesis), and (b) in Heterodontosauridae, showing medial mobility of the lower jaws.

More recently, the ways in which these herbivores chewed their food and how these jaw mechanisms evolved have been the focus of considerable research by D. B. Norman, D. B. Weishampel, A. W. Crompton, and J. Attridge. These studies have been based not only on comparisons of ornithopod skulls and teeth, but also on computer analyses of cranial mobility that might translate into special kinds of movement between chewing teeth.

What emerges from these studies is yet again more ornithopod diversity, this time at the level of feeding and foodstuffs. In the most primitive ornithopods, the very front of the cornified beak was relatively narrow and lacked teeth, suggesting a somewhat selective cropping ability. Iguanodontians, by contrast, lose their front teeth, broaden their snouts, and even develop a strongly serrate margin to their rhamphotheca. These animals were not selective feeders; instead, they hacked at and severed leaves and branches without much regard for what they were taking in. Whereas basal ornithopods were careful nibblers, most iguanodontians were lawn-mowers.

Once these gulp-fulls of leaves had passed the rhamphotheca into the mouth, all ornithopods chewed their food. Yet how they solved the problem of combining bilateral occlusion (where the teeth meet on both sides of the jaws at the same time) with chewing is one of the most intriguing aspects of dinosaur feeding, for both heterodontosaurids and euornithopods evolved different solutions to this problem, solutions that both parallel those "invented" by ungulate mammals (such as sheep or horses) but remain uniquely distinct from them and from each another.

On the basis of their skull architecture, patterns of tooth wear, and computer modeling, we know that heterodontosaurids chewed by combining vertical movement of the lower jaws with a slight degree of rotation of the mandible about their long axes (Figure 10.10b). In this way, they were able to move their upper and lower teeth in a transverse direction and thus break up the bits of plant food that the tongue had placed between them. Naturally, the fleshy cheeks prevented most of the food from falling out of the corners of the mouth.

Euornithopods, on the other hand, evolved a distinctly different pattern of skull movement in order to solve the problem of having bilateral occlusion and still chewing from side to side. Instead of loosening up the lower jaws to rotate about their long axes, euornithopods mobilized their upper jaws. This kind of mechanism, which Norman called pleurokinesis, involved a slight rotation of portions of the upper jaw, especially the maxilla (the bone that contains the upper teeth), relative to the snout and skull roof (Figure 10.10a). When the upper and lower teeth were brought into contact on both right and left sides, the upper jaws rotated laterally and the opposing surfaces of the teeth sheared past one another to break up plant food in the mouth. Unlike humans, in which the bones of the skull are solidly fused and locked together, an adaptation such as this requires flexibility at the joints between bones of the skull. In hadrosaurids, the complex occlusal surfaces afforded by the development of a dental battery would have made short work of virtually all foliage. Like the situation in heterodontosaurids, pleurokinesis represented an important advance for euornithopods, providing them the ability to chew a variety of plant foods, including those with a great deal of fiber.

As in all of the other ornithischians that have been discussed, once the food was properly chewed, it was swallowed and quickly passed into a capacious gut, which was present in all ornithopods and appears to have been relatively larger in the absolutely larger iguanodontians. Between the extensive chewing of food in the mouth and fermentation in the large gut, it is very likely that all ornithopods were well suited for a subsistence diet of low-quality, high-fiber vegetation.

Social behavior

From the time of their discovery, ornithopods of all kinds have attracted a good deal of attention, particularly for their oddly appearing ornamentation. The apparently outlandish crests on the heads – many of them hollow and highly chambered – of hadrosaurids, the tusks of heterodontosaurids, and the lumps on the forehead of *Ouranosaurus*, have called out for an explanation. It is safe to say that virtually all of these features – like those odd bumps and horns of ceratopsians, and for that matter the antlers of deer and horns of cows and antelope – hint at sophisticated social behavior.

Hadrosaurids have attracted the most attention, in large part because they clearly stand out from the crowd with their wild headgear. Once thought to relate to the aquatic habits of the group (see Box 10.1) or to the olfactory (sense of smell) function of the nasal cavity, much of the discussion about the functional significance of hadrosaurid ornamentation now centers on combat, display, and their reproductive consequences. In 1975, J. A. Hopson suggested that the unusual cranial features – principally involving the nasal cavity – that we see in hadrosaurids probably evolved in the context of social behavior among members of the same species. In particular, Hopson regarded the special cranial features in hadrosaurids as indicative of either intra- or interspecific aggression, more especially in the case of both solid and hollow crests in visual and vocal display. In order for crests to function as good signals to convey information about what species, what sex, and even rank an individual might be, they must be both visually and vocally distinctive. Only then can they be regarded as promoting successful matings by informing the consenting adults.

So how are we ever to make sense of these suggestions about unfossilizable behavior? Hopson made five predictions that link the fossil record of hadrosaurids to the social behaviors he anticipated were driven by sexual selection. First, to interpret incoming display information, hadrosaurids must have had both good hearing and good vision. These are qualities that cannot be measured directly in extinct vertebrates, but all hadrosaurids have large eye sockets, often with sclerotic rings that would have encircled the outer region of the eye. In all cases, eye size was quite large and so sight must have been reasonably acute. Similarly, we

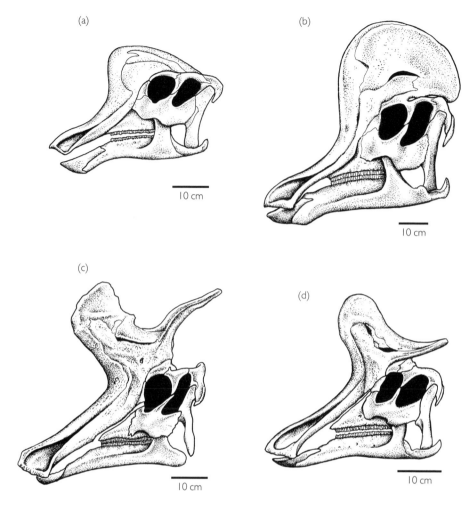

(a)

(b)

(c)

(d)

Figure 10.11. Growth and sexual dimorphism in lambeosaurine hadrosaurids. (a) Juvenile and (b) adult *Corythosaurus*. (c) Male and (d) female *Lambeosaurus*.

have evidence of the hearing via preserved middle and inner ear structures, also indicative of reasonable hearing across a wide range of frequencies in these animals.

Secondly, if the crest serves for visual display and as a vocal resonator, then its shape need not necessarily closely follow the shape of the cavities contained within. That is, the external shape of the crest may have been as important as its internal structure if it was to act in visual display. Again, this prediction is upheld by hadrosaurid fossils: in virtually all cases, the profile of the crest is much more elaborate or extensive than the walls of the internal plumbing.

If crests acted as visual signals (prediction 3), then they should be species specific in size and shape, and they should also be sexually dimorphic. This is amply upheld in large part thanks to studies by P. Dodson on the growth and development in lambeosaurine hadrosaurids (Figure 10.11). Using a variety of statistical techniques, Dodson was able to show that crests become most prominent when an animal approached sexual maturity. In addition, he demonstrated that each lambeosaurine species was dimorphic, particularly in terms of crest size and shape. Could these "morphs" be male and female? It certainly fits well with Hopson's prediction.

The last two predictions have to do with hadrosaurids in time and space. When several species occur together in the same area, they should exhibit great differences in the shape of their crests. Sameness would create a great deal of confusion among closely related hadrosaurids living in the same place, but distinctiveness in display structures would prevent such confusion, an obvious advantage during breeding season. Are crests more distinctive as the number of hadrosaurids living together goes up? The answer is "Yes." At Dinosaur Provincial Park in Alberta, Canada, where the number of hadrosaurids that have been found in the Dinosaur Park Formation (and thus thought to have lived together) is high, there are three distinctively crested lambeosaurines, one solid-crested form (*Prosaurolophus*) and two other species of hadrosaurine, each distinctive in its own right. In contrast, elsewhere where hadrosaurid diversity is lower, the variety of flamboyant head-dresses is decreased.

The last prediction, that crests should become more distinctive through time as a consequence of sexual selection (see Chapter 8), is not at all well supported. Hopson depended on the older, small-crested *Prosaurolophus* and the younger, large-crested *Saurolophus* being closely related, but this no longer seems to be the case (see below). In addition, lambeosaurine crests arguably become less distinctive over time.

If these supracranial contraptions were used for species recognition, intraspecific combat, ritualized display, courtship, parent–offspring

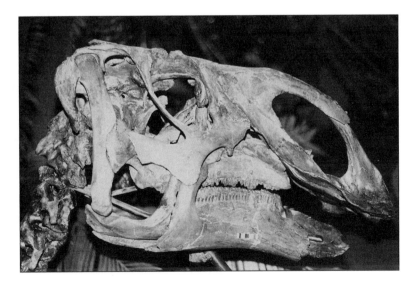

Figure 10.12. *Brachylophosaurus*, a solid-crested hadrosaurid from western North America.

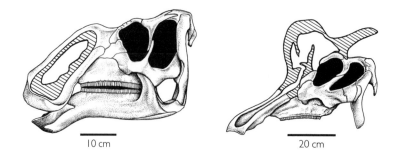

10 cm   20 cm

Figure 10.13. The circumnarial depression (indicated by cross-hatched region) which may have supported an inflatable flap of skin in hadrosaurines such as *Gryposaurus* (a). Highly modified nasal cavity housed within the hollow crest on top of the head of *Lambeosaurus* (b).

communication, and social ranking, the accentuated nasal arch and stout cranial crests seen in *Gryposaurus*, *Maiasaura*, and *Brachylophosaurus* were probably used for broadside or head-pushing during male–male combat (Figure 10.12). Hopson suggested that inflatable flaps of skin covered their nostrils and surrounding regions (Figure 10.13); these would have been blown up and used for visual display, as well as to make some noise – a kind of Mesozoic bagpipe. In *Prosaurolophus* and *Saurolophus*, this sac would have extended onto the solid crest that extended above the eyes, while in *Edmontosaurus*, where the nasal arch is not accentuated nor is there a crest, the complexly excavated nostril region may have housed an inflatable sac (Figure 10.14). With such an exceptional development of sacs around the nostrils and up and down the crest, ritualized combat with accompanying vocal and visual display became the norm.

When it came to display, none did it better than the lambeosaurines. In these animals, the hollow crests perched atop the head must have provided for instant recognition. This could have been achieved visually and by low honking tones produced in the large resonating chamber within the crest (Figures 10.13 and 10.15). Either way, by sight and/or through vocal cacophony, the crests of lambeosaurines would have functioned well as species-specific display organs.

Here we have a compelling case for hadrosaurid social behavior, but what of other ornithopods? Although the results are not as conclusive, it appears that the evolution of canine-like teeth of heterodontosaurids may have had something to do with intraspecific display and combat. Thulborn and R. E. Molnar independently suggested that, since these teeth are present only in mature "males," they would have been used not only in gender recognition, but also for intraspecific combat, ritualized display, social ranking, and possibly even courtship. A modern analogue is the tusked tragulids, living artiodactyls related to deer. Similarly, the development of a jugal boss in heterodontosaurids might also be interpreted as a form of visual display.

Figure 10.14. *Edmontosaurus*, a flat-headed hadrosaurid from the western USA.

Likewise, the low, broad bumps on top of the head of *Ouranosaurus* and the arched snout of *Muttaburrasaurus* and *Altirhinus* may well have similar behavioral significance (Figure 10.16). Perhaps these bumps aided individuals in the recognition of members of the same species, or members of the opposite sex. Or perhaps they were used in ritualized head-butting contests. *Ouranosaurus* was also equipped with extremely high spines on the vertebrae, which formed a high, almost sail-like ridge down its back. Like the case of *Stegosaurus* (see Chapter 6), it is possible that these long spines were covered with skin and used as a radiator or solar panel, to warm up or cool down. Alternatively (and not mutually

Figure 10.15. *Corythosaurus*, a hollow-crested hadrosaurid from the Late Cretaceous of western Canada. (Photograph courtesy of the Royal Ontario Museum.)

exclusively), they may have had a display function, providing the animal with a greater side profile than it would otherwise have had.

Display behavior in many ornithopods begins to make even more sense when considered with other aspects of their lifestyles. Consider communication between adults and between grown-ups and juveniles. There are several examples of single-species bonebeds – for example, *Dryosaurus, Iguanodon, Maiasaura, Hypacrosaurus*, and others – that support the notion that these animals were not only common but may have formed herds of

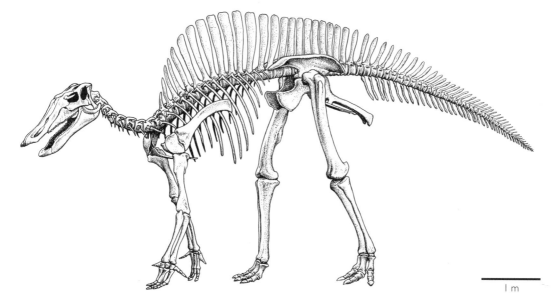

1 m

Figure 10.16. The Early Cretaceous iguanodontian *Ouranosaurus* from Niger.

r-strategist, a claim that is based on the inferred precocial nature of these dinosaurs. In contrast, *Maiasaura*, *Hypacrosaurus*, and perhaps other hadrosaurids that had nest-bound hatchlings requiring parental care, all appear to have been altricial and thereby K-strategists. Gazing elsewhere among ornithopods, it appears that precocity may be primitive for at least Euornithopoda, while altricial behavior probably evolved for the first time within the clade sometime prior to the origin of Hadrosauridae.

Whatever the broader meaning of these changes might be, we – and ornithopods – cannot escape from the effects of family. For ornithopods such as *Orodromeus*, parenting must have been easy – no provisioning or protection of the kids. But the toll to be paid was reduced survival of these offspring – wherever it was that they wandered off to. From a hadrosaurid perspective, however, it was a good thing to take care of the kids. For no matter how loud, squawky, and hard to handle these hatchlings might have been, Mom and Dad played a direct part in increasing their survival.

The evolution of Ornithopoda

Ornithopoda is defined as all the descendants of the common ancestor of a monophyletic clade that includes Heterodontosauridae and Euornithopoda (Figure 10.18) within this clade. Heterodontosauridae and Hadrosauridae are themselves monophyletic. Also containing a host of

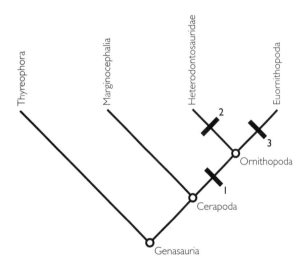

Figure 10.18. Cladogram of Genasauria, emphasizing the monophyly of Ornithopoda. Derived characters include: at **1** pronounced ventral offset of the premaxillary tooth row relative to the maxillary tooth row, crescentic paroccipital processes, strong depression of the mandibular condyle beneath the level of the upper and lower tooth rows, elongation of the lateral process of the premaxilla to contact the lacrimal and/or prefrontal; at **2** high-crowned cheek teeth, denticles on the margins restricted to the terminal third of the tooth crown, caniniform tooth in both the premaxilla and dentary; at **3** scarf-like suture between postorbital and jugal, inflated edge on the orbital margin of the postorbital, deep postacetabular blade on the ilium, well-developed brevis shelf, laterally swollen ischial peduncle, elongate and narrow prepubic process.

basal forms such as *Thescelosaurus*, *Hypsilophodon*, *Gasparinisaura*, and *Agilisaurus*, as well as Dryosauridae and forms sequentially more-closely related to Hadrosauridae, Ornithopoda is diagnosed on the basis of a number of derived features, among them pronounced ventral offset of the premaxillary tooth row relative to the maxillary tooth row, crescentic paroccipital processes, strong depression of the mandibular condyle beneath the level of the maxillary and dentary tooth rows, and elongation of the lateral process of the premaxilla to contact the lacrimal and/or prefrontal bones.

Where do ornithopods reside among the dinosaurs? They and margino-cephalians form a monophyletic group called Cerapoda. Successively larger groups in the hierarchy are Thyreophora and *Lesothosaurus* (see introductory text to Part II: Ornithischia) to form Ornithischia. The next larger inclusive node on the cladogram is Saurischia, which is the basal division within Dinosauria.

Given the high diversity (and phylogenetic complexity) of ornithopods, our aim in this section is to provide a basic overview of the phylogeny of this important group of ornithischians. We will therefore discuss the general shape of the ornithopod cladogram, noting the major divisions and the features that support these clades. Much of what follows is based on a wealth of discussion on ornithopod phylogeny, including the 1984 studies by P. C. Sereno, Norman, and A. Milner and Norman, as well as more recent work by Sereno, Norman, and Weishampel and colleagues.

Heterodontosauridae

Basally, Ornithopoda are divided into Heterodontosauridae and Euornithopoda. The first of these groups, Heterodontosauridae is defined as all the descendants of the common ancestor of *Heterodontosaurus* and *Lanasaurus*. Basally in the history of this clade, which also contains *Lycorhinus* and *Abrictosaurus*, these heterodontosaurids evolved high-crowned teeth, each bearing a chisel-shaped crown ornamented with denticles. In addition, and the principal basis for the name "heterodontosaurid," a large canine-like tooth is present in both upper and lower jaws. These "canines" are not the true canine teeth that characterize mammals.

Euornithopoda

Euornithopoda constitutes the remaining ornithopod clade (Figure 10.19). Defined as all the descendants of the common ancestor of *Agilisaurus* and Iguanodontia, euornithopods are characterized by features of the orbit (eye socket: scarf-like suture between postorbital and jugal, an inflated edge on the orbital margin of the postorbital bone), the pelvis (deep postacetabular blade on the ilium, well-developed brevis shelf, laterally swollen ischial peduncle, elongate and narrow prepubic process, tab-like obturator process on the ischial shaft), and femur (deep pit on the femoral shaft adjacent to the fourth trochanter).

This large euornithopod clade consists of a host of often relatively small, agile ornithopods such as *Hypsilophodon*, *Agilisaurus*, and *Gasparinisaura*, as well as a few somewhat larger, more robust forms (*Parksosaurus*,

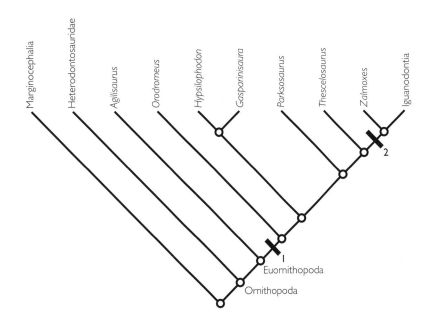

Figure 10.19. Cladogram of basal Euornithopoda, with more distant relationships with Heterodontosauridae and Marginocephalia. Derived characters include: at **1** subcircular external antorbital fenestra, distal offset to apex of maxillary crowns, strongly constricted neck to the scapular blade, ossification of sternal ribs, hypaxial ossified tendons in the tail; at **2** rectangular lower margin of the orbit, widening of the frontals, broadly rounded predentary, dentary with parallel dorsal and ventral margin, absence of premaxillary teeth, 10 or more cervical vertebrae, 6 or more sacral vertebrae, presence of an anterior intercondylar groove, inflation of the medial condyle of the femur.

Thescelosaurus), and Iguanodontia, residence of such dinosaurian luminaries as *Camptosaurus* (Figure 10.20), *Iguanodon*, and the hadrosaurids. Many of the smaller basal euornithopods had been placed in a group called Hypsilophodontidae, thought at that time to be a natural (monophyletic) group. However, with the discovery of new material, the recognition of new taxa, and the use of cladistic analyses to test the relationships of both new and earlier-known forms, Hypsilophodontidae has not stood up to scrutiny and therefore is now abandoned. In its place is the serial arrangement of *Agilisaurus*, *Hypsilophodon*, *Gasparinisaura*, *Thescelosaurus*, *Zalmoxes*, and other less well-known taxa more closely related to Iguanodontia. The evolution of these taxa successively entails modifications of the facial skeleton (reduction in size and rounding of the external antorbital fenestra, loss of contact between the lacrimal and premaxilla, reduction in length and repositioning of the palpebral), teeth (development of asymmetrical crowns, distinct cingulum), ribcage (ossification of sternal ribs), the development of ossified hypaxial tendons, and shoulder (distinct neck on scapula, angular deltopectoral crest).

How the many other basal members of Euornithopoda fit into this phylogeny remains to be seen. They may "shoe-horn" in nicely or they may

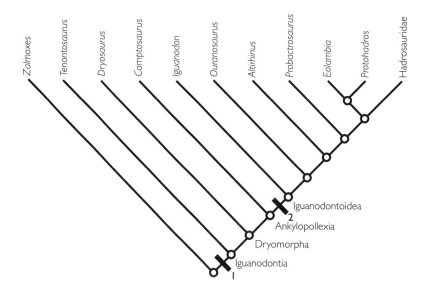

Figure 10.20. Cladogram of basal Iguanodontia. Derived characters include: at **1** premaxilla with a transversely expanded and edentulous margin, reduction of the antorbital opening, denticulate margin of the predentary, deep dentary ramus; loss of sternal rib ossification, loss of a phalanx in digit III of the hand, compressed and blade-shaped prepubic process; at **2** strong offset of premaxilla margin relative to the maxilla, peg-in-socket articulation between maxilla and jugal, development of a pronounced diastema between the beak and mesial dentition, mammillations on marginal denticles of teeth, maxillary crowns narrower and more lanceolate than dentary crowns, closely appressed metacarpals II–IV, deep triangular fourth trochanter, deep extensor groove on femur.

radically reorganize what we've just described. For now, however, we must regard *Atlascopcosaurus*, *Leaellynasaura*, *Qantasaurus*, *Notohypsilophodon*, *Fulgurotherium*, *Drinkeria*, and *Yandusaurus*, as euornithopods without a more exact phylogenetic home.

Iguanodontia

By far, Iguanodontia is the bushiest of ornithopod clades, claiming not only the very diverse Hadrosauridae, but also a variety of more basal forms (Figure 10.20). Iguanodontians uniquely share the following features: transversely expanded and toothless premaxilla, smoothly round oral margin of the predentary, deep dentary with parallel dorsal and ventral borders, loss of a phalanx in the third digit of the hand, and a compressed, blade-shaped prepubic process.

Tenontosaurus is the most basal iguanodontian ornithopod; remaining iguanodontians are called Dryomorpha after the most basal member, *Dryosaurus*. Climbing slightly higher, we encounter the more restrictive clade called Ankylopollexia – the "fused thumbs." These forms share a number of derived features, among them relatively close packing of the teeth, upper teeth with a prominent ridge on their outer side, fusion of the wrist bones, and the formation of the spiked thumb. Within this clade, *Camptosaurus* (Figure 10.21) is the most basal, while *Lurdusaurus* is the next most derived.

Figure 10.21. The Late Jurassic iguanodontian *Camptosaurus* from the Western Interior of the USA. (Photograph courtesy of the Royal Ontario Museum.)

The next included clade – Iguanodontoidea – includes species of *Iguanodon* and all more derived iguanodontians and is diagnosed by numerous changes in the skull and teeth that relate to the feeding mechanics of this group, as well as modifications of the hand, pelvis, femur, and foot. Contained within Iguanodontoidea, successively are *Iguanodon*, *Ouranosaurus*, *Altirhinus*, *Probactrosaurus*, *Eolambia* (which may be closely related to *Altirhinus*), *Protohadros*, and Hadrosauridae. *Jinzhousaurus* and *Nanyangosaurus*, both from the Early Cretaceous of China, also appear to be closely related to Hadrosauridae, although their exact positions are not yet clear.

Hadrosauridae
Until the 1990s, the monophyly of Hadrosauridae had long been assumed, but had not been demonstrated by cladistic analysis. One of the first of these recent forays into hadrosaurid phylogeny was conducted by Horner, who provocatively suggested a diphyletic origin for Hadrosauridae, with lambeosaurines deriving from an *Ouranosaurus*-like iguanodontian, and hadrosaurines from a more *Iguanodon*-like iguanodontian. These conclusions spawned a number of more recent cladistic analyses, including those by Horner himself, all of which now strongly support the monophyly of Hadrosauridae (Figure 10.21), defined as that clade of ornithopod dinosaurs consisting of the most recent common ancestor of *Telmatosaurus* and *Parasaurolophus*, plus all the descendants of this common ancestor. Diagnostic of Hadrosauridae are numerous modifications of the feeding apparatus (including the develop-ment of a dental battery and modification of the jaw joint and its support), and changes in the shoulder and knee joints.

Two major clades – Lambeosaurinae and Hadrosaurinae – constitute most of Hadrosauridae (Figure 10.22) and they are each other's sister-group,

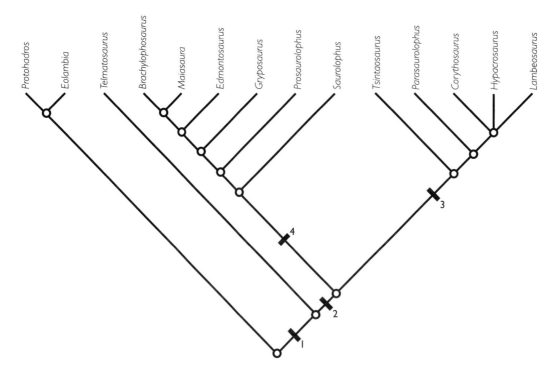

Figure 10.22. Cladogram of Hadrosauridae. Derived characters include: at **1** three or more replacement teeth per tooth position, posterior extension of the dentary tooth row to behind the apex of the coronoid process, absence of the surangular foramen, absence or fusion of the supraorbital to the orbit rim, long coracoid process, dorsoventrally narrow proximal scapula, very deep, often tunnel-like intercondylar extensor groove; at **2** absence of the coronoid bone, reduction in surangular contribution to coronoid process, double-layered premaxillary oral margin, triangular occiput, eight or more sacral vertebrae, reduced carpus, fully open pubic obturator foramen, absence of distal tarsals II and III; at **3** maxilla lacking an anterior process but developing a sloping dorsal shelf, groove on the posterolateral process of the premaxilla, low maxillary apex, a parietal crest less than half the length of the supratemporal fenestrae; at **4** presence of a caudal margin on the circumnarial fossa.

collectively dubbed Euhadrosauria. Immediately outside Euhadrosauria but within Hadrosauridae are *Telmatosaurus* and probably *Bactrosaurus*, *Claosaurus*, and *Tanius*. Euhadrosaurian hadrosaurids are united by several changes in the skull (among them, absence of a coronoid bone, modification of the oral margin of the premaxilla, triangular form to the back of the skull), eight or more sacral vertebrae, and alteration of the hand, pelvis, and ankle.

Features uniting Lambeosaurinae, the clade of "hollow-crested" hadrosaurids, are all associated with modifications of the skull having to do with the evolution of the nasal cavity into a supracranial crest. Most basal among lambeosaurines is *Tsintaosaurus*, which had first been considered a hadrosaurine, but is now known to possess key lambeosaurine features. Lambeosaurines above *Tsintaosaurus* – *Parasaurolophus*, *Corythosaurus*, *Hypacrosaurus*, and *Lambeosaurus* – are united by additional changes in the supracranial crest and an increase in height of the neural spines of the back vertebrae. Among the four last-mentioned lambeosaurines, relationships include *Parasaurolophus* as the basal taxon and *Corythosaurus*, *Hypacrosaurus*, and *Lambeosaurus* sharing closest, but still not yet fully resolved relationships.

The phylogenetic relationships of hadrosaurines are currently not well supported. The clade itself is united by a single feature of the expression of the external nares on the snout. The basal hadrosaurine appears to be *Lophorhothon*, with a succession of yet higher clades that includes one formed of *Naashoibitosaurus*, *Saurolophus*, and an Argentinean form referred to *Kritosaurus*, and the other consisting of successively more exclusive relationships among *Prosaurolophus*, *Gryposaurus*, *Edmontosaurus*, *Brachylophosaurus*, and *Maiasaura*.

In addition, there is a host of hadrosaurids that – for one reason or another, mostly having to do with their incomplete preservation – are yet unresolved at the highest reaches of the iguanodontian tree. These include *Gilmoreosaurus*, *Tanius*, *Jaxartosaurus*, *Aralosaurus*, *Barsboldia*, and *Nipponosaurus* from Asia, *Secernosaurus* from South America, and *Hadrosaurus* and *Kritosaurus* from North America. As always, further discoveries and careful research will most likely provide us with that important bit of information to place these wayward ornithopods in their phylogenetic context.

Ornithopods meet history: a short account of their discovery

In 1822, Gideon and Mary Ann Mantell discovered peculiar-looking teeth from the Wealden beds of the Tilgate Forest in what is now West Sussex, England. These teeth, later given the name *Iguanodon* (*iguana* – lizard; *don* – teeth) by Gideon Mantell in 1825, provided humans with their first inkling of some of the behemoth denizens of the Mesozoic. From that time forward, *Iguanodon* fossils began to accumulate in Mantell's and other collections – teeth, jaws, limb bones, and vertebrae, and a number of English paleontologists, among them T. H. Huxley, R. Owen, and Mantell himself, set about the task of studying them. Out of their work came a picture of dinosaurs, ornithopods among them, as gigantic (estimated length in excess of 35 m), pachydermoid reptiles that walked on all fours in the most lumbering fashion. In keeping with this picture, a spike-like bone found with *Iguanodon* remains was fitted to its nose, which gave the dinosaur a profile like a modern rhinoceros.

Nearly a quarter century later and some 6000 km across the Atlantic Ocean, Joseph Leidy of the Academy of Natural Sciences of Philadelphia obtained some large bones from a quarry in Upper Cretaceous marls in nearby Haddonfield, New Jersey. These fossils, named *Hadrosaurus* (*hadro* – stout) by Leidy in 1858, created quite a stir in the mid-1800s, in part because they settled some important early issues about dinosaur biology. For example, Leidy used the disparity in length between the fore- and hindlimb to argue that *Hadrosaurus* (and by inference *Iguanodon*) walked predominantly on its hindlimbs. But even more important – and influential – *Hadrosaurus* was the first dinosaur to be mounted and exhibited to the public. It is said that attendance sky-rocketed at the museums where the 10 m long *Hadrosaurus* was displayed; dinosaurs have always been big draws.

Thereafter through the turn of the nineteenth century, there was to be a flurry of activity on both sides of the Atlantic. In 1869, Huxley and P. Matheron, respectively, were to name two relatively small (2–3 m long) ornithopods: *Hypsilophodon* (*hypsi* – high; *loph* – crest) from the same beds as *Iguanodon*, but from the Isle of Wight off the southern coast of England, and *Rhabdodon* (*rhabdo* – ribbon) from the Late Cretaceous of southern France.

However important these discoveries were to become, almost everything was eclipsed in 1878 by the recovery of complete and multitudinous (an unbelievable 31, to be exact) *Iguanodon* skeletons from a coal seam some 300 m beneath the small mining town of Bernissart in southern Belgium. Over the next seven years, Louis Dollo, curator of paleontology at the Institut Royal de Science Naturelle de Belgique in Brussels, described and interpreted this Bernissart bonanza (Figure 10.23). His studies focused on the new anatomical information provided by the material, on the taxonomic composition of the assemblage (two species of *Iguanodon*), on the size and posture of the body (7–10 m long; bipedal posture much like that of *Hadrosaurus* from New Jersey), and on the functional significance of particular skeletal elements (the spike went on the hand, not the nose.).

Turning back to North America, it was the opening of the Western Interior of the USA to paleontological exploration that prompted further discoveries during the last three decades of the 1800s. In short order, discovery in 1872 of new hadrosaurid material from Late Cretaceous age beds of Kansas (subsequently dubbed *Claosaurus* (*klao* – break) by O. C. Marsh in 1890) was followed by the 1885 announcement of *Camptosaurus* (*kamptos* – flexible), a new, medium-sized (5–7 m long) ornithopod dinosaur from the Upper Jurassic Morrison Formation of Wyoming. Nine years later, in 1894, Marsh again described a new ornithopod, this time a smaller contemporary of *Camptosaurus*, which he named *Dryosaurus* (*dryos* – oak).

The close of the nineteenth century marked a major turnover in dinosaur paleontologists; E. D. Cope died in 1896 and Marsh in 1899 (see Box 6.2). The new generation of dinosaur researchers of the first quarter of the twentieth century was to make an outstanding series of discoveries, particularly in newly explored regions of North America, but also in Asia, Africa, and South America. The first discoveries of the 1900s, however, were made not here, but in Transylvania, in the foothills of the Southern Carpathian Mountains. Here, dinosaur fossils found on the estate of the Nopcsa family were described by F. Nopcsa in 1900 as *Limnosaurus* (*limnos* – lake). Because this name had been used previously for a fossil crocodile, *Limnosaurus* was renamed *Telmatosaurus* (*telmat* – swamp) by Nopcsa in 1903.

Thereafter, the first quarter-century of the 1900s saw a mushrooming of ornithopod discoveries. Of the 11 new kinds of ornithopod dinosaurs, all but three were collected from the small, but rich, badlands

Figure 10.23. Several death-posed *Iguanodon*, the great beast of Bernissart, Belgium.

of Upper Cretaceous rocks along the Red Deer River of Alberta, Canada, during the Great Canadian Dinosaur Rush. Floating on log rafts and dragging horse-drawn wagons, B. Brown excavated a host of now-classic hadrosaurids, including *Saurolophus* (*lophus* – crest) in 1912, *Hypacrosaurus* (*hypakros* – highest, referring to spines) in 1913, *Corythosaurus* (*koryth* – crown) in 1914, and *Prosaurolophus* (*pro* – before) in 1916. Close behind were two Canadian dinosaur paleontologists, L. M. Lambe and W. A. Parks. Lambe's discoveries include *Gryposaurus* (*grypos* – hooked), named in 1914, and *Edmontosaurus* (from Edmonton), named in 1920, while Parks christened *Parasaurolophus* (*para* – near) in 1922 and *Lambeosaurus* (after Lambe) in 1923. He also described a new ornithopod (as a new species of *Thescelosaurus*; see below), which in 1937 C. M. Sternberg renamed *Parksosaurus* (after Parks).

Elsewhere in the world, only three other ornithopods were described over this 20 year period. Prior to his exceptionally rewarding travels in Alberta, Brown spent time collecting ornithopod dinosaurs in the region of the San Juan Basin, New Mexico. From there, he described a new hadrosaurid, *Kritosaurus* (*kritos* – separate), in 1910. *Thescelosaurus* (*theskelos* – astonishing), a basal euornithopod, was discovered in Upper Cretaceous rocks of Wyoming by C. W. Gilmore in 1913, while *Lycorhinus* (*lykos* – wolf; *rhinus* – snout) – originally thought to be a synapsid but now known to be a heterodontosaurid ornithopod – was collected from Lower Jurassic rocks of South Africa and described by S. H. Haughton in 1924.

The second quarter-century produced a host of discoveries from elsewhere in the world, principally Asia and Australasia. *Tanius* (named for H.-C. Tan, the Chinese geologist who discovered its remains), a flat-headed hadrosaurid from China, was described by C. Wiman in 1929. F. von Huene, renowned dinosaur paleontologist from Tübingen, Germany, announced *Fulgurotherium* (*fulgur* – lightning; *therion* – beast, referring to

Lightning Ridge) from Lower Cretaceous rocks of New South Wales, Australia, in 1932; unfortunately it was regarded as a theropod dinosaur until the 1980s, when it was re-examined by R. E. Molnar. It is now properly regarded as a non-iguanodontian ornithopod.

With the 1930s came the first descriptions of ornithopod dinosaurs from the Central Asiatic Expeditions. *Bactrosaurus* (*baktron* – club, referring to vertebrae) and *Mandschurosaurus* (from Manchuria) – two hadrosaurids from the Late Cretaceous of the Inner Mongolian region of China – were described by Gilmore in 1933. The Inner Mongolian *Mandschurosaurus* was renamed *Gilmoreosaurus* in 1979 by M. K. Brett-Surman in honor of Gilmore's efforts to understand these first hadrosaurids from the Gobi Desert.

Although not so well known as the Gobi discoveries, there are dinosaurs from elsewhere in Asia. *Nipponosaurus* (Nippon is the Japanese name for their country), the first and most famous hadrosaurid from Japan, was named by T. Nagao in 1936, while *Jaxartosaurus* (from the Jaxartes River, Kazakhstan) was described by A. N. Riabinin in 1939. Upper Cretaceous beds of the Kazakhstan desert were to produce other ornithopod dinosaurs, as we will see.

To round out the first half of the twentieth century, R. S. Lull and N. E. Wright presented the first summary and revision of any of the ornithopod groups, this time for North American hadrosaurids, in 1942. Their work covered hadrosaurid anatomy, paleoecology, and taxonomy. Like similar earlier treatments of ceratopsians, stegosaurs, and theropods, this work was to become a landmark study for generations to come.

The years since 1950 have been marked by a healthy mixture of discovery, reflection, and revision on a worldwide scale. Beginning in Asia, studies published in the early 1950s by A. K. Rozhdestvensky introduced a new species of *Saurolophus*, this time from Mongolia, and a new iguanodontian from the Early Cretaceous, also from Mongolia. Originally named *Iguanodon orientalis*, this form is now known as *Altirhinus*, thanks to research by Norman published in 1998. Also from Asia came *Tsintaosaurus* (from Tsintao), described by C.-C. Young in 1958. This Late Cretaceous hadrosaurid from Shandong, China, is truly one of the most unusual, sporting a unicorn-type horn from the top of its skull.

Elsewhere, work continued on the dinosaurs collected at the beginning of the century in Tendaguru, Tanzania (see Box 11.1). In 1955, W. Janensch began publishing work on his new ornithopod *Dysalotosaurus* (*dysalotos* – uncatchable, later to be referred to *Dryosaurus*).[3] And back in North America, a new hadrosaurid from the Late Cretaceous of Alberta was described as *Brachylophosaurus* (*brachys* – short) by Sternberg in 1953, while further south and to the east W. Langston Jr

---

3 There is a double entendre in this fossil's name. The name *Dysalotosaurus* – ("uncatchable lizard") is sometimes thought to be a reference to its gracile, sleek, morphology, but in fact the species name, *lettowvorbecki*, suggests the real intent behind the name. A World War I German general, Paul Emil von Lettow-Vorbeck, stationed in Tanzania, proved uncatchable to pursuing British and South African armies. In 1919 (just after the end of Word War I), the unrepentant German paleontologist, H. Virchow, in naming this Tanzanian dinosaur, celebrated this fact with its name.

christened *Lophorhothon* (*rhothon* – nose), a new hadrosaurid from the Late Cretaceous of Alabama.

In 1962, A. W. Crompton and A. J. Charig announced a new, and quite peculiar form, *Heterodontosaurus* (*heteros* – different; *odont* – tooth). This creature with tusks and molar-like teeth from the Early Jurassic of South Africa has been in and out of skirmishes over its affinities within Ornithischia, but now appears to be secure as an ornithopod. Most importantly for early ornithopods, *Heterodontosaurus* is now known from virtually complete skulls and an exquisite skeleton.

Later that decade, results of the Sino-Soviet Paleontological Expedition in China became available. The first ornithopod to come from these efforts, *Probactrosaurus* was described by Rozhdestvensky in 1966. Hailing from the Early Cretaceous of Inner Mongolia, China, *Probactrosaurus* has featured widely in discussions of the ancestry of Hadrosauridae and has recently been redescribed by Norman. Hadrosaurids also received a new member, in the form of *Aralosaurus* (from the Upper Cretaceous Aral region of Kazakhstan), described by Rozhdestvensky in 1968.

Finally, from the close of the 1960s throughout the early 80s, P. M. Galton began what can only be called a *tour de force* on ornithopod osteology, taxonomy, and phylogeny. A virtual one-man-show, Galton was eventually to take on: reconstructions of the limb musculature, locomotor mechanics, jaw mechanics, and feeding in several euornithopods; taxonomic revisions and redescriptions of *Parksosaurus*, *Hypsilophodon*, *Thescelosaurus*, and *Dryosaurus*; the naming of new taxa (*Othnielia* (for Othniel Charles Marsh), from the Late Jurassic of Colorado, Utah, and Wyoming, *Valdosaurus* (*valdus* – Weald, from the Wealden deposits) from the Early Cretaceous of England and Niger, and *Bugenasaura* (see below); and the paleobiogeographical and evolutionary patterns in the entire ornithopod group.

Although Galton's work on such a diverse array of ornithopods dominated from the end of the 1960s until the advent of cladistic approaches to phylogeny reconstruction (Chapter 3), a host of discoveries throughout this interval added important new information on ornithopod diversity. Indeed, the 1970s were to see more discoveries than even during the most prolific times of the Canadian Dinosaur Rush. J. H. Ostrom started the ball rolling in 1970 with *Tenontosaurus* (*tenon* – tendon), an Early Cretaceous ornithopod from Montana whose affinities have been of some contention, but which is now regard a basal iguanodontian. Halfway around the world was discovered *Shantungosaurus*, a gigantic hadrosaurid described in 1973 by S. Hu from the Late Cretaceous of Shandong, China.

Two new heterodontosaurids were also announced in 1975. The first, by C. E. Gow, was named *Lanasaurus* (*lana* – wooly; named to honor Harvard University's A. W. Crompton, whose nickname is "Fuzz"). Like *Heterodontosaurus*, *Lanasaurus* also comes from the Early Jurassic of South Africa. The second heterodontosaurid, named *Abrictosaurus* (*abriktos* – awake) by Hopson in 1975, hails from similar Lower Jurassic rocks of both

South Africa and neighboring Lesotho. Unlike *Heterodontosaurus*, these two new forms are known from only fragmentary skull material.

Further to the north on the African continent, we shift from heterodontosaurids to iguanodontians, for in 1976 P. Taquet announced the peculiar, high-spined *Ouranosaurus* (in Nigerian, *ourane* – brave). This ornithopod, from the Early Cretaceous of Niger, bears only a modest similarity to *Iguanodon* and has been pivotal to some of the more recent discussions of ornithopod phylogeny.

The new ornithopods of 1979 came shotgun-style from three of the four corners of the globe: from China, Argentina, and the USA. From the Middle Jurassic of Sichuan, China, the new basal euornithopod *Yandusaurus* (Chinese: *yan* – salt; *du* – capital), was described by X.-L. He. *Secernosaurus* (*secerno* – divide), the first hadrosaurid from the Southern Hemisphere, was announced by Brett-Surman from Upper Cretaceous beds of Rio Negro, Argentina. And Horner and R. Makela stunned the world with *Maiasaura* (*maia* – good mother), a hadrosaurid from the Late Cretaceous of Montana, whose remains included adult and hatchling specimens provided the first inkling of parental care in dinosaurs.

The unleashing of new ornithopods slowed down only slightly during the 1980s. In North America, *Zephyrosaurus* (Zephyros, Greek god of the west wind) was an Early Cretaceous basal euornithopod described by H.-D. Sues in 1980, while *Orodromeus* (*oros* – mountain; *dromeus* – runner), another basal euornithopod, was named by Horner and Weishampel in 1988. Both species come from Montana. In Asia, we were introduced to *Barsboldia* (named for Mongolian paleontologist R. Barsbold by T. Maryańska and H. Osmólska in 1981), a hadrosaurid from the Late Cretaceous of Mongolia, and *Gongbusaurus* (Chinese, *gong* – worker; *bu* – board; referring to Board of Works), a Late Jurassic euornithopod from Sichuan and Xinjiang provinces, China described by Z.-M. Dong and colleagues in 1983. Australia had a boom decade for ornithopods during the 1980s, with *Muttaburrasaurus* (from Muttaburra, described by A. Bartholomai and R. E. Molnar in 1981), an iguanodontian from the Early Cretaceous of Queensland, and *Atlascopcosaurus* (for Atlas Copco Co., which supplied excavation equipment) and *Leaellynasaura* (for Leaellyn Rich, who helped in the discovery), two euornithopods from the Early Cretaceous of Australia that were described in 1989 by T. H. Rich and P. Vickers-Rich. Ten years later, Rich and Vickers-Rich were to announce the existence of another basal euornithopod from the Early Cretaceous of Australia – *Qantassaurus* (named for Qantas, the Australian airline) – in 1999.

As important as it is to find new taxa, other studies tackled important problems of ornithopod paleobiology and evolution. P. Dodson and Norman separately began to create inroads into these problems in 1980 by redescribing ornithopod anatomy and questioning aspects of ornithopod systematics (Dodson on *Camptosaurus* and *Tenontosaurus*, Norman on *Iguanodon*). Also in 1984, the time was right for full-blown cladistic interpretations of ornithopod phylogeny: Norman and Sereno published cladistic analyses that included ornithopod dinosaurs and both researchers followed up these earlier studies with more detailed work in 1986.

The 1990s continued the march to find new ornithopods and to understand their place in the Mesozoic world. By mid-decade, we had *Agilisaurus* (*agili* – agile), *Drinker* (named in honor of Edward Drinker Cope), and *Bugenasaura* ("large cheeked lizard"), all basal euornithopods, the first described by Q. Pang from the Middle Jurassic of Sichuan, China, the second by R. T. Bakker and co-workers from the Late Jurassic of Wyoming, and the last by Galton from the Late Cretaceous of the western USA. Three hadrosaurids were also christened over this same time. Two come from the Late Cretaceous of New Mexico – *Anasazisaurus* ("Anasazi lizard") and *Naashoibitosaurus* ("Naashoibito lizard"; both described by A.-P. Hunt and S. Lucas) and the third discovered from Upper Cretaceous rocks of Siberian Russia, named *Amurosaurus* (for the Amur River, which forms part of the boundary between Russia and China) by Y. Bolotsky and S. Kurzanov.

By the close of the century, another five ornithopods were to be named. Two were non-iguanodontian ornithopods from the Late Cretaceous of Argentina: *Gasparinisaura* (named for Zulma Gasparini, Argentinean specialist on extinct crocodilians; *Gasparinisaura* was originally thought to be a basal iguanodontian) named by R. A. Coria and L. Salgado, and *Notohypsilophodon* (*noto* – southern) described by Martinez. The rest, basal iguanodontians all, have had much to say about the phylogeny of this group, particularly as relates to hadrosaurid origins. These include: *Eolambia* ("dawn lambeosaur") from the Early Cretaceous of Utah, named by J. J. Kirkland; *Protohadros* ("first hadrosaur") from the early Late Cretaceous of Texas, named by Head; and *Lurdosaurus* ("heavy lizard"), a massive form that lived with *Ouranosaurus* in the Early Cretaceous of what is now Niger, described by P. Taquet and D. A. Russell.

This first decade of the new millennium has hardly started and ornithopod discoveries still abound. China is once again proving its richness: *Charonosaurus* (named for Charon, the boatman in Greek mythology who ferried the souls of the dead across the River Styx to Hades), a lambeosaurine described by P. Godefroit and colleagues from the Late Cretaceous of northern China; *Nanyangosaurus* (named for Nanyang in Henan Province, where this iguanodontian was discovered) and *Jinzhousaurus* (named for Jinzhou, the large area in western Liaoning Province), both from the Early Cretaceous of China and both close to hadrosaurid origins, the former named by X. Xu and colleagues and the latter by X. Wang and Xu. Another ornithopod, *Jeholosaurus* ("Jehol lizard") comes from the Lower Cretaceous rocks in China that have yielded the feathered dinosaurs (see Chapter 14); it was described by Xu and colleagues.

Two other ornithopods – from Europe – have been described so far in the first years of the 2000s. *Draconyx* ("dragon claw"), a basal iguanodontian described by O. Mateus and M. Antunes, comes from the Late Jurassic of Portugal. And *Zalmoxes* (named for the Dacian deity *Zalmoxes*), known since the time of Nopcsa as *Rhabdodon*, is the sister-group to

Iguanodontia described by Weishampel and colleagues from the Late Cretaceous of Romania.

The last of the so far known millennial ornithopods is *Planicoxa* ("flat hipbone"), an iguanodontian from the Early Cretaceous of Utah, described by T. DiCroce and K. Carpenter.

Important readings

Dodson, P. 1975. Taxonomic implications of relative growth in lambeosaurine hadrosaurids. *Systematic Zoology*, **24**, 37–54.

Galton, P. M. 1974. The ornithischian dinosaur *Hypsilophodon* from the Wealden of the Isle of Wight. *Bulletin of the British Museum (Natural History) Geology*, **25**, 1–152.

Hopson, J. A. 1975. The evolution of cranial display structures in hadrosaurian dinosaurs. *Paleobiology*, **1**, 21–43.

Horner, J. R. 1984. The nesting behavior of dinosaurs. *Scientific American*, **250**, 130–137.

Horner, J. R., Weishampel, D. B. and Forster, C. A. 2004. Hadrosauridae. In Weishampel, D. B., Dodson, P. and Osmólska, H. (eds.), *The Dinosauria*, 2nd edn. University of California Press, Berkeley, pp. 438–463.

Lull, R. S. and Wright, N. E. 1942. Hadrosaurian dinosaurs of North America. *Geological Society of America Special Paper 40*, pp. 1–242.

Norman, D. B. 1980. On the ornithischian dinosaur *Iguanodon bernissartensis* from the Lower Cretaceous of Bernissart (Belgium). *Institut Royal de Science Naturelle de Belgique, Mémoire*, **178**, 1–103.

Norman, D. B. 1986. On the anatomy of *Iguanodon atherfieldensis* (Ornithischia: Ornithopoda). *Bulletin, Institut Royal de Science Naturelle de Belgique, Science de la Terre*, **56**, 281–372.

Norman, D. B. 2004. Basal Iguanodontia. In Weishampel, D. B., Dodson, P. and Osmólska, H. (eds.), *The Dinosauria*, 2nd edn. University of California Press, Berkeley, pp. 413–437.

Norman, D. B., Sues, H.-D., Coria, R. A. and Witmer, L. M. 2004. Basal Ornithopoda. In Weishampel, D. B., Dodson, P., and Osmólska, H. (eds.), *The Dinosauria*, 2nd edn. University of California Press, Berkeley, pp. 393–412.

Ostrom, J. H. 1961. Cranial morphology of the hadrosaurian dinosaurs of North America. *Bulletin of the American Museum of Natural History*, **122**, 33–186.

Taquet, P. 1976. Géologie et paléontologie du gisement de Gadoufaoua (Aptien du Niger). *Cahiers de Paléontologie*, pp. 1–191.

Weishampel, D. B. 1984. The evolution of jaw mechanisms in ornithopod dinosaurs. *Advances in Anatomy, Embryology and Cell Biology*, **87**, 1–110.

PART III

Saurischia: predators and giants

First coined by Cambridge University's H. G. Seeley in 1887, Saurischia originally consisted of Sauropodomorpha (Chapter 11) and its sister-taxon Theropoda (Chapters 12, 13, and 14). Seeley imagined saurischians as having a different primitive archosaurian ("thecodont") ancestor from that of ornithischians (Chapter 5), and thought of dinosaurs as a heterogeneous group of advanced archosaurs. A modern view includes Sauropodomorpha and Theropoda as well as a few primitive taxa that appear to be neither sauropodomorphs nor theropods within Saurischia. Saurischians include both the smallest of dinosaurs and the super-giants, as well as the most agile of predatory dinosaurs and the most ponderous plant-eaters.

Despite such disparate membership, Saurischia is monophyletic, defined as all Dinosauria closer to *Allosaurus* than to *Stegosaurus* (Figure III.1). The clade is diagnosed by more than a dozen derived features (Figure III.2), including modifications of the external nares, elongation of the rearward neck vertebrae (contributing to a relatively long neck), the development of accessory articulations between the dorsal vertebrae (the so-called hyposphene–hypantrum articulations), expanded transverse processes of the sacral vertebrae, twisted and enlarged thumb, and modifications of the pelvis. All in all, this is a well-supported dinosaurian clade.

Until recently, saurischians neatly divided into theropods and sauropodomorphs *à la* Seeley. However, recent discoveries of new taxa and better-preserved material of earlier-known taxa have provided a new

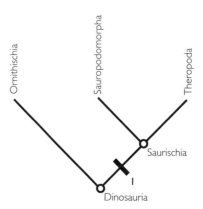

Figure III.1. Cladogram of Dinosauria, emphasizing the monophyly of Saurischia. Derived characters include: at I fossa expanded into the forward corner of the external naris, the development of a subnarial foramen, a concave facet on the axial intercentrum for the atlas, elongation of the centra of forward cervical vertebrae, hyposphene–hypantrum articulation on the dorsal vertebrae, expanded transverse processes of sacral vertebrae, loss of distal carpal V, twisting of the first phalanx of manual digit I, well-developed supracetabular crest, and restriction of the medioventral lamina of the ischium to the proximal third of the bone.

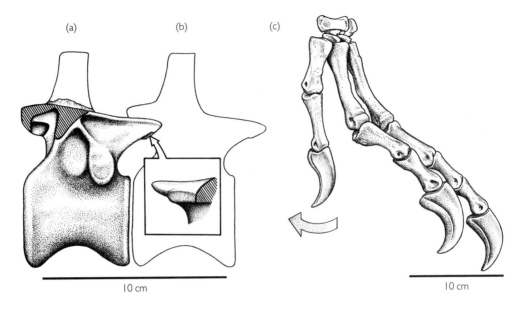

Figure III.2. (a) Dorsal lateral vertebrae of *Herrerasaurus* indicating the extra hyposphene–hypantrum articulations. (b) Hypantrum in medial view. (c) Twisted thumb (digit I of the hand).

window on the complexity of saurischian origins. These include *Herrerasaurus*, *Eoraptor*, *Staurikosaurus*, *Saturnalia*, and *Guaibasaurus*.

When J. A. Gauthier first assessed the phylogenetic relationships of theropods, it was necessary to establish who among remaining dinosaurs were the closest relatives to Theropoda. He identified Sauropodomorpha as the sister-group of Theropoda. No problem there; this union is the familiar Saurischia. And as we all know, Ornithischia is the sister-group to Saurischia. However, two relatively poorly known forms, *Staurikosaurus* and *Herrerasaurus* (Figure III.3a and b), were problematic. Because they were originally thought to be theropods and therefore dinosaurs, Gauthier's Dinosauria included not only Saurischia and Ornithischia, but also these two lesser well-known forms as basal members of the clade (Figure III.4a).

With new material of *Herrerasaurus* and the discovery of *Eoraptor* (Figure III.3c), P. C. Sereno and collaborators in the 1990s suggested that these two forms, plus *Staurikosaurus* were actually basal members of Theropoda (Figure III.4b). As a result, Dinosauria reverted solely to Saurischia and Ornithischia. With the subsequent discovery of *Guaibasaurus* and *Saturnalia* (Figure III.5), and new analyses of all these forms conducted by Max Langer, the situation again became more intricate and controversial. Dinosauria stayed the same, but Saurischia revealed more complexity than before. While theropods share a more recent common ancestor with sauropodomorphs, this relationship does not include *Herrerasaurus* and *Staurikosaurus* (together forming Herrerasauridae), *Eoraptor*, and *Guaibasaurus*. Instead, Saurischia is divided into two main groups: Herrerasauridae and a so-far unnamed

(a)

(b)

(c)

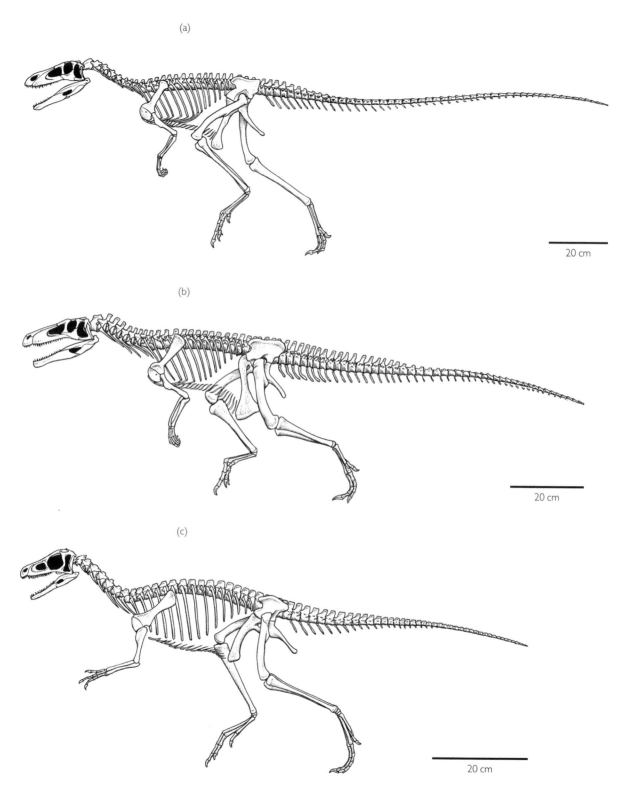

Figure III.3. (a) *Staurikosaurus*, (b) *Herrerasaurus*, (c) *Eoraptor*.

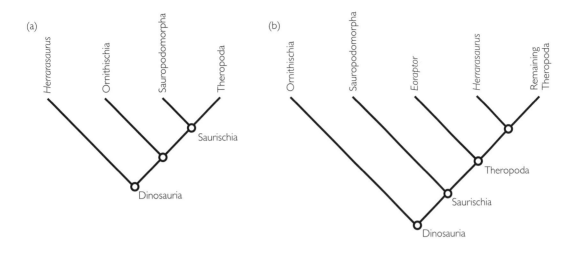

Figure III.4. (a) Gauthier's interpretation of basal saurischian relationships; (b) Sereno's interpretation of basal saurischian relationships.

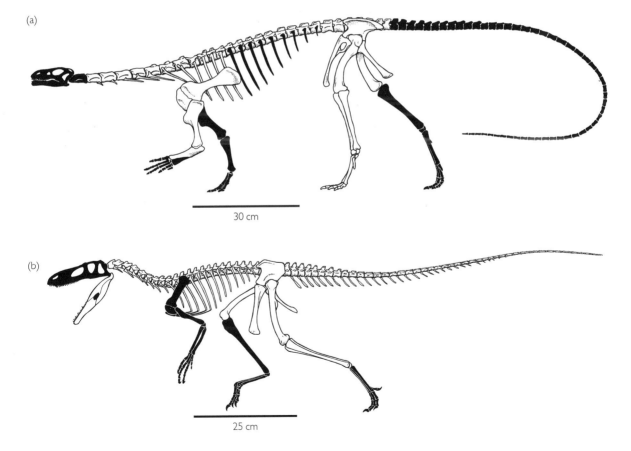

Figure III.5. (a) *Saturnalia*, (b) *Guaibasaurus*. Solid black areas are where lack of data (missing bones) have been supplemented by similar features in related dinosaurs.

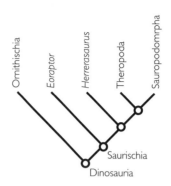

Figure III.6. Cladogram of Saurischia.

clade comprising *Eoraptor*, *Guaibasaurus*, *Saturnalia*, Theropoda, and Sauropodomorpha (Figure III.6). The features establishing these relationships are hard to come by because of the incomplete nature of the material of several of these basal saurischians, so this chunk of dinosaurian evolutionary history could well be unstable in the future.

Important readings

Gauthier, J. A. 1986. Saurischian monophyly and the origin of birds. *Memoirs of the California Academy of Sciences*, **8**, 1–55.

Langer, M. 2004. Basal Saurischia. In Weishampel, D. B., Dodson, P. and Osmólska, H. (eds.) *The Dinosauria*, 2nd edn. University of California Press, Berkeley, pp. 25–46.

Langer, M. A., Abdala, F., Richter, M. and Benton, M. J. 1999. A sauropodomorph dinosaur from the Upper Triassic (Carnian) of southern Brazil. *Comptes Rendus de l'Académie des Sciences, Paris, Sciences de la Terre et des Planètes*, **329**, 511–517.

Sereno, P. C. 1999. The evolution of dinosaurs. *Science*, 284, 2137–2147.

Sereno, P. C., Forster, C. A., Rogers, R. R. and Monetta, A. M. 1993. Primitive dinosaur skeleton from Argentina and the early evolution of Dinosauria. *Nature*, **361**, 64–66.

Sereno, P. C. and Novas, F. E. 1992. The complete skull and skeleton of an early dinosaur. *Science*, **258**, 1137–1140.

Figure 11.1. *Diplodocus*, one of the best-known sauropodomorphs, from the Late Jurassic of the Western Interior of the USA.

CHAPTER 11

Sauropodomorpha: the big, the bizarre, and the majestic

Sauropodomorphs appear to be the popular embodiment of the consummate dinosaur. First, many of them were *extremely* large. Secondly, they had a long neck and tail suspended from a rotund body supported on four lumbering legs. Thirdly, they are depicted as having nearly no brain power. And, finally, they are extinct. Isn't that what dinosaurs are all about?

But for us size and majesty might be a better way to describe sauropodomorphs. These extraordinary dinosaurs pushed the extremes of terrestrial body size – to the tune of 75,000 kg and possibly more (Figure 11.1). In doing so, they taxed biomechanical and physiological design – weight-support, neural circuitry, respiration, digestion, and the like – to the limit. It isn't easy being huge.

What does it take to be building-sized: to grow to up to 40 m long and tower 6 m at the shoulder? Take the shape of your basic sauropod, and add to it a long neck constructed by a complex system of girders and air pockets to maximize lightness and strength. At the end place a tiny skull – silly, until we realize that only an idiot would design a large, heavy skull at the end of an extremely long neck. The skull itself is unusual: it has relatively simple, peg-like or spatulate teeth, and nostrils that, instead of residing at the tip of the snout, have a phylogenetic tendency to migrate upward, toward the top of the head (Figure 11.2). Add to this four pillar-like limbs that would do a Greek temple proud. The limb bones are composed of bone denser than that found in the upper parts of the skeleton; again a sophisticated adaptation placing the weight and strength in the skeleton where it was most needed. Sauropodomorphs lived from the beginning of dinosaur history (Late Triassic) until its denouement (the close of the Cretaceous). Over this long interval, sauropodomorphs managed to walk or be carried to nearly all points of the compass, from Europe to Africa, to Asia, to North America, to South America, to Australia, to Antarctica (Figure 11.3).

Many sauropods were big, and so you'd expect that their bones would be easily found. The record of individual sauropodomorphs, however, can be very poor. Except for a few prosauropods (some of which are known from complete skeletons), most sauropodomorphs are known from incompletely preserved material, commonly missing heads, parts of tails, and feet. P. Dodson noted this enigma: why are animals as large and presumably durable as sauropodomorphs not better preserved in the fossil record? In answer, he has suggested that, in animals as large as the largest sauropods, preservation of a complete skeleton is balanced against the vagaries of both sedimentation and erosion. Perhaps many of the sauropodomorphs were just too large to be easily buried. It is one thing to bury a dog carcass and quite another to bury a whale. This matches the observation that those body parts most likely to be lost before burial – the tail, feet, and skull – are just the elements commonly missing when sauropodomorphs are collected.

The dinosaurs that look like "brontosaurus" are but one part of Sauropodomorpha (*saurus* – lizard; *pod* – foot; *morpho* – form); the other consists of a relatively short-lived clade called prosauropods (*pro* – before;

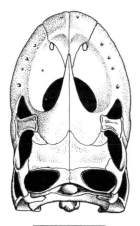

10 cm

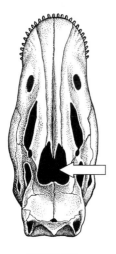

10 cm

Figure 11.2. Dorsal view of the skull of (a) *Brachiosaurus* and (b) *Diplodocus*. Note the dorsally placed external nares, especially in *Diplodocus* (arrow).

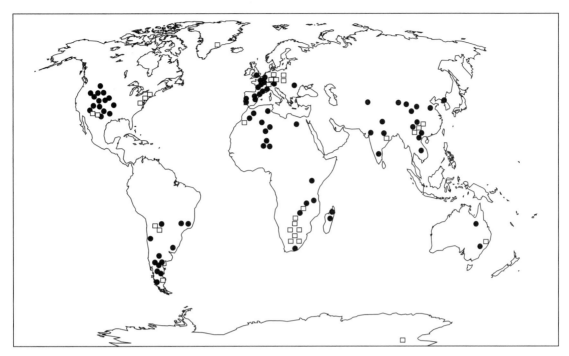

Figure 11.3. Global distribution of Sauropodomorpha. Prosauropoda indicated by solid circles; Sauropoda indicated by open squares.

Figure 11.4). Prosauropods are first known from the Late Triassic, from many corners of the world – North America, Asia, Europe, Africa, and South America. Later sauropodomorph history, that of the sauropods (Figure 11.5), is rich in evolutionary ebb and flow, including right up to the last moments of the Cretaceous. By some accounts, sauropodomorphs spawned well over 100 different species over this interval.

Sauropodomorphs are split roughly one-third to two-thirds between prosauropods and sauropods. Within prosauropods are two principal groups, anchisaurians and plateosaurians, with few more primitive forms at the base of the anchisaurian + plateosaurian clade. Sauropods

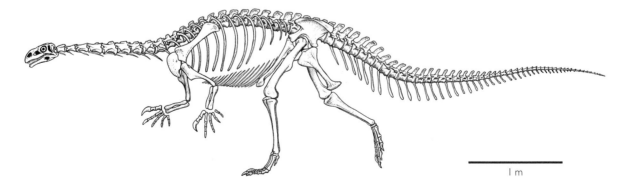

1 m

Figure 11.4. Left lateral view of the skull and skeleton of *Plateosaurus*.

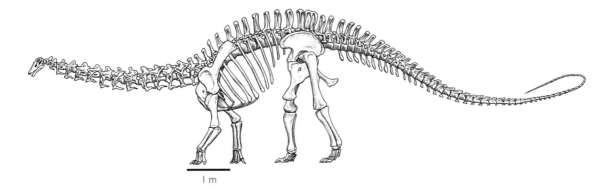

| m

Figure 11.5. Left lateral view of the skull and skeleton of *Apatosaurus*.

on the other hand consist of eusauropods, which comprise nearly all but a few primitive forms, and neosauropods, which consist of diplocoids and macronarians. This latter group contains a host of well-known sauropods (*Camarasaurus* and *Brachiosaurus*, among them), as well a many others.

Lifestyles of the huge and ancient

When we find the remains of these magnificent animals, they come from a myriad of different environments. All are known from floodplain sediments to channel and even eolian sandstones. In the Upper Jurassic Morrison Formation of the Western Interior of the USA, where sauropods are known in great abundance and where their sedimentary context has been carefully studied by Dodson, A. K. Behrensmeyer, and R. T. Bakker, sauropod remains are distributed across a broad spectrum of sedimentary environments, from those interpreted as lakes, rivers, and floodplains. By way of comparison with the distribution of the carcasses of large terrestrial mammals such as elephants and rhinos living today in places like the Serengeti of eastern Africa, it appears that sauropods were similarly able to range across all available habitats found in the ancient landscapes represented by the Morrison Formation.

In addition to such catholic utilization of living space, these animals may have had to cope with long dry seasons during the year. Sedimentological and paleontological evidence suggests that the ancient environment of the Morrison Formation was subject to seasonal aridity. Such annual droughts may have been severe enough to have forced sauropods into large-scale migration to more lush areas. We will have more to say about migration when considering sauropod herding.

Was seasonal aridity typical of sauropod habitats? At Tendaguru (Box 11.1) in southeastern Tanzania, as well as in northern Texas at the famous Glen Rose trackway sites (Figure 11.6) and in Maryland where the remains of the brachiosaurid *Astrodon* have been uncovered, there is strong geological evidence that these environments were once close to the sea and potentially quite humid. Perhaps these were some of the conditions that sauropods found most congenial.

In the swamps? But wait – doesn't everybody know that sauropods frequented swamps, where their great bulk was buoyed up by the water? In

BOX 11.1

Tendaguru!

Tendaguru! The name strikes like the boldness of a movie marquee: the hinterland of Tanzania on the eastern coast of Africa. This spot, which today is monotonously formed of broad plateaus blanketed by dense thorn trees and tall grass thick with tsetse flies, was formerly the site of perhaps the greatest paleontological expedition ever assembled, and much – thousands of millennia – before that the place where dinosaurs came to die.

Let's go back to 1907, when Tanzania was part of German East Africa. This was the era of massive western European colonialism in Africa. With the widespread colonialism came scientists. And to then German East Africa came paleontologists in search of fossils.

The fossil wealth of Tendaguru was first discovered in 1907 by an engineer working for the Lindi Prospecting Company. Word spread quickly, ultimately to Professor Eberhard Fraas, a vertebrate paleontologist from the Staatliches Museum für Naturkunde in Stuttgart, who happened to be visiting the region. So excited was he at the prospect of collecting dinosaurs after his visit to Tendaguru that he took specimens back to Stuttgart (including what was eventually to be called *Janenschia*) and more especially started drumming up interest among other German researchers to continue field work in the area.

It was W. Branca, director of the Humboldt Museum für Naturkunde in Berlin, who was the first to seize upon the opportunity presented to him by Fraas. Yet before mounting an expedition of the kind demanded by Tendaguru, Branca had to tackle the problem of its financial backing. By seeking support from a great many sources, he received more than 200,000 deutschmarks – a fortune for the time – from the Akademie der Wissenschaften in Berlin, the Gesellschaft Naturforschender Freunde, the city of Berlin, the German Imperial Government, and almost a hundred private citizens.

With money, materials, and supplies in hand, the Humboldt Museum expedition set off for Tendaguru in 1909. For the next four field seasons, it was

Figure B11.1.1. Werner Janensch, the driving force behind the extraordinarily successful excavations at Tendaguru, Tanzania.

bonanza time. Under the leadership of moustached and jaunty Werner Janensch (Figure B11.1.1) for three of these seasons (Hans Reck took charge in the fourth season), these years were to see possibly the greatest dinosaur collecting effort in the history of paleontology. The first season involved nearly 200 workers, mostly natives, laboring in the hot sun as they dug huge bones out of the ground. During the second season, there were 400 workers and in the third and fourth seasons 500 workers. By the end of the expedition's efforts, some 10 km^2 of area was covered with huge pits, attesting to the diligence and hard work of these laborers.

But there's more. Many of these native workers brought their families with them, transforming the

(continued on p. 234)

BOX 11.1 *(cont.)*

dinosaur quarries at Tendaguru into a populous village of upwards of 900 people. With all these people, water and food were a severe problem. Not available locally, water had to be brought in, carried on the heads and backs of porters. And with the vast quantities of food that had to be obtained for workers and their families, and the pay for work carried out in the field, it is not surprising that the funds amassed by Branca disappeared at a great rate.

Still, the rewards were great indeed. Over the first three seasons, some 4,300 jackets were carried back to the seaport of Lindi – a four-day walk away, a trip made 5,400 times by native workers each with the fossils balanced on his or her head and back – to be shipped from there to Berlin.

Overall, work at Tendaguru involved 225,000 man-days and yielded nearly 100 articulated skeletons and hundreds of isolated bones. When finally unpacked and studied, what a treasure-trove: in addition to ornithischians (*Kentrosaurus*, *Dryosaurus*) and theropods (*Elaphrosaurus*), and a

pterosaur as well, the Tendaguru expeditions claimed not only two new kinds of sauropod (*Tornieria* and *Dicraeosaurus*), but also new material of *Barosaurus* and the finest specimen of *Brachiosaurus* ever found.

The Humboldt Museum never went back to Tendaguru after 1912. In 1914, the World War I erupted and, with the Treaty of Versailles, German East Africa became British East Africa. This shift in the continuation of European colonialism brought British paleontologists to Tendaguru in 1924, under the direction of W. E. Cutler. This team from the British Museum (Natural History) hoped to enlarge the quarried area and retrieve some of the left-over spoils from the German effort. From 1924 to 1929, the British expedition had its ups and downs, finding more of the kinds of dinosaurs discovered earlier, but suffering some severe health problems including malaria, from which Cutler died in 1925. There has been no significant paleontological effort at Tendaguru since.

Figure 11.6. Five parallel trackways of Late Jurassic age, Morrison Formation, Colorado, USA. Tracks are thought to have been made by a diplodocid. (Photograph courtesy of M. Lockley.)

this way, they could have remained deeply submerged, breathing with only their nostrils poking out of the water. Such an idea may have originated as a consequence of R. Owen's (1842) mistaken notion of the affinities of *Cetiosaurus* (remember, he thought it to be an exceedingly large marine crocodile). It was later argued that sauropods were too big and heavy to have lived on land; that their legs would not have supported them; and that, like a whale, they needed the buoyancy of water to permit them to breathe. It was said that the long tails were used for swimming. Finally, it was suggested that the nostrils at the top of the skull were like the snorkel on a submarine, allowing the animal to breathe while remaining fully submerged (and presumably hidden and protected). Although as early as 1904 E. S. Riggs argued coherently that these animals were terrestrial, sauropods have persistently been viewed as having a fully aquatic existence, with their bodies submerged 6 m or more below the surface of the water and their necks craning to keep the head, and hence the nostrils, just above the water's surface.

The idea of aquatic sauropods began unraveling in the 1950s and especially during the 1960s and 70s. In 1951, K. A. Kermack examined the barometric consequences of the lungs of a submerged sauropod. The air that these animals had to breathe would obviously be at atmospheric pressure (i.e., 1 atmosphere of pressure). However, because the thorax (in vertebrates, the part of the body between the neck and stomach) – and hence the lungs – is under a column of water some 6 m or more deep, this part of the sauropod's respiratory anatomy is under nearly 2 atmospheres pressure (water pressure goes up 1 atmosphere of pressure every 9.8 m). This pressure at depth would tend to collapse the thorax, pushing whatever air was in the lung up and out of the body. How the next breath might be taken is hard to say, since the lungs would have to be expanded against high pressures, well beyond any experienced in vertebrates today. Unless sauropods had exceedingly powerful chest muscles, these animals certainly would have been unable to breathe in. For this reason, Kermack argued, it is perhaps better to envision sauropods as terrestrial animals.

Sauropods clambered out of the water for good with Bakker's studies in the late 1960s and 70s. This research emphasized the sturdy, pillar-like construction of the legs and feet of sauropods, surely strong enough for walking – at least slowly – on land. Additionally, Bakker pointed to the narrow, slab-sided thorax – very unlike the amphibious hippopotamus, but like rhinos and elephants – as evidence of terrestrial habits. Not since these studies have sauropods been submerged up to their nostrils and we hope that they never will again.

So what were the long necks for, if not for keeping the head above water? From a land-based perspective, sauropod necks have been likened to that of giraffes, inviting the inference that Mesozoic giants fed in tall trees. For some sauropods such as *Cetiosaurus*, *Diplodocus*, and *Apatosaurus*, such feeding habits do not appear possible, but for others, such as *Brachiosaurus*, it may have been true. While mounting a partial skeleton of *Cetiosaurus* at the Leicesterhire Museum in England, J. Martin experimented with the neck vertebrae to see what sort of restrictions there were on its range of motion, concluding that this Middle Jurassic sauropod was unable to upwardly flex its neck as far as had been thought previously. A dozen years after Martin's work, Kent Stevens and Mike Parrish demonstrated – using computer modeling – that similar limitions in the cervical region are also found in diplodocids (*Diplodocus* and *Apatosaurus*). Their research, manipulation of a three-dimensional virtual model of the neck vertebrae, provided the feeding envelope (maximal vertical and horizontal range of head movement) of these two sauropods, again indicating that they generally browsed for vegetation within 2 or 3 m of the ground.

For *Brachiosaurus*, things may have been different. Not only was the neck very long, but the front limbs were longer than those in the rear (Figure 11.7). With this "extra boost," its head could apparently be raised to a height of 13 m, providing the opportunity to feed on foliage to which virtually no one else had access. But animals like *Brachiosaurus* had to pay a price for such posture. Now that its head was perched so high, its brain (the relatively smallest among dinosaurs) would have towered about 8 m above its heart. In contrast to the condition noted by Kermack for a submerged sauropod, under terrestrial conditions, the heart would have to pump blood at very high pressure to get it to the brain. Otherwise, the animal would have no recourse but to faint when it lifted its head.

What sort of blood pressure are we talking about to pump blood from the heart to the head of a sauropod? Remember that, when you go to the doctor and your blood pressure is taken, it comes in two numbers, both of which are measures of pressure in terms of millimeters of mercury, abbreviated mmHg. The first, the systolic pressure (usually

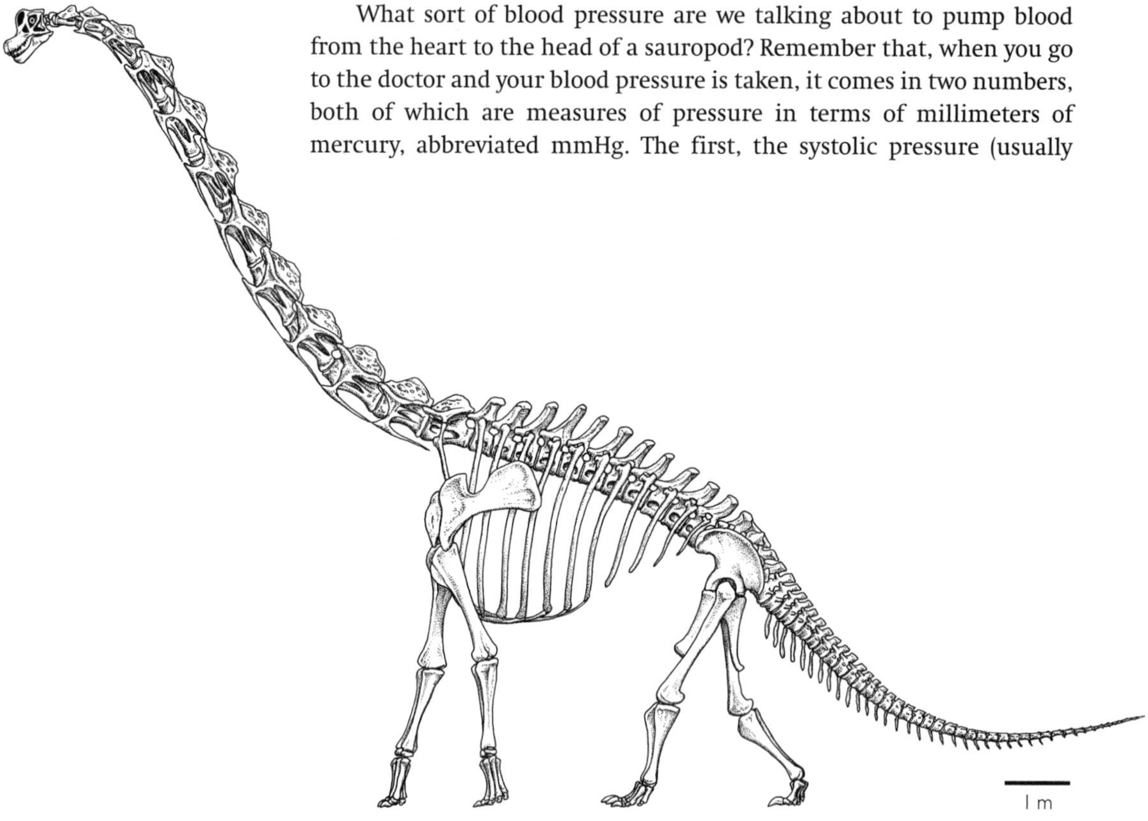

1 m

Figure II.7. Left lateral view of the skull and skeleton of *Brachiosaurus*.

Figure 11.8. Systolic blood pressures compared: (a) a sauropod (approximately 630 mm), (b) a giraffe (320 mm), and (c) a human (150 mm).

around 130 mmHg in healthy people), indicates the pressure your blood is experiencing when the heart contracts. It is during this part of heart activity that blood is shot out to the extremities of the body; hence, it is at its highest pressure. In contrast, during heart relaxation, blood pressure (diastolic) is at its minimum. In healthy people, this diastolic pressure ranges between 70 and 80 mmHg.

We can estimate what kind of systolic pressure would be necessary to get blood from the heart to the head in a full-grown *Brachiosaurus*. To push blood through the arteries up its 8.5 m long neck (or thought of another way, to counter the pressure of an 8.5 m column of blood), the heart of a *Brachiosaurus* must pump with a pressure exceeding 630 mmHg. Such blood pressures would have been much larger than those in any living animal. Most mammals – including ourselves – have pressures of 110 to 150 mmHg and even giraffe have only about 320 mmHg. It would indeed take a very muscular heart – some estimate one weighing as much as 400 kg – to do the pumping (Figure 11.8). It's hard to imagine how the biomechanics of such a gigantic pump would have been feasible and a recent study by R. S. Seymour and H. B. Lillywhite suggests that it would have been impossible.

However, another aspect of body posture also influences our inferences about the heart and blood pressure in sauropods. Initially suggested by J. B. Hatcher in 1901, and revived by Bakker in 1971, *Diplodocus* and the like may have gained access to foliage at high levels in the trees by adopting a tripodal posture, rearing up on their hindlimbs and using their tails as a "third leg" (Figure 11.9). This posture should sound familiar; we considered it in Chapter 6 when considering the possibility that stegosaurs could also do it. Similarly, the center of gravity was positioned just in front of the hips, making it biomechanically rather easy to rear upward in its hindlimbs.

Recall that many sauropods have distinctive, well-developed, Y-shaped neural arches. Here is the anatomical function of this feature: the head and neck of the animal were supported by a strong and taut ligament cradled between the bifurcations of the neural spines of the dorsal and cervical vertebrae in diplodocids (apparently also in *Camarasaurus*). So raising the head and neck would not have been a constant muscular struggle; much of the weight would have been taken up by this ligament (Figure 11.10). In tripodal posture, these dinosaurs would have had to pay the same price as *Brachiosaurus* (which itself was probably not able to rear up): elevated blood pressure and a large heart to produce it.

So it comes down to the sometimes conflicting issues of blood pressure, heart size, the bony construction of neck vertebrae, and head height as we infer how sauropods held their necks. Held high, the pressure necessary to get blood from the heart to the head would have been extreme, necessitating a cardiac pump that may itself been bio-

Figure 11.9. A sauropod reconstructed in a tripodal posture, using the tail as a "third leg."

Figure 11.10. Anterior neck vertebrae in *Diplodocus*. The neural spines are bifurcated, and are thought to have held a ligament supporting the neck, the nuchal ligament (shown in black) running from the head, down the neck, and beyond.

mechanically impossible. In diplodocids and cetiosaurs at least, the vertebral column restricted elevation of the neck. And finally, many of these sauropods had bifurcations of the cervical neural spines that would have accommodated very thick longitudinal ligaments attaching to the back of the head. These ligaments had no possibility of contracting – that job is only for muscles. Instead, what they would have been good for was resisting downward bending of the head and neck, much like a very taut rubber band. In this way, the neck was more likely to have been held horizontally, because it was supported there by the long ligament between the neck and head. And a horizontal neck would have placed the head at approximately the same level as the heart, with no outrageously high blood pressures required. Still these animals may have occasionally reared up on their hindlimbs – during intraspecific combat, when feeding on highly placed leaves in tall trees whose lower foliage had been depleted, and perhaps during procreation.

Before leaving the subject of the physiological consequences of the extraordinarily long necks of sauropods, we turn to the nature of respiration. For prosauropods and especially sauropods, the trachea (wind-pipe) would have been exceptionally long, approximately the same length as those arteries carrying blood from the heart to the brain. The trachea is necessary to bring oxygen into contact with the alveoli in the lungs, those physiologically active sites where oxygen is transmitted to the blood and where carbon dioxide is passed back to the air.

The key words here are "physiologically active": the alveoli are and the trachea and most of the brochial passageways are not. In animals that pass air bidirectionally into and out of the lungs (i.e., during inhalation and exhalation) rather like a bellows (mammals, lizards, crocodilians, and snakes), the trachea creates physiological dead-space: some portion of the inspired air never reaches the alveoli. Its simply brought into the respiratory system and returned without being involved in oxygen–carbon dioxide exchange. By contrast, birds have a unidirectional air-flow, whereby inhaled air passes across the physiologically active tissues of the lung (in this case, called faveoli), nearly all the oxygen is absorbed into the bloodstream, and the now oxygen-depleted air is run through a series of pulmonary sacs around the lungs and back into the trachea for exhalation. Obviously, the avian system wrings more oxygen out of the air than the bidirectional, bellows-style lungs found in mammals and other tetrapods.

But what of sauropodomorphs? Here we take a clue from both birds and giraffes. In animals that have long necks, the problem of physiological dead-space would probably be acute if they didn't have uni-directional air flow. Long-necked bidirectional breathers like giraffes circumvent the problem of dead-space by having an inordinately narrow trachea: dead-space is reduced by limiting the surface area of the trachea. In fact, it is thought by some that giraffe may be the longest-necked animals capable of combining bellows-style lungs and a very long trachea. That being the case, it may be that sauropodomorphs, especially if these giant forms had an elevated metabolism (see Chapter 15), may have had unidirectional, avian-style lungs in order to eliminate the problems associated with all of that physiological dead-space engendered by the very long trachea. Given the elongation of the neck region in Saurischia as a whole, isn't it possible that these unidirectional lungs should be described as saurischian and not "avian"?

Feeding

Sauropodomorphs evidently were an important and diverse group of well-adapted herbivorous dinosaurs that almost certainly fed from the crowns of quite tall trees. Still, by comparison with modern herbivorous mammals and especially with other heribvorous dinosaurs such as hadrosaurids, the skull does not appear to be particularly well-designed for powerful chewing (Figures 11.11 and 11.12). Yes, there is a ventrally offset jaw joint in animals such as *Plateosaurus*, *Coloradisaurus*, *Apatosaurus*, and even *Camarasaurus* (see introductory text to Part II Ornithsichia), but overall the skull is relatively small and lightly built, with little room for jaw muscles. Most sauropod teeth have simple crowns: triangular, spatulate, or slender and pencil-like (Figure 11.13). There is even a tendency in the clade to limit the teeth to the front of the jaws. In all cases, there is nothing like the full arcades or dental batteries seen in other dinosaurian herbivores (see Chapters 9 and 10).

P. M. Barrett and G. R. Upchurch have studied in great detail the evolutionary pattern of sauropod feeding by linking various aspects of jaw mechanics and body form with phylogeny. They observed that sauropods

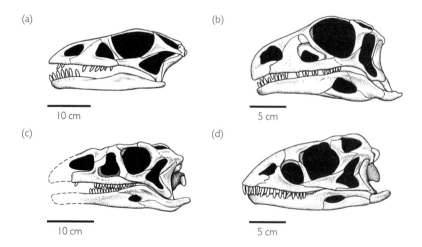

(a)

10 cm

(b)

5 cm

(c)

10 cm

(d)

5 cm

Figure 11.11. Left lateral view of the skull of (a) *Anchisaurus*, (b) *Coloradisaurus*, (c) *Lufengosaurus*, and (d) *Yunnanosaurus*.

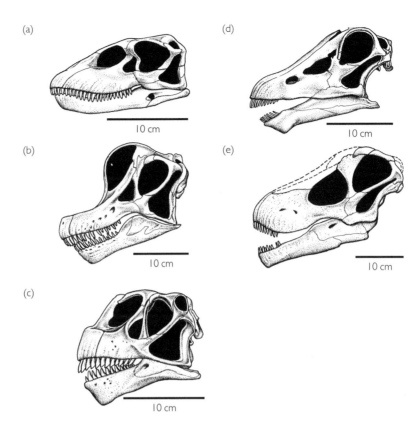

Figure 11.12. Left lateral view of the skull of (a) *Shunosaurus*, (b) *Brachiosaurus*, (c) *Camarasaurus*, (d) *Diplodocus*, and (e) *Nemegtosaurus*.

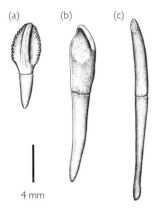

Figure 11.13. Teeth in selected sauropodomorphs. (a) Leaf-shaped prosauropod tooth of *Plateosaurus*; (b) spatulate tooth of sauropod *Camarasaurus*; (c) pencil-like tooth of *Diplodocus*. The lower part of each tooth is the root.

lengthened their forelimbs over the primitive condition seen in prosauropods and other saurischians. Not far from the base of the sauropod evolutionary tree is an increase in body size and the development of spatulate teeth, the latter viewed as a means for cropping coarser vegetation.

By the time we reach more derived sauropods, the snout broadens, the lower jaw strengthens, and wear indicating front and rearward movement of the jaws is found on the teeth. Several major changes in the jaw system can be identified with further sauropod evolution. Among neosauropods, that great clade of sauropods that includes camarasaurs, brachiosaurs, and titanosauroids (camarasauromorphs) on the one hand and diplodocoids on the other, the skull shows additional strengthening (closure of the antorbital fenestra). Camarasauromorphs generally show a shortening and elevation of the skull, indicating a more powerful biting force, but otherwise retain the feeding features seen in other, more primitive neosauropods. In contrast, the peg-like teeth arranged at the front of the jaws (forming a tooth comb), apical tooth wear (instead of longitudinal), and elongation of the snout in diplodocoids indicate an abandoning of fore–aft jaw movement and a shift to exploitation of a more delicate food source in these animals.

In sum, sauropods evolved a variety of ways to grab onto, and begin to prepare, food in the mouth. Tooth form and especially tooth wear indicate that these gigantic herbivores either nipped and stripped foliage, unceremoniously delivering a succulent bolus to the gullet without much modification in the mouth. It is doubtful that many of these animals fed selectively, given the size of the jaws, the nature of the dentition, and body size, although diplodocoids may have been relatively more selective than other sauropods.

Swallowing sped the bolus down its long travel along the esophageal canal, whereupon it entered the abdomen, and in particular the gizzard. This muscular chamber, sitting just ahead of the glandular part of the stomach, is thought to have been developed in both prosauropods and sauropods; enclosed within it would have been a collection of gastroliths, rather large and smoothly polished stones that acted as a gastric mill. Contraction of the walls of the gizzard churned the gastroliths, thereby aiding in the mechanical breakdown of food as it passes further along in the gut. Gastroliths are known – albeit rarely – in both prosauropods and sauropods (including the discovery of in-place gastropods with the skeleton of *Seismosaurus*).

In all sauropodomorphs, the gut must have been capacious, even considering the forward projecting pubis (in contrast to all ornithischians, which rotated the pubis rearward to accommodate an enlarged gut; see Chapter 7). J. O. Farlow envisioned these animals as having an exceptionally large fermentation chamber in the hindgut that would have housed endosymbionts; that is, bacteria that lived within the gut of the dinosaur. The endosymbionts would have chemically broken down the cell walls of the plant food, thereby liberating their nutritious contents. Considering the size of the abdominal cavity in sauropodomorphs, these animals probably fed on low-quality food items (i.e., foliage with high fiber content; see Chapter 16); perhaps they also had low rates of passage of food through the gut in order to ensure a high level of nutrient extraction from such low-quality food. Still, these huge animals with their comparatively small mouths must have been constant feeders to acquire enough nutrition to maintain themselves. The digestive tract of a sauropod had to have been a non-stop – if slow-speed – conveyor belt.

As we earlier discussed, sauropodomorphs were the tallest browsers for their time – and for all time in many cases. Prosauropods like *Euskelosaurus* and *Plateosaurus* were able to feed at up to 3 m above the ground, particularly if they assumed a tripodal posture. This was tall for the Late Triassic and Early Jurassic. Likewise, some later sauropods may have been able to do so as well, at least until they felt faint. And the elongate necks seen in all members of Sauropodomorpha certainly extended vertical feeding ranges. Such ranges have been estimated to have been up to four or five stories. In sauropodomorphs capable of rearing up on their hindlimbs to feed, it is likely that the hand manipulated leaves and branches to the mouth or possibly assisted in balancing while the animal craned for foliage just out of reach.

From their anatomy, we have good reason to suspect that sauropodomorphs browsed at high levels, but were there tall trees for them to browse on? Yes, indeed; the Late Triassic had its share of tall plants, the likes of which *Plateosaurus* and *Massospondylus*, among other prosauropods, fed upon. These include ferns, conifers, seed plants, cycads, and ginkgoes. During the Jurassic, considered by some the hey-day of sauropod evolution, a great variety of conifers, with fewer kinds of ginkgoes, cycads, ferns, and horsetails, constituted the tall plants available to a browsing *Omeisaurus*, *Diplodocus*, or *Seismosaurus* (see Chapter 16). And for the Cretaceous sauropodomorphs such as *Saltasaurus*, *Quaesitosaurus*, and *Alamosaurus*, there were the emergent angiosperms, some of which probably reached tall-tree height before the end of the period (see Chapter 16).

However, beyond recognizing this potential link between sauropodomorphs and their fodder, we have very little direct evidence of their diets. It has been suggested that a pile of carbonaceous material, old stems, bits of leaves, and other plant material collected from the abdominal region of a sauropod skeleton found in the Morrison Formation of Wyoming constitutes fossilized gut contents. Interestingly, this skeleton also had a packet of gastroliths in its belly region. Another potential example comes to us from the Upper Jurassic of Utah. Consisting of sections of small twigs and branches, it contains neither leaves nor carbonized residues. If either of these collections of plant hash are to be believed as stomach contents, they give the impression that sauropod digesta was rather coarse and fibrous.

Locomotion

The early history of locomotion in Sauropodomorpha is consistent with the primitive condition for all dinosaurs: bipedality (see Chapter 5). In the most primitive of prosauropods (e.g., *Thecodontosaurus*, *Anchisaurus*, *Ammosaurus*), the forelimbs are shorter than the hindlimbs and the trunk region is relatively short, suggesting that these animals walked principally on their hindlimbs rather than on all fours. However, the largest and most derived of prosauropods (among them *Riojasaurus* and *Blikanasaurus*) appear to have become fully quadrupedal.

In all cases, prosauropods appear to have been quite slow, perhaps the slowest of all bipedal dinosaurs. R. A. Thulborn calculated that most moved around at no more than 5 km/h, about the average walking speed of humans. Whether such an inference is true might be judged by checking it against prosauropod footprints and trackways. The few prosauropod trackways about which we are confident come from Lower Jurassic and Early Cretaceous rocks of the American southwest and from Upper Triassic sediments of the southern Africa country of Lesotho.[1]

What do these tracks tell us about prosauropod locomotion? Unfortunately, not much about walking or running rates. Nevertheless, there is information to be gleaned about limb posture from these

---

1 On the face of it, there is an abundance of so-called prosauropod prints throughout the world. However, most of these – despite claims in the literature – were not made by prosauropods at all. Instead, their makers were theropods, ornithopods, primitive ornithodirans, or even crocodylomorphs.

trackways. First, they all come from animals walking quadrupedally; none indicates the bipedal stance or locomotion that is suggested for primitive prosauropods on the basis of skeletal information. Furthermore, when walking on all fours, the prosauropod print-maker had a rather broad trackway, with the oval prints of the hindfoot turned outward from the midline. In keeping with the rearward-positioned center of gravity, the imprints of the hands are smaller and somewhat shallower than the feet. Interestingly, the large thumb claw appears to have made a mark in the ground only when the hand sank deeply into the substrate. Otherwise, it was clearly held high enough to clear the surface.

For sauropods, both skeletons and trackways reveal a great deal more about locomotion in these animals than they do for prosauropods. Again, we look to Thulborn's estimates of speed based on skeletal information. In 1990, he calculated maximal speeds for such sauropods as *Brachiosaurus*, *Diplodocus*, and *Apatosaurus*; all ranged between 20 and 30 km/h, a reasonable clip for animals the size of a house and weighing in excess of three to ten elephants.

Probably more to the point, though, sauropods probably walked a good deal slower most of the time, perhaps at rates of 20 to 40 km/day. These slower rates are based on sauropod trackways known in great quantities from the Jurassic and Cretaceous of North and South America, Europe, Asia, and Africa. Remarkable footprint assemblages tell us not only about the presence of particular track-makers in the area, but also about the details of walking and/or running, some of the subtleties of the animals' behavior, and of course the abundance of the animals as they paraded across the ground.

Sauropod trackways tend to be quite narrow and their prints immense. The hand is horseshoe shaped. In contrast, the foot is ovoid and there is an indication of toe claws and a heal pad. Most significantly, relatively few trackways include a tail-drag mark, providing strong evidence that many sauropods carried their immense tails clear of the ground.

Social behavior

The generally slow rate of their progression is certainly consistent with the long-distance movements we suggested earlier for migratory sauropods. This particular aspect of sauropod – indeed more generally of sauropodomorph – social behavior has long been of interest to dinosaur paleobiologists. F. von Huene first suggested mass movement of large groups (should we call them herds – or flocks?) of *Plateosaurus* from the more easterly highlands of Europe during the Late Triassic to account for the large, virtually monospecific bonebed at Trossingen, Germany. Since that time, numerous other prosauropods have been discovered in single-species mass accumulations. Likewise, the many now-famous mass accumulations in the Morrison Formation in the USA, the Tendaguru Beds of Tanzania, the Lower Jurassic sauropod sites of India, and most recently the Middle Jurassic extravaganzas of Sichuan, China, together with the vast sauropod footprint assemblages described earlier, all speak loudly to the existence of gregariousness of at least some sauropods, including *Shunosaurus*, *Diplodocus*, and *Camarasaurus*. Sauropods living in large groups must have been capable of wreaking severe damage on local vegetation, either by stripping away all the foliage they could reach or by trampling into the ground all of the shrubs, brush, and trees that might have been in the way. If many kinds of sauropod actively depleted their food source, as P. Dodson and colleagues suggested in 1980, then it is equally likely that these herds had to migrate elsewhere for a bite to eat. In contrast, there were many other sauropods, including *Brachiosaurus* and *Haplocanthosaurus* from the Morrison Formation, and *Opisthocoelicaudia* from Mongolia, that are not so numerous. Is it possible that they lived a more solitary existence, perhaps remaining behind as the likes of *Shunosaurus* or *Camarasaurus* left them in the sunset?

Beyond gregariousness, our discussion of sauropodomorph behavior culminates with the time-honored subjects of defense and reproduction. Defense in sauropods is obvious: large size confers the supreme deterrent against an attack. But for many prosauropods and of course the young of any of these animals, protection via size was far off in both ontogenetic or phylogenetic time. Thus for the chief weaponry in these animals we look to the large and trenchant thumb claw that is especially well developed in prosauropods of any age and even in young and vulnerable adult sauropods (despite reduction of the rest of the digits of the hand).

Sauropodomorph growth and development

Until recently, we knew next to nothing about the early occasional discoveries of subadult specimens, including a few relatively complete skeletons of both prosauropods and sauropods. But nothing like the great hadrosaurid nesting grounds of Montana. Indeed, Bakker, on the basis of pelvic structure and the rarity of eggs attributable to sauropods, suggested that (some? all?) sauropods gave birth to live young. The point was at least debatable until the late 1990s.

In November 1997, an expedition led by Luis Chiappe from the Natural History Museum of Los Angeles County, Lowell Dingus from the American Museum of Natural History in New York, and Rodolfo Coria from the Carmen Funes Museum in Plaza Huincal, Argentina, went to the rugged Upper Cretaceous badlands of central Argentina to look for dinosaur remains, including – if luck was with them – important specimens of the some of the rarest of all vertebrate fossils, early birds. Luck was with them, but not of the avian kind. Instead, their searches led them to another most rare of fossil occurrences: a sauropod nesting ground. This site, known as Auca Mahuevo ("Auca more eggs"), consists of a massive nesting ground covering more than a square kilometer and littered with tens of thousands of large unhatched eggs. Upon further investigation, four layers of eggs were uncovered and, in each layer, the eggs were organized into clusters of between 15 and 34 eggs, thought to represent individual nests or clutches. Most spectacular, a high proportion of these eggs contained embryonic skeletons. And with some of these embryos are the impressions of embryonic skin.

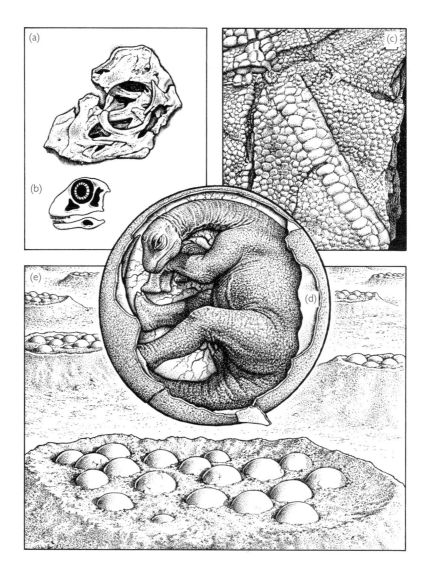

Figure 11.14. Titanosaurian remains from the Auca Mahuevo locality of Patagonia, Argentina. (a) Titanosaur skull (fossil); (b) reconstructed skull; (c) titanosaur skin impression (fossil); (d) reconstructed egg/embryo; and (e) schematic field of nests.

The importance of Auca Mahuevo is at least three-fold. First, it is not at all common for eggs to preserve their original embryonic contents. Yet here was an unambiguous association of eggs with embryos. Upon examining the tiny skeletons inside the egg, the investigators were able to determine that they were the first known embryonic remains of sauropods. All appear to belong to the same species, but which species could not be confidently determined. In several eggs, nearly complete embryonic skulls indicate that these sauropods were probably titanosaurians (Figure 11.14). Thus the "who are they?" question has been partially solved and provided the Rosetta stone for determining the affinities of similar kinds of empty eggs known from elsewhere in the world.

Secondly, the incredible geographical extent of the nesting horizons certainly speaks for colonial nesting and implies gregarious behavior in these titanosaurians. Clearly we have several enormous colonies, to which mothers would return on a regular basis. On the basis of such complex social structure, we might ask whether these titanosaurids engaged in parental care. Because there is no fossil evidence of adults preserved at Auca Mahuevo, investigators have inferred that the females probably left the site after laying their eggs, although it is possible that they may have communally guarded the whole nesting area from its periphery. How else to account for the existence of large numbers of nesting sites, which must have represented a temptingly rich food resource for contemporary predators?

Finally, these titanosaurian embryos provide the opportunity for discovering the details of early stages of sauropod growth and development. This perspective on dinosaur life histories is still in its infancy. From the microscopic structure of the embryonic bone, we can estimate the early growth stages of the embryos and compare these rates with those obtained from juveniles and adults. From these stages, clutch size, and other aspects of reproductive biology, we can perhaps begin to explore why sauropods may have become as large as they did. Did they prolong the length of time that they devoted to growth? Or did they become so large because their growth rate increased? Perhaps it was some of both. In any event, the availability of embryonic titanosaurids provides an avenue of research that bears directly on the early stages of dinosaur growth.

Beyond the Argentinean bonanza, what do we known about them or about the general aspects of sauropod reproduction, growth, and life histories? Sex in these animals assuredly involved coupling between a tripodal male and a quadrupedal female; however, beyond this most basic of poses all else remains speculative.

There are as yet no prosauropod eggs (although it is possible that a nest of six eggs – with embryonic material – retrieved from the Lower Jurassic of South Africa may have been laid by *Massospondylus*). For sauropods, the story is less desperate: although they are very rare, in at least one case eggs have been associated with a particular sauropod *Hypselosaurus* from the Upper Cretaceous of southern France and others from Mongolia and India have also been inferred to have been laid by unspecified sauropods. Laid in linear pairs, these eggs may also have been covered by mounds of vegetation to keep them at optimal temperature and humidity.

Once hatched, sauropodomorphs apparently grew at very high rates. Whereas once thought to be slow growing (they were after all only "reptiles" and exceptionally large ones at that, so that only growth rates from modern lizards and turtles – not birds – were used to extrapolate to sauropod-sized animals), new studies of the microscopic structure of sauropod bone indicates rapid and continuous rates in both prosauropods and sauropods. Rather than imagining animals taking about 60 years to reach sexual maturity and having a longevity of perhaps 200–300 years, A. Dunham and colleagues suggested that it should take about 20 years or less for a sauropod (and probably for a prosauropod as well) to become sexually mature. Similarly, lifespans for these animals were probably on the order of 100 years.

Whatever their individual longevities, sauropodomorphs were the largest terrestrial life forms of their times and indeed of all time. Yet their evolutionary history is not entirely one of becoming bigger. In 1914, F. Nopcsa recognized that many of the Late Cretaceous dinosaurs from Transylvania that he was studying, including the titanosaurian *Magyarosaurus dacus*, were much smaller than those known from elsewhere in the world. Nopcsa viewed these smaller forms as dwarfs living on islands. Eighty-five years later, C.-M. Jianu and D. M. Weishampel had another look at Nopcsa's hypothesis, on the basis of a better sauropod fossil record and using statistical and phylogenetic approaches. Their work suggested that, among adult neosauropods, *Magyarosaurus* appears to be represented by the smallest individuals (probably on the order of 5–6 m in length), that *Magyarosaurus* was more similar to subadults than to adults of other taxa, and, that this "juvenile" morphology may constitute dwarfing by paedomorphosis (i.e., the retention of juvenile characteristics into adulthood) in *Magyarosaurus*. It may yet be possible to explore whether this paedomorphosis was carried out by changes in growth rates, late beginning of growth, or early cessation of growth, now that samples of perinatal sauropods are becoming available and our understanding of bone histology is much more extensive.

So it looks like sauropodomorphs were fast-growing, yet slow-paced, giants of the Mesozoic. From prosauropods to sauropods, their high-browsing skills were never challenged throughout their reign. Today, they continually surprise and inspire with their towering qualities and the biomechanical and evolutionary consequences of such great size.

The evolution of Sauropodomorpha

Sauropodomorpha is a diverse and long-lived clade, containing two major groups of dinosaurs – Prosauropoda and Sauropoda (Figure 11.15). Sauropodomorpha is easily diagnosed by more than a dozen derived features. Among these are a relatively small skull (about 5% of body length), deflected front end of the lower jaw (the lower front end of the mandible is angled rearward), elongate, peg-like (lanceolate) teeth with coarsely serrated crowns, at least 10 neck vertebrae that form a very long neck, dorsal and caudal vertebrae added to the front and hind end of the sacrum, an enormous thumb equipped with an enlarged claw, a large hole in the pubis near the hip socket, the obturator foramen, and an elongate femur.

Sauropodomorpha shares a close phylogenetic relationship with Theropoda, and thereafter with a few basal taxa (among them *Eoraptor* and *Herrerasaurus*), altogether forming the monophyletic clade known as Saurischia. The next more distant relationship is with Ornithischia; relationships further afield fall outside of Dinosauria (see Chapter 5).

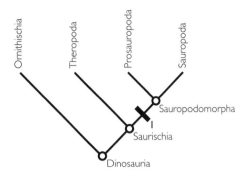

Figure 11.15. Cladogram of Dinosauria emphasizing the monophyly of Sauropodomorpha. Derived characters nclude: at **1** irelatively small skull (about 5% of body length), deflected front end of the lower jaw, elongate lanceolate teeth with coarsely serrated crowns, at least 10 neck vertebrae that form a very long neck, dorsal and caudal vertebrae added to the front and hind end of the sacrum, enormous thumb equipped with an enlarged claw, a very large obturator foramen in the pubis, and an elongate femur.

Prosauropoda

While some dinosaur paleontologists have argued that prosauropods are not a natural group, the consensus now seems to be that they are indeed monophyletic (Figure 11.16). Thus Prosauropoda can be defined as the common ancestor of *Thecodontosaurus* and *Plateosaurus* and all the descendants of this common ancestor. The features that diagnose this large clade include several modifications of the upper and lower jaws, elongation of the back vertebrae and changes in their articulations with each other, and modifications of the hand including a 45° twisting of the large thumb claw.

Recent phylogenetic work by P. M. Galton and P. Upchurch indicates that Prosauropoda can be subdivided into two major groups, with a few falling outside these subclades. These latter include *Thecodontosaurus*, a 2.5 m long, fully bipedal herbivore from the Late Triassic of Great Britain that has a lightly built skull with jaw joint set only slightly beneath the level of the tooth rows. Remaining prosauropods are diagnosed by having a separate opening in the braincase for the middle head vein, elongation of the second neck vertebra, an acetabulum that is complete open medially, and other aspects of the pelvis and hindfoot. This so-far unnamed clade includes Anchisauria and Plateosauria. Anchisaurians include two smaller clades, one consisting of *Anchisaurus* itself and *Ammosaurus*, and the other of *Riojasaurus*, *Melanorosaurus*, *Camelotia*, and *Lessemsaurus*; these taxa are united by having reduced prefrontal and frontal bones in the skull, at least five premaxillary teeth, elongate forelimb, straight femur, and changes in the pelvis and foot. Plateosaurians, on the other hand, include the majority of prosauropods, among them *Yunnanosaurus*, *Massospondylus*, *Lufengosaurus*, *Euskelosaurus*, and *Plateosaurus* (listed here in successively closer relationship). At their base, plateosaurians have elongate cervical vertebrae, reorganization of the sacrum, and modification of the wrist.

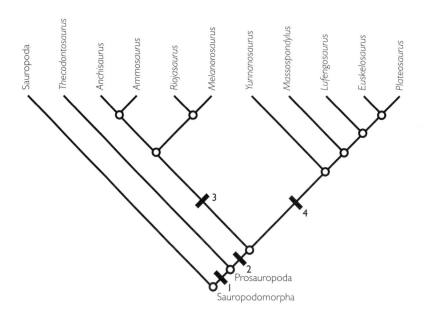

Figure 11.16. Cladogram of a monophyletic Prosauropoda. Derived characters include: at **1** lateral lamina on the maxilla, strap-like ventral process of the squamosal, ridge on lateral surface of the dentary, elongate posterior dorsal centra, distal carpal I transversely wider than metacarpal I, phalanx I on manual digit I with a proximal heel, a 45° twisting of the large thumb claw; at **2** separate opening for vena cerebralis media above the trigeminal foramen, axis centrum that is three times longer than high, short and robust metacarpal I, an acetabulum that is complete open medially, subtriangular distal end of ischium, increased robustness of metatarsals II and III; at **3** prefrontal length approximately that of the frontal, frontal excluded from supratemporal fossa, at least five premaxillary teeth, forelimb length greater than 60% hindlimb length, straight femoral shaft, fourth trochanter is displaced to the caudomedial margin of the shaft, hour-glass-shaped proximal end of metatarsal II; at **4** long retroarticular process on the lower jaw, longest postaxial cervical centrum at least three times as long as high, a dorsal vertebra added to the sacrum, proximal carpals present, large obturator foramen.

Sauropoda

Currently Sauropoda is defined as the most recent common ancestor of *Blikanasaurus* and *Saltasaurus* and all the descendants of this ancestor (Figure 11.17). This clade is supported by more than a dozen unique features, including important biomechanical modifications of the cervical vertebrae, increase in forelimb length, as well as changes in the elbow, forefoot, hindlimb proportions, and hindfoot that relate to increasing weight support.

Evolution within Sauropoda has only recently been evaluated using cladistic approaches, thanks to the independent investigations of Upchurch and J. Wilson. Their most recent works are consistent with each other. First, they indicate that sauropods consist of several primitive taxa (among them *Blikanasaurus*, *Vulcanodon*, and *Kotasaurus*) on the one hand and Eusauropoda on the other. Eusauropods, diagnosed by many features, including a broadly rounded snout and other changes in the

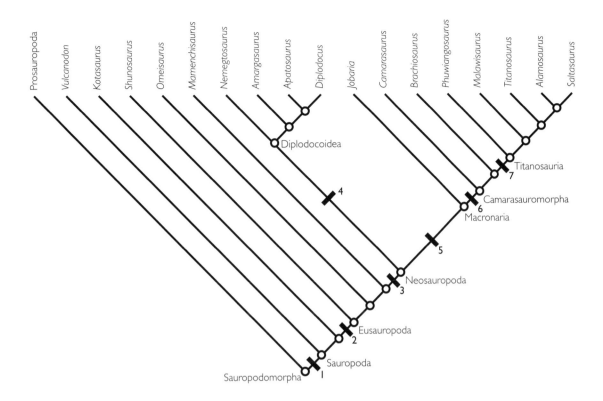

Figure 11.17. Cladogram of Sauropoda, with more distant relationships with Prosauropoda and Theropoda. Derived characters include: at **1** special laminar system on forward cervical vertebrae, forelimb length greater than 60% of hindlimb length, triradiate proximal end of ulna, subrectangular distal end of radius, length of metacarpal V greater than 90% that of metacarpal III, compressed distal end of ischial shaft, reduced anterior trochanter on femur, femoral shaft elliptical in horizontal cross-section, tibia length less than 70% of femur length, metatarsal III length less than 40% of tibia length, proximal end surfaces of metatarsals I and V are larger than those of metatarsals II, III and IV, metatarsal III length is more than 85%, metatarsal V length of ratio is 0.85 or higher; at **2** broadly rounded snout, caudal margin of external naris that extends behind the posterior margin of the antorbital fenestra, lateral plate on premaxillae, maxillae, and dentaries, loss of the anterior process of the prefrontal, frontals wider than length, wrinkled tooth crown enamel, most posterior tooth positioned beneath antorbital fenestra, at least 12 cervical vertebrae, neural spines of the cervical vertebrae that slope strongly forward, dorsal surface of sacral plate at the level of dorsal margin of ilium, block-like carpals, metacarpals arranged in U-shaped colonnade, manual phalanges wider transversely than proximodistally, two or fewer phalanges for manual digits II–IV, strongly convex dorsal margin of ilium, loss of the anterior trochanter of femur, lateral muscle scar at mid-length of fibula, distally divergent metatarsals II–IV, three phalanges on pedal digit IV, ungual length greater than 100% of metatarsal length for pedal digit I; at **3** subnarial foramen on premaxilla–maxilla suture, preantorbital fenestra in base of ascending process of maxilla, quadratojugal in contact with maxilla, pedal digit IV with two or fewer phalanges; at **4** subrectangular snout, fully retracted external nares, elongate subnarial foramen, reduction of angle between midline and premaxilla–maxilla suture to 20° or less, most posterior tooth rostral to antorbital fenestra; at **5** greatest diameter of external naris greater that of that of orbit, subnarial foramen found within the external narial fossa; at **6** nearly vertical dorsal premaxillary process, splenial extending to mandibular symphysis, acute posterior ends of pleurocoels in anterior dorsal vertebrae, metacarpal I longer than metacarpal IV; at **7** prominent expansion of rear end of sternal plate, very robust radius and ulna.

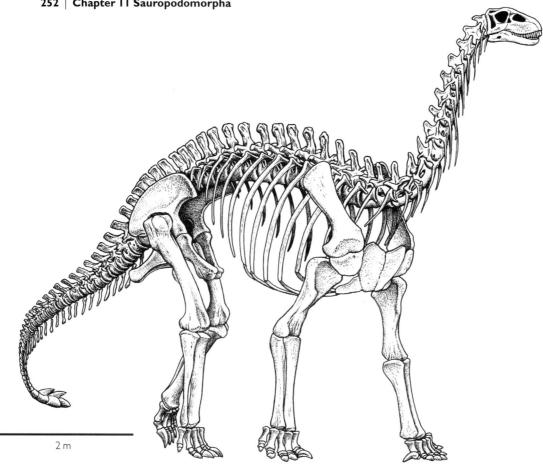

2 m

Figure 11.18. The eusauropod *Shunosaurus*, from the Middle Jurassic of Sichuan Province, China.

facial skeleton and palate, increase in the number of cervical vertebrae, and modifications of the forefoot, pelvis and hindlimb, are a more resticted set of sauropods, consisting of *Shunosaurus* and all remaining taxa. *Shunosaurus* itself was a 9 m long sauropod from the Middle Jurassic of China known from several nearly complete skeletons (Figure 11.18). Its skull is relatively long and low, vaguely reminiscent of the prosauropod condition (nostrils in front of the snout, many small and spatulate teeth).

Numerous taxa are present at the base of the eusauropods, including *Patagosaurus*, *Cetiosaurus*, *Omeisaurus*, and *Mamenchisaurus*. At the top of the stack of these basal eusauropods is *Lourinhasaurus*, a Late Jurassic form from Portugal. It represents the sister-taxon to a more exclusive clade known as neosauropods. Neosauropoda, diagnosed as having a second opening in front of the antorbital fenestra (called the preantorbital fenestra), modification of the cheek region, and changes in the ankle, is split into diplodocoids on the one hand and macronarians on the other.

Diplodocoidea   Diplodocoids are diagnosed as having a subrectangular snout, fully retracted external nares, and reduction of the upper temporal

BOX 11.2

The decapitation of "*Brontosaurus*"

With the discoveries of dinosaurs in the Western Interior of the USA during the late nineteenth century, box-car loads of brand new, but often incomplete, sauropod skeletons were shipped back east to places like New Haven and Philadelphia. It was Yale's O. C. Marsh who described one of these new sauropods as *Apatosaurus* in 1877. With further shipments of specimens and more studies, Marsh again named a "new" sauropod in 1879 – *Brontosaurus*.

Years went by and – thanks to the burgeoning popularity of many kinds of dinosaur – the public came to know the name *Brontosaurus* much better than it did the earlier-discovered *Apatosaurus*. Nevertheless, there was the suspicion by many sauropod researchers that *Apatosaurus* and *Brontosaurus* were the same kind of sauropod. In fact, this case was made in 1903 by E. S. Riggs. Since then, most sauropod workers have regarded *Brontosaurus* as synonymous with *Apatosaurus*. If *Apatosaurus* and *Brontosaurus* are two names for the same sauropod, the older, name, *Apatosaurus*, should be applied to this Late Jurassic giant.

But the more interesting story is not in the names, but in the heads. Again we go back to Marsh. Lamenting in 1883 that his material of "*Brontosaurus*" (now *Apatosaurus*) had no head, he made his best guess as to the kind of skull this animal had: one like *Camarasaurus*. And it was thus that *Apatosaurus* donned the short-snouted profile of its Morrison cohort.

Enter H. F. Osborn, then curator of vertebrate paleontology and powerbroker of the American Museum of Natural History, and W. J. Holland, curator of fossil vertebrates at the Carnegie Museum of Natural History and equally stalwart in his pursuit of "getting it right" about sauropods. Contemporary dinosaur researchers in the early part of the twentieth century, these two skirmished over the issue of whose head should reside on the neck of *Apatosaurus*. Osborn followed Marsh and had his mount of this majestic sauropod topped with a camarasaur head, while Holland was strongly persuaded that *Apatosaurus* had a more *Diplodocus*-like head (based on a somewhat removed yet associated skull found near an otherwise quite complete skeleton at what is now Dinosaur National Monument). But Holland gained no adherents and his mount of *Apatosaurus* in the Carnegie Museum remained headless in defiance of Osborn's dogma. After Holland's death, however, the skeleton was fitted with a camarasaur skull, almost as if commanded by Osborn himself.

Whose head belongs to whom was finally resolved in 1978 by D. S. Berman and J. S. McIntosh. Through some fascinating detective work on the collection of sauropod specimens at the Carnegie Museum, these two researchers were able to establish that *Apatosaurus* had a rather *Diplodocus*-like skull – long and sleek, not blunt and stout as had previously been suggested. As a consequence, a number of museums that display *Apatosaurus* skeletons celebrated the work of Berman and McIntosh (and Holland) by conducting a painless head transplant – the first ever in dinosaurian history.

fenestra, among other features of the skull. This clade is presently known from *Nemegtosaurus* and *Quaesitosaurus* from the Late Cretaceous of Mongolia, *Rayososaurus* and *Amargosaurus* from the Early Cretaceous of Argentina, *Dicraeosaurus* from the Late Jurassic of Tanzania, and finally Diplodocidae.

The best known by far among diplodocoids (and even among sauropods) are the diplodocids. United by having skid-shaped hemal arches beneath the tail vertebrae, deeply cleft V-shaped neural spines in the shoulder region, and ischia that are expanded at their ends, this group of sauropods is perhaps better known than any other. Here we find *Diplodocus*, from the Late Jurassic of the western USA (see Figure 11.1). Completely known down to its toe bones (including new evidence of spiky skin), this 27 m long sauropod has a very giraffe-like skull, with an elongate snout that houses pencil-like teeth along its very front margin. The nostrils are located on the very top of the skull. And the tail is drawn out into in a series of long cylindrical vertebrae that together forms a "whiplash," thought to serve in defense against predators.

Other diplodocids are slightly less impressive in size, but no less intriguing. The 21 m long *Apatosaurus* (also from the Late Jurassic of the western USA) is known best either as the dinosaur with the head transplant or by its incorrect name "*Brontosaurus*" (Box 11.2). At about 25 m in length, *Barosaurus* hales from the Late Jurassic of the western USA and from Tanzania – as does *Brachiosaurus* – providing important information on intercontinental distribution patterns among dinosaurs.

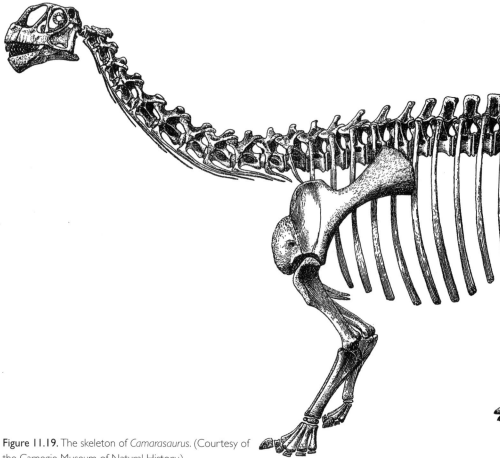

Figure 11.19. The skeleton of *Camarasaurus*. (Courtesy of the Carnegie Museum of Natural History.)

Macronaria The other half of Neosauropoda is Macronaria, which have enlarged external nares and a subnarial foramen lying within the external narial fossa. Several taxa (among them *Bellusaurus* and *Jobaria*) form a small basal macronarian clade, which is the sister-group to camarasauromorphs. As the name implies, Camarasauromorpha in the shape of camarasaurs includes not only *Camarasaurus* (Figure 11.19), but also *Haplocanthosaurus* and titanosauriforms. The skulls of these taxa have external nares with a forward margin that are elevated and vertical, modified pleurocoels (hollow cavities in the vertebrae) in the back region (Figure 11.20), and an increase in the height of the forefoot. *Camarasaurus* is one of the more familiar of all sauropods. Known from the Late Jurassic of the western USA, it is the most common of all North American sauropods. Perched on the end of its 18 m long body, its head was short, high, and powerfully built, and equipped with fewer teeth than we've yet seen in sauropods. Its vertebrae had very deep pleurocoels and, dorsally, the neural spines at the back of the neck were divided by a U-shaped cleft.

Haplocanthosaurus is another camarasauromorph from the Late Jurassic of the western USA. Although not nearly so well known as *Camarasaurus*, it appears to be the sister-group to Titanosauriformes. This latter clade, defined as the most recent common ancestor of *Brachiosaurus* and *Saltasaurus*, has its neural arches toward the front half of the centrum for the middle tail vertebrae rather than in the middle and an ilium that is rounded in front, as well as other features.

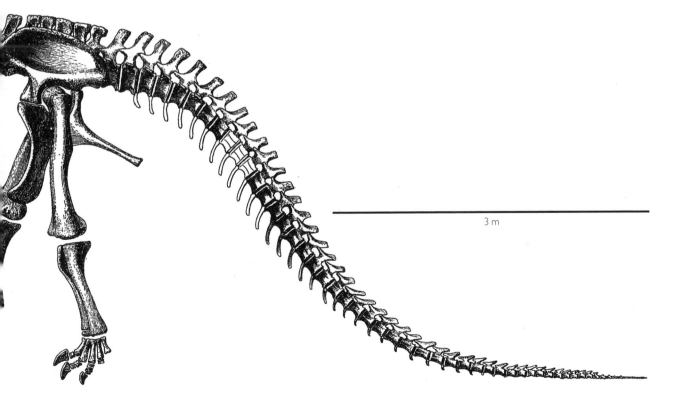

3 m

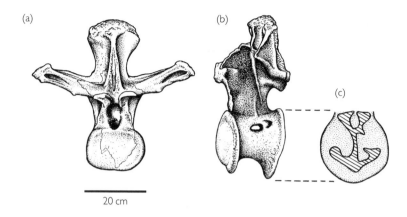

Figure 11.20. (a) Front view and (b) left lateral view of one of the back vertebrae of *Brachiosaurus*, with pleurocoels indicated in (c) cross-section.

Brachiosaurids are those titanosauriforms with an elongate humerus with an enlarged muscular crest. *Brachiosaurus* itself (see Figure 11.7) captured several decades-worth of people's imaginations as the largest land-living animal of all time (measuring 23 m long and weighing in excess of 50,000–60,000 kg). Now supplanted by the likes of *Supersaurus* and *Seismosaurus*, *Brachiosaurus* nevertheless is still by far the best known of all of these earthly giants.

Titanosauria

The next nested group within Titanosauriformes is Titanosauria, which is defined by having a prominent expansion of the rear end of the sternal plate and an extremely robust radius and ulna. Titanosaurians consist of a succession of ever-more closely related forms – *Phuwiangosaurus*, *Austrosaurus*, *Malawisaurus*, *Gondwanatitan*, *Titanosaurus*, *Lirainosaurus*, *Alamosaurus*, *Pelegrinisaurus*, *Opisthocoelicaudia*, and *Saltasaurus*.

There are many named titanosaurians, but we'll concentrate on several of the better-known forms. *Malawisaurus* is one of the more recent discoveries among titanosaurians. The best-known dinosaur from the small central African country of Malawi, the Early Cretaceous *Malawisaurus* is a modestly large sauropod (9 m in length) known from a good portion of the skeleton, including skull elements, teeth, vertebrae, and limb elements. On the other hand, *Titanosaurus*, the namesake of the group, is best known from the Late Cretaceous of India. What we know of *Titanosaurus* are its tail vertebrae and most of the limb bones, all indicating a 12 m long sauropod.

Virtually all titanosaurians hail from the Southern Hemisphere (i.e., Gondwana, which early-on included India). *Alamosaurus*, by contrast, is an enigmatic sauropod (even though it is known from a quarter of an articulated skeleton) from the Late Cretaceous of the western USA (Figure 11.21). From this and other potentially related material (referred to *Alamosaurus* but possibly representing other titanosaurians), it is clear that *Alamosaurus* was quite a large animal, probably measuring up to 21 m in length.

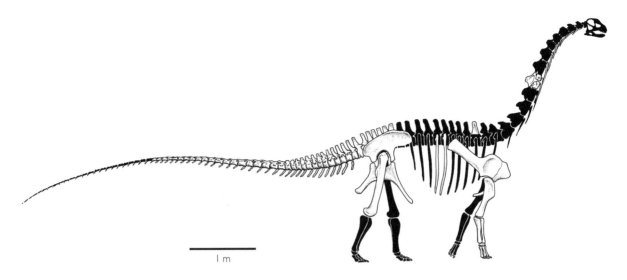

1 m

Figure 11.21. *Alamosaurus.* Unknown parts of the skeleton are shaded in black.

Finally among titanosaurians, we have *Opisthocoelicaudia* from the Late Cretaceous of Mongolia, and *Saltasaurus* from roughly contemporary rocks in Argentina. The former, a 12 m long sauropod, has a deep midline cleft in the neural spines of the vertebrae, as well as other unusual skeletal features (six sacral vertebrae, very short tail, massive and relatively short forelimbs, greatly reduced foot). Unfortunately, the skull of *Opisthocoelicaudia* is not yet known. *Saltasaurus* is presently known from disarticulated and somewhat incomplete material. However, this stocky limbed form, about the same size as *Opisthocoelicaudia*, has provided scientists with more than the usual skeletal equipment. For from *Saltasaurus*, we have unambiguous evidence that these animals were covered with a pavement of bony dermal armor called osteoderms. Now also known in *Malawisaurus*, *Agustina*, *Magyarosaurus*, and *Ampelosaurus*, these globular and button-like osteoderms apparently covered the animal's back and provided more protection (in addition to size alone) against predatory attacks.

Sauropodomorphs meet history: a short account of their discovery

Although the first-discovered sauropodomorphs post-date the discovery of theropods and ornithopods, they were well on hand when R. Owen coined the term Dinosauria for their inclusion. In 1836, H. Riley and S. Stutchbury announced their discovery of *Thecodontosaurus* (*theke* – socket; *dont* – tooth), a relatively small (2.5 m long) animal from Upper Triassic strata of western England and Wales. Despite its early discovery and study, it was not added to Owen's Dinosauria until 1870 (by T. H. Huxley). Discoveries from the same region, described by K. A. Kermack in 1984, have added new and important anatomical and phylogenetic information on this prosauropod.

A year after the announcement of *Thecodontosaurus*, H. von Meyer described what was to become the best known of all Late Triassic and Early Jurassic sauropodomorphs, *Plateosaurus* (*plateos* – flat). Originally discovered in Upper Triassic rocks of northern Bavaria, but – like *Thecodontosaurus* – not included in Owen's Dinosauria until 1855 (at von Meyer's hands), abundant *Plateosaurus* material is now known from throughout the rest of Germany, as well as from France and Switzerland. And in 1841, the year before he coined the name Dinosauria, Owen announced *Cetiosaurus* (*keteios* – whale-like), the first gigantic dinosaur to be discovered, in this case from Middle Jurassic strata of England. Owen originally thought that *Cetiosaurus* was an exceptionally large crocodile, and it took another 30 years before J. Phillips brought it back into the sauropod fold.

Although there were some very important sauropodomorph discoveries following Owen's christening of Dinosauria (his own recognition of *Massospondylus* (*masso* – massive; *spondyl* – spool, referring to the spool-shaped centra of the vertebrae) from the Early Jurassic of southern Africa in 1854, P. A. Johnston's naming of the Early Cretaceous sauropod *Astrodon* (*astro* – star) from Maryland, Huxley's christening of *Euskelosaurus* (*eu* – true; *skele* – limb) – another prosauropod from the Late Triassic of southern Africa in 1866, and P. Matheron's 1869 description of the second known sauropod *Hypselosaurus* (*hypselos* – high) from the Late Cretaceous of France), it wasn't until the opening up of the western frontier of the USA to scientific exploration that the first sauropodomorph boom began.

As has been the case so many times with the history of new dinosaur discoveries, it was the material from the Western Interior of the USA that cut the Gordian knot of sauropod anatomy and evolution. And again the two major figures in these discoveries were E. D. Cope and O. C. Marsh (see Box 6.2). From the late 1870s to the close of the nineteenth century, these two workers furiously tapped into the riches of the Late Jurassic Morrison Formation from such classic localities as Como Bluff, Wyoming, and Morrison and Garden Park, Colorado. They collected a huge variety of new dinosaurs, among them the sauropods *Camarasaurus* (*kamara* – chamber), *Apatosaurus* (*apato* – trick or false; so-named because the tail bones appeared to Marsh more like those of a lizard than of a dinosaur), *Diplodocus* (*diplo* – two; *docus* – spar or beam), *Barosaurus* (*barys* – heavy) and best known of all "*Brontosaurus*" (*bronto* – thunder; but see Box 11.2). Although at times eclipsed by more recent collections, these discoveries were immensely important in providing for the first time anatomical information from complete skeletons and often skulls of these gigantic Mesozoic herbivores.

Elsewhere in the world, other sauropodomorphs were also beginning to see the light of day. Owen's successor at the British Museum (Natural History), R. Lydekker, described *Titanosaurus* (named for Titan, the giants of Greek mythology) in 1877. This large, but poorly known – even to today – sauropod from the Late Cretaceous of India is now claimed to come from France, Spain, Madagascar, and Laos, making it perhaps the most widely known of all sauropods. Over this period, Lydekker was also responsible for making known important sauropod material from South America and Africa. Meanwhile, Marsh was busy working closer to his own backyard than his more western research. In 1885, he announced the prosauropod *Anchisaurus* (*anchi* – near), and another – *Ammosaurus* (*ammos* – sand) – in 1891. Both are known from the Early Jurassic of the northeastern USA, with an additional record of *Ammosaurus* from the same time in Arizona.

The beginning of the twentieth century was marked by continued work in the Western Interior of the USA, particularly by the American Museum of Natural History working again at Como Bluff and nearby Bone Cabin Quarry in Wyoming, the Field Museum of Natural History working in western Colorado, and the Carnegie Museum in Wyoming, Colorado, and especially Utah. In Utah, a particularly rich quarry was discovered by E. Douglass in 1909 and worked by the Carnegie Museum until 1923, thereafter by the Smithsonian Institution, and then by the University of Utah. It is now run by the U.S. Park Service and is famous as Dinosaur National Monument.

Overall, the American Museum (New York) and Carnegie Museum (Pittsburgh) efforts have yielded one of the greatest collections of sauropods on earth. The best of the bounty includes several sauropod skeletons that were subsequently mounted for exhibition in a number of museums (*Apatosaurus* in the American Museum,[2] in the Carnegie Museum, and in the Field Museum (Chicago), in the Yale Peabody Museum (New Haven), and at the University of Wyoming (Laramie), *Camarasaurus* in the Carnegie Museum and in the Smithsonian Institution (Washington, DC), and *Diplodocus* in the Carnegie Museum, the Smithsonian Institution, and Denver Museum of Natural History). Equally important to our story are the two discoveries of new sauropods that came from these expeditions: *Haplocanthosaurus* (*haplos* – single; *akantha* – spine), described by J. B. Hatcher in 1903 from the Morrison Formation (Upper Jurassic) of Colorado, and *Brachiosaurus* (*brachion* – arm), also described by E. S. Riggs in 1903. This latter sauropod, also from the Upper Jurassic Morrison Formation of Colorado, is now known from elsewhere in the western USA and, more importantly, from along the eastern coast of Africa in what is now Tanzania.

All the while these sauropod discoveries were taking place in the USA, European workers were continuing to study the odd scrap of prosauropod and sauropod material as it became available. Some of these specimens appeared to represent previously unknown kinds of dinosaur. For example, in 1908, von Huene announced a new prosauropod which he called *Sellosaurus* (*sellos* – saddle) from rocks that date to the Late Triassic near the southern German town of Trossingen, till then (and since) famous for the manufacture of Hohner harmonicas (attention, all you blues harp players).

At the same time as all these early twentieth century efforts were taking place in the USA and Europe, huge quantities of sauropod bones, along with those of other dinosaurs, were being unearthed in what is

---

2 This most-famous of museums recently eclipsed its own mount of *Apatosaurus* with an exciting, free-standing mount of a *Barosaurus* mother rearing up to protect its baby against a marauding *Allosaurus*. The head of the mother reaches up three stories into the air (see Figure 1.7).

now the eastern African country of Tanzania. It was here, in the Tendaguru Hills, that one of the most truly awesome, intensive dinosaur expeditions ever took place (Box 11.1). When the dust from the digging had settled, the Tendaguru expeditions claimed not only a wealth of ornithischian and theropod dinosaurs, but also two new kinds of sauropod (*Tornieria* (named for German paleontologist G. Tornier), a newly recognized sauropod named by Sternfeld in 1911, and *Dicraeosaurus* (*dikraios* – bifurcated), described by W. Janensch in 1914), as well as new material of *Barosaurus* and the finest specimen of *Brachiosaurus* ever found – now mounted and peering into the fourth floor balcony of the Humboldt Museum für Naturkunde.

While the Berlin contingent of dinosaur paleontologists was reveling in the wealth of Late Jurassic bones in Africa, others were busy back home. Beginning in 1911 and continuing sporadically until 1932, the Staatliches Museum für Naturkunde and the Museum für Geologie und Paläontologie at the University of Tübingen excavated a small hill outside the southern town of Trossingen (remember *Sellosaurus* and the Hohner harmonicas?). This excavation, in Upper Triassic rocks, was to yield one of the largest mass accumulations of any prosauropod, in this case of *Plateosaurus*. Important from an anatomical perspective as well as from what it might tell us about *Plateosaurus* behavior, evolution, and taphonomy, the Trossingen site has since been studied by von Huene and M. Sander.

From the 1920s until the close of the 1930s, sauropodomorph discoveries came about in a rather piecemeal fashion. In 1922, C. W. Gilmore described a new Late Cretaceous sauropod from New Mexico (now also known from Utah and Texas) that he dubbed *Alamosaurus* (named for the Ojo Alamo Formation of New Mexico). Two years later, S. H. Haughton described *Melanorosaurus* (*melanos* – black; *oros* – mountain) from the Late Triassic of South Africa. Finally, in 1929, C. Wiman described new sauropod from the Late Jurassic of Shandong, China. Originally named *Helopus* (*helo* – marsh; *pus* – foot), this name had already been used for a bird, so in 1956 A. S. Romer provided this animal with a new name, *Euhelopus*, which has been used thereafter. And closing out the decade, von Huene published his large and very valuable tome on sauropods (and other dinosaurs) from the Cretaceous of South America.

The 1930s saw three important sauropod milestones: the publication of yet another von Huene monograph, this one – released in 1932 – summarizing a wealth of taxonomic and anatomical observations on saurischian dinosaurs, in the process describing a new sauropod from the Late Cretaceous of Transylvania that he named *Magyarosaurus* (the Magyars were central Asian invaders of what is now Hungary more than 1,000 years ago). Shortly thereafter and an ocean and continent away, B. Brown and colleagues from the American Museum of Natural History began working the Upper Jurassic (Morrison Formation) Howe Quarry, in north-central Wyoming. This site has yielded such an abundance of dinosaur bones – among them one of the best collections of sauropods – that even today have yet to be fully studied and described.

Finally, the close of the decade saw the opening of the richest of all dinosaur fossil fields: those of China. From the late 1930s onward, a vast array of new sauropodomorphs has been described. C.-C. Young began the avalanche in 1939 with his description of *Omeisaurus* (named for Mount Emei, a sacred mountain in Sichuan Province), a primitive sauropod from the Late Jurassic. Shortly thereafter (1941), there was the Early Jurassic *Lufengosaurus* (named for Lu-Feng, a locality in Yunnan Province) followed by the contemporary *Yunnanosaurus* (for Yunnan Province, where this dinosaur was discovered) described in 1942. Twelve years later, Young described the remarkable long-necked *Mamenchisaurus* (named for Mamenchi Ferry at Jinshajiang, Sichuan Province) from the Late Jurassic.

While new sauropodomorph specimens continued to be uncovered throughout the mid-1900s, it was not until the late 1960s that new kinds of prosauropods and sauropods began to be discovered again. The first of these, a prosauropod named *Riojasaurus* (for Rioja Province), was described from Upper Triassic strata of Argentina by Bonaparte in 1969. From Mongolia, Nowinski described the Late Cretaceous sauropod *Nemegtosaurus* (named for the Nemegt Formation of Mongolia; see Figure 11.12e) two years later. The coming of 1972 saw the discovery of two new sauropodomorphs: the so-far poorly known prosauropod *Azendohsaurus* (named for the town of Azendoh, near the fossil deposits in the Atlas Mountains of Morocco), described by J. M. Dutuit from the Late Triassic, and the primitive sauropod *Vulcanodon* (*vulcan* – volcano) described by M. A. Raath from the Early Jurassic of Zimbabwe. Shortly thereafter (1975), S. L. Jain and colleagues announced their discovery of *Barapasaurus* (*bara* – big; *pa* – leg; based on several Indian languages) from the Early Jurassic of India. Together with *Vulcanodon*, this primitive form has provided interesting insights into the base of sauropod phylogeny.

The close of the 1970s brought us *Opisthocoelicaudia* (*opistho* – hind; *coel* – hollow; *caud* – tail; the name acknowledges concave rear end of the tail), a sauropod from Upper Cretaceous strata of Mongolia described by M. Borsuk-Bialynicka in 1977. At the same time but some 18,000 km away in Argentina, J. F. Bonaparte and colleagues announced four new sauropodomorphs: *Coloradisaurus* (named for the Los Colorados Formation from which it was recovered; see Figure 11.11b), a prosauropod described in 1978 from the Upper Triassic of La Rioja Province, another Upper Triassic prosauropod – *Mussaurus* (*mus* – mouse, known from one of the smallest articulated skeletons, albeit as a hatchling) – also from the Late Triassic but from Santa Cruz Province, and two new sauropods. The first, *Patagosaurus* (named for the Patagonia region of Argentina), was described in 1979 from the Middle Jurassic of Chubut Province, while the second, the Late Cretaceous *Saltasaurus* (named for Salta Province, Argentina, where this sauropod was found), was described in 1980.

Following the important revisionary work on the prosauropod *Massospondylus* by M. R. Cooper in 1981, the 1980s belong to central and eastern Asia and to the USA. S. Kurzanov and A. Bannikov began the Asian advance with their description of a new Late Cretaceous sauropod from

Mongolia, which they named *Quaesitosaurus* (*quaesitus* – abnormal or uncommon) in 1983. But soon the story turns to China and the work of Z.-M. Dong and colleagues: *Shunosaurus* (from Shuno, an old name for the Sichuan region of China), and *Datousaurus* (from Malay *datou* – chieftain), described, respectively, in 1983 and 1984 from the Middle Jurassic of Sichuan Province, China. This region has continued to yield abundant Middle and Upper Jurassic sauropod material to this day, so much so that we can continue to expect some profound discoveries from here in the future.

On the other side of the Pacific Ocean, discoveries of new sauropods were again coming into prominence. Working in the Morrison Formation of Colorado, J. A. Jensen announced three new sauropods in 1985, an unprecedented number since the heyday of the late 1800s and early 1900s: *Supersaurus* (*super* – above), *Ultrasauros* (*ultra* – beyond), and *Dystylosaurus* (*di* – two; *stylos* – beam). Not only was this a surprising number of new sauropods, but they were huge, the largest of all land-living animals known until then.

The year 1985 also brought us two new prosauropods: *Camelotia* (alluding to the legendary Camelot, near where the dinosaur was found), a prosauropod from the Late Triassic of England described by Galton, and *Blikanasaurus* (named for Blikana Mountain, South Africa), described by Galton and J. van Heerden from the Late Triassic of Lesotho. A year later, Bonaparte announced a new sauropod, *Lapparentosaurus* (named for A. F. de Lapparent, noted French dinosaur paleontologist and Jesuit priest, whose greatest works were published in the mid-1900s) from the Middle Jurassic of Madagascar. Before the close of the decade, two new taxa were added to the sauropod roster: *Aeolosaurus* (named for Aeolus, Greek god of the winds and also alluding to the windiness of Patagonia) from the Late Cretaceous of Argentina, named by J. Powell in 1987, and *Kotasaurus* (named for the Kota Formation) from the Early Jurassic of India, identified by P. Yadagiri in 1988.

The 1990s and continuing into the present century have been a bonanza of discoveries of new sauropodomorphs. Dong was first off the line in 1990 with his description of *Bellusaurus* (*bellus* – fine), a sauropod known principally from a number of juvenile skeletons collected from Middle-to-Upper Jurassic strata in north central China. That same year, Z. Bai and colleagues described a new prosauropod from the Lower Cretaceous of Yunnan, China, which they called *Yimenosaurus* (lizard from Yimen).

The following year, sauropodomorphs pushed yet again at the edges of gigantism with the announcement of *Seismosaurus* (*seismos* – earthquake), a new and gargantuan sauropod from the Morrison Formation of New Mexico described by D. D. Gillette. This dinosaur, thought to be on the order of 35 m long, is the largest animal yet known to have walked on land. Indeed, it may be nearly as large as a land-dwelling flesh-and-blood creature can become. The same year saw the descriptions of two new mid-to-Late Cretaceous Argentinian sauropods, *Andesaurus* (named for Andes Mountains of Argentina) described by J. Calvo and Bonaparte, and *Amargasaurus* (named for Amarga Canyon in Neuquen Province,

Argentina), a bizarre form with exceedingly long bifurcated neural spines described by Salgado and Bonaparte. Finally, R. Wild provided a new name – *Janenschia* (named for W. Janensch, leader of the Tendaguru expeditions) – for one of the Late Jurassic sauropods from Tendaguru, Tanzania, in 1991.

Thereafter, new sauropodomorphs have been almost literally spilling out of the rocks. *Jainosaurus* (named for Indian vertebrate paleontologist S. L. Jain), a sauropod named by A. Hunt and co-authors in 1994, comes from the Upper Cretaceous of India. Two new sauropodomorphs – the prosauropod *Jingshanosaurus* (from Jingshan), named by Y. Zhang and Z. Yang from the Lower Cretaceous of China, and the sauropod *Ampelosaurus* (*ampelo* – vineyard), christened by J. Le Loeuff from Upper Cretaceous beds of southern France – were announced in 1995. Another pair of sauropods, both from Argentina, were named in 1996: *Rayososaurus* (named after the Rayoso Formation from which it was collected) from the Early Cretaceous by Bonaparte and *Pellegrinisaurus* (named for Lake Pellegrini) from the Late Cretaceous by Salgado.

Although 1997 witnessed the discovery of only one new sauropod – *Hudiesaurus* (*hudie* – butterfly, referring to the shape of the vertebrae) named by Dong from the Late Jurassic of China – and 1998 saw two new forms (*Gongxianosaurus* (named for Gongxiang County) from the Early Jurassic of China, named by X.-L. He and co-workers and *Lourinhasaurus* (named for the Lourinha Formation) from the Late Jurassic of Portugal by P. Dantas and collaborators), 1999 was a banner year for sauropodomorph discovery. Twelve new sauropodomorphs were named in that year alone. Three come from Argentina: *Tehuelchesaurus* (named for the Tehuelche tribe) named by T. H. Rich and colleagues from the Middle Jurassic, *Lessemsaurus* (named for Don Lessem) from the Late Jurassic by Bonaparte, and *Agustinia* (named for Agustin Martinelli) also by Bonaparte from the Early Cretaceous. Another two hail from Brazil. The first is a sauropod named *Gondwanatitan* (Titan from Gondwana) from the Late Cretaceous named by A. Kellner and S. de Azevedo, while the other – *Saturnalia* (Saturnalia is the Roman winter solstice festival) named by M. Langer and co-workers – comes from the Late Triassic of Brazil and may not be a sauropodomorph at all, but instead the sister-group to the sauropodomorph clade itself.

Africa was host to three important sauropod discoveries in 1999. *Atlasaurus* (named for the Atlas Mountains by M. Monbaron and co-workers) comes from the Middle Jurassic of Morocco, while *Jobaria* (named for Jobar, a creature of Tuareg mythology) and *Nigerosaurus* (named for Niger), both named by Sereno and collaborators, are known from the Early Cretaceous of Niger.

Europe, Asia, and North America complete the remainder of sauropodomorphs discovered in the last year of the twentieth century. Europe produced *Dinheirosaurus* (lizard from Porto Dineiro) named by Bonaparte and Mateus from the Late Jurassic of Portugal, and *Lirainosaurus* (*lirain* – slender, in Basque) from the Late Cretaceous of Spain, named by J. L. Sanz and colleagues. Half a world away, the sauropod *Tangvayosaurus* (named for Tang Vay, Laos) was discovered by R. Allain and co-workers from the Early Cretaceous of Laos. And finally,

V. Tidwell and co-authors announced *Cedarosaurus* (named for the Cedar Mountain Formation) from the Early Cretaceous of Utah.

Although just beginning, the twenty-first century has so far given us another dozen new sauropodomorphs from around world. The majority come from Asia: *Isanosaurus* (named for the Isan locality) from the Late Triassic of Thailand, named by E. Buffetaut and colleagues, *Chuanjiesaurus* (named for the Chuanjie Formation) from the Middle Jurassic of China, described by X. Fang and co-workers, *Jiangshanosaurus* (Jiangshan County is in Zhejiang Province, China) from the Early Cretaceous of China, named by F. Tang and colleagues, and *Huabeisaurus* (named for the Huabei locality) from the Late Cretaceous of China, named by Q. Pang and Z. Cheng.

From Africa come another three sauropods: *Tendaguria* from the Late Jurassic named by Bonaparte and colleagues after the famous Tendaguru locality in Tanzania from which it comes. *Paralititan* (tidal Titan) from the Late Cretaceous of Egypt named by M. Smith and co-workers, and *Rapetosaurus* (rapeto – mischievous giant) also from the Late Cretaceous but this time from Madagascar, named by K. Curry Rogers and C. Forster. Europe sports two new sauropodomorphs, a prosauropod from the Late Triassic of Germany named *Ruehleia* (named for Hugo Rüle von Lilienstern) by Galton, and a Late Jurassic–Early Cretaceous sauropod from Spain named *Losillasaurus* (named for the Losilla locality) by M. Casanovas and co-authors. Two new sauropods, both Early Cretaceous in age, come from the USA: *Sauroposeidon* (Poseidon was the Greek god of, among other things, earthquakes) from Oklahoma, named by M. Wedel and collaborators, and *Venenosaurus* (poison lizard) from Utah, described by Tidwell and co-authors. Finally, Salgado and C. Azpilicueta described *Rocasaurus* (named after General Roca city) from the Late Cretaceous of Argentina.

We have now "traveled" through Sauropodomorpha, discussing along the way their wildlife paleobiology, phylogenetic relationships, and the history of their discovery. Although we have wended our way through what amounts to nearly 150 different species, there is still a great deal of work to do on sauropodomorphs, the vast majority of which were the largest animals ever to have lived on land.

Important readings

Alexander, A. McN. 1989. *Dynamics of Dinosaurs and other Extinct Giants*. Columbia University Press, New York, 167pp.

Galton, P. M. 1990. Basal Sauropodomorpha – prosauropods. In Weishampel, D. B., Dodson, P. and Osmólska, H. (eds.), *The Dinosauria*, 1st edn. University of California Press, Berkeley, pp. 320–344.

Galton, P. M. and Upchurch, P. 2004. Prosauropoda. In Weishampel, D. B., Dodson, P. and Osmólska, H. (eds.), *The Dinosauria*, 2nd edn. University of California Press, Berkeley, pp. 232–258.

Seymour, R. S. and Lillywhite, H. B. 2000. Hearts, neck posture and metabolic intensity of sauropod dinosaurs. *Proceedings of the Royal Society of London*, series B, **2167**, 183–187

Upchurch, P., Barrett, P. M. and Dodson, P. 2004. Sauropoda. In Weishampel, D. B., Dodson, P. and Osmólska, H. (eds.), *The Dinosauria*, 2nd edn. University of California Press, Berkeley, pp. 259–321.

CHAPTER 12

Theropoda I:
Nature red in
tooth and claw

Figure 12.1 (see page 265). The king of the tyrant lizards, *Tyrannosaurus rex*.

*T*yrannosaurus *rex* rules. An awesome amalgam of huge teeth, sinewy haunches, and scimitar claws (Figure 12.1), the "king of the tyrant lizards" has exerted a compelling hold over generations of fans since its discovery in 1902. *Tyrannosaurus*, however, didn't spring out of nowhere; instead it is only one of many kinds of meat-eating dinosaurs, some so incredible that they appear to be inspired by nightmares. These include the notorious *Velociraptor* (the "raptor" that terrorized its way through all three *Jurassic Park* movies), a host of ostrich-mimicking ornithomimosaurs, the brutish, horned *Carnotaurus*, long-snouted spinosauroids, and outlandish oviraptorosaurs (Figures 12.2 and 12.3).

Grouped together as Theropoda (*thero* – beast; *pod* – foot), the group created in 1881 by O. C. Marsh, these dinosaurs have had a long evolutionary history extending back to the Late Triassic and, as we shall see in Chapter 13, going all the way to the present. Included are a great variety of forms, among them ceratosaurs, carnosaurs, tyrannosauroids, therizinosauroids, ornithomimosaurs, oviraptorosaurs, troodontids, dromaeosaurids, and birds, as well as a number of other taxa that fit among these major theropod groups. These dinosaurs have been found on every continent, having lately been discovered in the heart of blustery and frigid Antarctica (Figures 12.4 and 12.5). In this chapter, we will concentrate on non-avian (i.e., non-bird) theropods, holding off on the relationships of animals like *Deinonychus* to such creatures as *Archaeopteryx*, *Sinornis*, and others until Chapter 13.

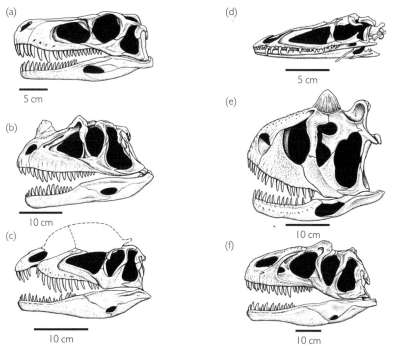

Figure 12.2. Left lateral view of the skull of (a) *Herrerasaurus*, (b) *Ceratosaurus*, (c) *Dilophosaurus*, (d) *Coelophysis*, (e) *Carnotaurus*, and (f) *Allosaurus*.

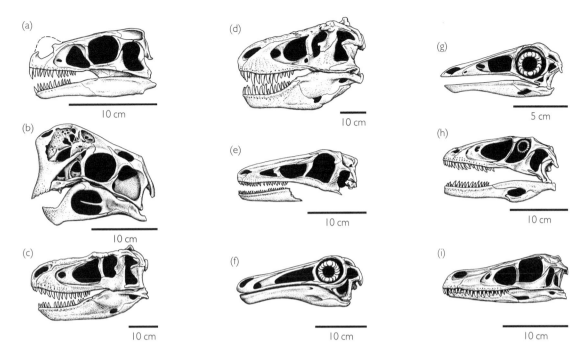

Figure 12.3. Left lateral view of the skull of (a) *Ornitholestes*, (b) *Oviraptor*, (c) *Albertosaurus*, (d) *Tyrannosaurus*, (e) *Saurornithoides*, (f) *Gallimimus*, (g) *Dromiceiomimus*, (h) *Deinonychus*, and (i) *Velociraptor*.

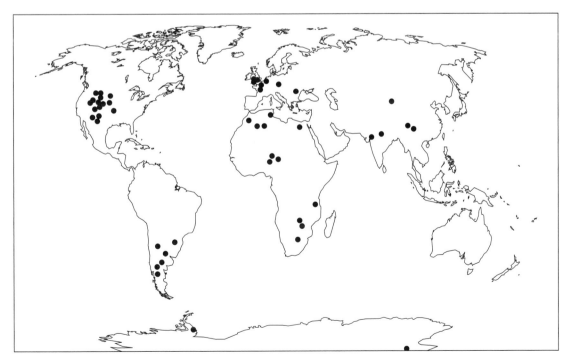

Figure 12.4. Global distribution of non-coelurosaurian Theropoda.

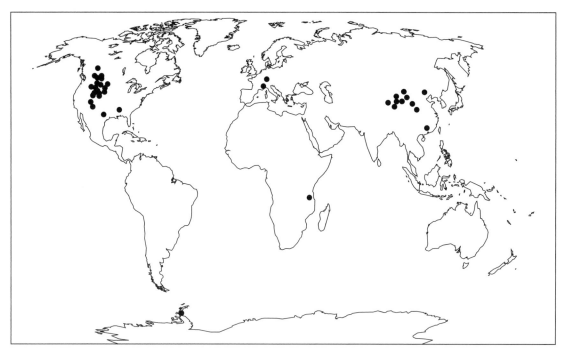

Figure 12.5. Global distribution of non-avian Coelurosauria.

(a)

(b)

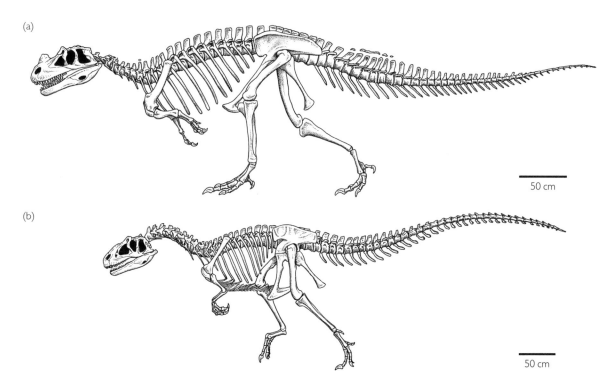

50 cm

50 cm

Figure 12.6. Left lateral view of the skull and skeleton of (a) *Ceratosaurus* and (b) *Allosaurus*.

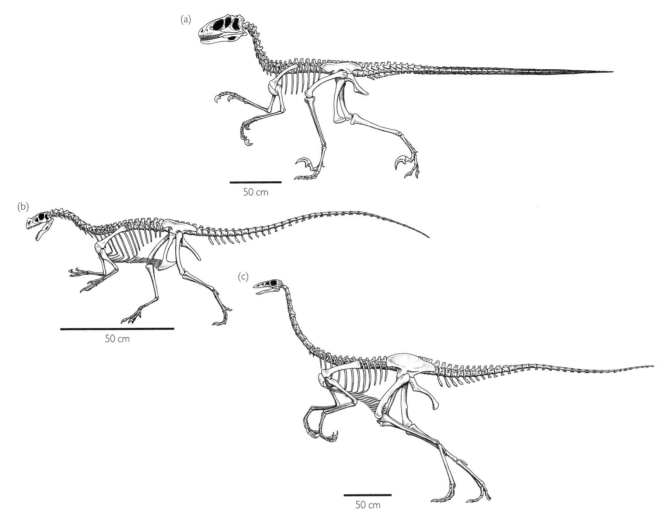

Figure 12.7. Left lateral view of the skull and skeleton of (a) *Deinonychus*, (b) *Ornitholestes*, and *Struthiosaurus* (c).

All theropods had a bipedal stance (Figures 12.6 and 12.7) and many are characterized as having sharp, commonly serrated teeth.[1] Beyond these two features, however, there was much variation: theropods came in various sizes and personalities. Some entirely lost their teeth, while others had teeth that looked like 15 cm long scimitars. And even though *Tyrannosaurus*, at about 13 m in length, has been a focal point of predatory dinosaurs, the majority of theropods were much smaller. A few were only a meter or so long. Some of the largest developed very short and stout forelimbs with only two stubby fingers, while smaller theropods sported extremely long arms with large hands and slim fingers. Regardless, theropod evolution is thought to have been associated in one

---

1 The teeth of theropods appear to have been a primitive carry-over from earlier in ornithodiran history (see Chapter 5).

way or another with what the animals did best: track, attack, and feed. For all the non-avian theropods – the "conventional" theropods – appear to have been irredeemable carnivores through and through.

Theropod lives and lifestyles

Going their way

Imagine living in a world threatened by the largest terrestrial carnivore of all time and you have the Late Cretaceous of both western North America and central and eastern Asia. Yet beware. Living in fear of *Tyrannosaurus* or *Tarbosaurus* would not have been enough. As the enormous tyrannosaurids presumably devoured their way through herds of ceratopsians or hadrosaurids, small, agile troodontids and dromaeosaurids stalked the landscape in search of their next meals as well. Hunting in packs and armed with recurved razor blades on their feet, these prehistoric killing-machines must have efficiently un-zipped the bowels of their unfortunate victims.

Non-avian theropods have been collected from a broad range of depositional settings, from fluvial channels and overbank deposits to lacustrine environments and even eolian dunes. Habitat preference – if it existed – is not known. In most instances, skeletal remains are found in isolation and often in a disarticulated state. This is the usual condition for the preservation of any dinosaur specimen. However, several often-remarkable mass accumulations of single theropod species provide important insights into the conditions not only of death and burial but perhaps of the animal's biology as well. For the most part, these mass graveyards include both juveniles and adults. There is *Syntarsus* in Zimbabwe, South Africa, and Arizona. And in neighboring New Mexico, near Abiquiu in the northwestern part of the state, there is Ghost Ranch, one of the most profoundly rich sites for any theropod, this time yielding an extraordinary abundance – several hundred individuals – of *Coelophysis*. Ghost Ranch is rivaled only by the Cleveland-Lloyd Quarry of east-central Utah, from which literally tonnes of *Allosaurus* bones have been collected since 1927. What might be the significance of these mass burials? Could it have been that these theropods were gregarious, living in large family groups, which then perished in some sort of catastrophe? Or perhaps each accumulation represents a communal feeding site? Are these relics of pack-hunting? Whatever we might speculate produced these mass burials – from geological to biological factors – thus far, none of these ideas has yet received a detailed analysis.

Theropods as living organisms

No one would deny the predatory ability of theropods. Inherited from their archosaurian forebears, such behavior is amply demonstrated by their panoply of claws and teeth. But first let's step back a little from theropod weaponry to look at the animal as a whole.

Non-avian theropods range in size from less than a meter (*Microraptor*) to animals growing to upward of 15 m in length (*Tyrannosaurus, Carcharodontosaurus, Giganotosaurus*). In the case of *Tyrannosaurus* from the Late Cretaceous of North America, the largest of these creatures dwarfed their presumed prey by nearly 35% of body length. J. O. Farlow has argued that this disparity in body size is due to both physiological and ecological causes. The immense size of tyrannosaurids and a few other theropods may have been a product of the abundance and rapid turnover of food sources: the large herds of rapidly growing, abundantly reproducing hadrosaurids and ceratopsids thought to have lived concurrently with these predators. So, the ready and continuously replenishing food source may have allowed tyrannosaurids, at least, to reach their large size. The lack of this kind of abundant resource today, Farlow argued, may restrict the size of modern mammalian carnivores.

All theropods are obligatory bipeds, unable to walk or run on any-thing but their hind legs. The body was balanced directly over the pelvis, with the vertebral column held nearly horizontally. Evidence from the skeleton and trackways indicates that the hind legs were held close to the body. Theropod trackways are always narrow gauge, in some cases so narrow that one foot appears to have been placed ahead of the other, rather than along side it. The trackways, as well as skeletal material, also indicate that the foot was held in a digitigrade stance – that is, on the toes with the ball of the foot held high off of the ground – and that the digits themselves terminated in sharp claws.

Theropods (especially troodontids and ornithomimosaurs) have short femora compared to the great length of the rest of the hindlimb. This condition tends to emphasize powerful and long strides as the animal runs. Calculations of running speeds on the basis of hindlimb proportions (see Box 15.3) indicate that many of these animals were indeed quite rapid movers, probably clocking in at 40 to 60 km/h. Some footprint evidence bears these numbers out; for example, a trackway in Texas demonstrates that a theropod once thundered away at upward of 45 km/h. For these reasons as well as those of skeletal anatomy, it is clear that all theropods could easily pursue a range of both fast and agile prey.

How fast the likes of *Tyrannosaurus* and other large theropods ran has been controversial, to say the least. Some investigators, the British bio-mechanicist R. McNeill Alexander among them, have calculated, using limb proportions and models of leg motion, that large theropods were limited to walking at no more than 4 km/h. Others have suggested maximum running speeds of nearly twice this value, using similar approaches. J. R. Hutchinson and M. Garcia took a slightly different approach to infer running speeds in theropods. They modeled how much muscle mass would be needed for these animals to support their bodies at different walking and running speeds. For small theropods, their model indicated reasonably fast running – no problems there. But for an adult *Tyrannosaurus* to achieve even Alexander's estimate, it would have been near or above its maximum muscular capacity and to move at higher speeds it would have required so much leg musculature that it assuredly would have collapsed under its own weight. This being the case, it is doubtful that tyrannosaurs and most prob-ably other large theropods were capable runners or could reach high speeds.

Weaponry The hindlimb, and especially the sharply clawed feet, had important functions that extended far beyond running abilities. Here we speak of the delicious topic of prey dismemberment. Nearly all theropods

10 cm

Figure 12.8. Left foot of *Deinonychus* with its disemboweling second-toe claw.

carried a significant portion of their weaponry on their feet (Figure 12.8). Imagine having great razor blades on your feet and this will allow you to get an inkling of the damage that the hind claws of theropods might inflict on a subdued dinner. In dromaeosaurids and troodontids, the claw on the second digit of the foot was especially huge and sharp. Because of its joint with the rest of the toe, this scimitar-like claw is capable of a very large arc of motion. During normal walking and running, this weapon was retracted so as to protect it from abrasion and other damage from contact with the ground. But, when needed, it could be flexed back into a lethal position and, with the powerful kicking motion of the rest of the leg, this razor-bladed foot could slash its way into the belly of some hapless herbivore, disemboweling the animal in one rapid stroke.

Not all of a theropod's weapons were on its feet – not by a long shot. Commonly equally well developed were the powerfully built forelimbs, equipped with powerful, grasping hands (Figure 12.9). Death surely came as often and as forcefully at the hands as at the feet of many a theropod. How can it have been otherwise? All theropods evolved from a common ancestor that uniquely evolved an enlarged hand from the shorter-handed ornithodiran condition. In addition, the largest of digits (I, II, III) were capable of extreme extension, of great advantage in grabbing onto large prey. The next evolutionary step in theropod evolution – that of lengthening the fingers and capping them with a trenchant claw – makes a great deal of sense simply from a slashing point of view. Other reorganizations to this already modified hand include changes in the wrist and hand to allow for yet better grasping ability in avetheropods and later on in the development of the semilunate carpal bone. This modification provided a much greater-than-normal grasping ability, of obvious importance to great manual dexterity and the ability to sever flesh from bone.

The formidable armament of theropods was lethally coupled with exceptional balance. The nearly horizontal position of the vertebral

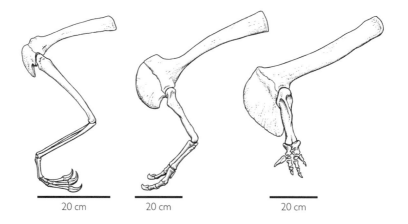

20 cm   20 cm   20 cm

Figure 12.9. Left forelimb of (a) *Struthiomimus*, (b) *Tyrannosaurus*, and (c) *Carnotaurus*.

column took advantage of the center of gravity being positioned near the hips. While this was obviously important to predatory dinosaurs both large and small, it probably was not enough for the kinds of agility we suspect was part of the daily routine of some of the small and highly aggressive theropods. For in animals such as dromaeosaurids, balance was everything. On the basis of their light-weight, yet powerfully built skeletons, these animals must have had an extraordinary degree of agility, balance, and leaping ability. Then there are the feet; with such weaponry, dromaeosaurids flung themselves at their fleeing prey, kicking with great accuracy with one of their dangerous feet while supporting the body with the other leg. Or possibly they even attacked with both feet simultaneously, their prey suffering sure death from the onslaught of twin-bladed damage.

The ability of theropods like *Deinonychus* and *Velociraptor* to engage in such acrobatics may well have come down to the design of their tail. One of the derived features of dromaeosaurids is an ability to pivot a stiffened tail just behind the pelvis. This stiffening would allow the rigid tail to move as a unit in any direction and thus it functioned as a dynamic counter-balancing device against the motions of the long arms and grasping hands. This must have been of great importance to a dromaeosaurid as it gripped with its large, powerful hands and struck out with one of its hind feet to dispatch a fleeing victim.

If grasping hands and long arms generally typify theropod skeletons, how does one explain the outrageously diminutive forelimbs of *Tyrannosaurus* and *Carnotaurus*? Not easily. For example, *T. rex* could not even reach its mouth with one of its hands. It is not at all clear why these animals independently miniaturized their arms. It has been suggested that perhaps the arms were short in order to balance an overly large head. This idea makes some biomechanical sense. In bipedal animals that have exceptionally large heads (like *Tyrannosaurus*, but not so much the case in *Carnotaurus*), increasing head size may require downsizing other aspects of the front half of the body to remain balanced with the back half at the hips. Still, there's the issue of what functions these dinky limbs may have had. The bones of the arms and hands are surprisingly stout, and there is good reason to believe that the shoulder and upper part of the limb were quite powerfully muscled. In fact, K. Carpenter and M. Smith have calculated that the arm of *Tyrannosaurus* may have been able to lift 300 kg. English paleontologist Barney Newman suggested that extending a small forelimb might be enough to catapult one of these large animals up from its presumed, belly-down, sleeping position. Whether tyrannosaurids slept belly-down remains speculation, and, whether getting up was a function of the forelimbs, it is clear from the robust limbs and large, stout claws at the tips of the fingers that the forelimbs aspired to greater purposes.

Jaws and teeth   As with just about any predatory animal, the head is impressive and terrifying. In the case of the big tyrannosaurids, the heads could be upwards of 1.75 m in length. In the case of smaller theropods, heads commonly have a brace of serrated blade-like teeth and can reveal acute visual capabilities and even hint at serious brain power. Furthermore it is from heads that we learn most about the details of choice of meals, killing techniques, and the mechanics of food ingestion.

In general, theropod skulls are rather primitive, reminiscent of those of many non-dinosaurian ornithodirans. By comparison with ornithopods, for example, they have few of the specializations that suggest much chewing took place. Still, because of the well-rounded occipital condyle and its articulation with the first part of the cervical vertebrae, it is thought that the skull had considerable mobility on the neck. So too do many of the joints of the skull with each other. Their appearance of mobility, however, may be a phantom; instead they may be constructed in such a fashion to lessen or dissipate the stresses passing through the skull as the animal bit, subdued, and dismembered its struggling prey.

Several recent studies have computer modeled the architecture of the theropod skull to understand the kinds of force that would pass through it during the acts of seizing and slicing prey. This burgeoning field of computer-based theropod jaw mechanics has produce a number of important studies, including those by D. Henderson and D. Weishampel, who used computer modeling to assess biomechanical convergence in the jaw systems of carnivorous archosaurs including theropods, and those by E. Rayfield, who used techniques of finite element analysis from engineering and combined it with three-dimensional computed tomography (CT scanning) of a skull of *Allosaurus* (Figure 12.10). Rayfield and colleagues concluded that *Allosaurus* used a high-impact slash-and-tear attack on its prey using its powerful neck muscles to drive the skull downward. When the head was retracted, the teeth sliced and tore flesh away before it was swallowed.

It is the sharply pointed, often recurved and serrate teeth in the upper and lower jaws that ultimately served the animal best in handling its prey. It has never been an issue that an animal equipped with sharply

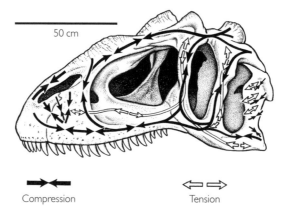

Figure 12.10. The skull of *Allosaurus* with its finite element analysis model, indicating the regions of stress that pass through it as a function of biting.

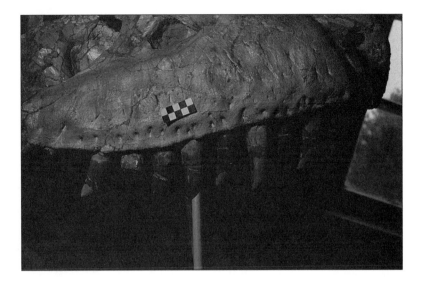

Figure 12.11. The impressive of *Tarbosaurus*, right lateral view, with scimitar-like teeth lining its jaws. Scale in centimeters.

keeled and pointed teeth was anything but a meat-eater (Figure 12.11). Yet it took till 1992 for someone to look at the details of theropod tooth construction and its significance to the cutting of flesh. Comparing different theropod teeth with an assortment of serrated steel saw blades, W. L. Abler concluded that the teeth of smaller theropods such as *Troodon*, with their prominently pointed serrations, functioned effectively in a "grip-and-rip" fashion, a bit like a serrated hacksaw blade. Tyrannosaurids, on the other hand, with their broad teeth and thickly rounded serrations, had weaker cutting ability. Based upon creases and striations in the teeth, however, Abler speculated that the teeth were subjected to complex, strong, and violent forces, such as might occur with a powerful, actively struggling prey. Beyond this, the precise function of tyrannosaurid teeth remains somewhat enigmatic (Box 12.1).

In addition, Abler observed pockets at the base of the serrations along the front and back of each tooth. These he suggested may have acted as havens for colonies of bacteria. As is the case in the living Komodo dragon (a large, predatory lizard), bacterial infections would have provided a double whammy to the victims of a theropod biting attack.

At least twice in their history, particular groups of non-avian theropods have drastically reduced or lost all of their teeth. Ornithomimosaurs lose their upper dentition first, while retaining only a few of the most forward lower teeth (this is the condition in the most primitive ornithomimosaur *Harpymimus*). All more-derived members of the clade lose even this dental modicum. In oviraptorosaurs, there is not even this brief glimpse of the transformation in the number of teeth in the jaws: all of these unusual-looking dinosaurs are toothless, both up and down, at the most primitive level of the history of the clade. Instead, between their shortened upper jaws is a pair of peg-like projections dead-center in middle of the palate.

BOX 12.1

Triceratops as spoils … or spoiled *Triceratops?*

In 1917, L. M. Lambe suggested that *Gorgosaurus* was not so much an aggressive predator, but instead maintained its sustenance by scavenging. The basis for his remarks was the absence of heavy wear on the teeth of this theropod – these animals therefore must have fed primarily on the softened flesh of putrefying carcasses. This interpretation has appeared on and off again in discussions of theropod diet and hunting behavior, frequently enough to be something like a cottage industry in anecdotal "knowledge" about these animals.

The notion of theropod scavenging rests on the assumption that tooth wear was usually absent and that carcasses were readily available. It is further bolstered by the lack of a convincing account of why the forelimbs of these animals are so small. We take each of these sources of support in turn. First, contrary to Lambe's claims, tooth wear is present on the teeth of nearly all large theropods. Not that that "proves" that tyrannosaurids and other large theropods had to have been active predators; both modern scavengers and active predators alike can have a high degree of wear on their teeth. The commonness of carcasses, putrefied or otherwise, was probably dependent on the nature of the season – dry, stressful seasons probably claimed their share of dead hadrosaurids, ceratopsians, sauropods, and the like. Interestingly, this potentially great contribution of carcasses would have been in the form of tough, dry bodies, sort of dinosaurian jerky (sun-dried meat) – not the kind of soft, predigested carrion postulated by Lambe. We suspect a supply of the soft variety would have been quite inadequate. Finally, those tyrannosaurid forelimbs may not be an explanatory burden to

those advocating active predation. These animals may well have enjoyed having extra appendages to slice, dice, and dismember their prey, but both G. S. Paul's theropods-as-sharks and Farlow and R. E. Molnar's suffocation hypothesis are fully consistent with a head-first attack style that, we believe, could have been carried out without the aid of long arms.

Recently, the suggestion of tyrannosaurids as carrion eaters has come from the observation that these theropods had surprisingly broad, bulbous teeth. Modern scavengers such as the hyena likewise have broad teeth, which are used to crush the bones of carcasses. This notion has been pooh-poohed by scientists who cannot imagine a dinosaur with the size and obvious carnivorous equipment of *T. rex* being a scavenger.

What is clear is that *Tyrannosaurus* teeth clearly exceed the size and shape that would be predicted by an allometric or scaling effect (see Chapter 15); that is, the teeth increased in size in a way that exceeds the size increase that might be predicted from its enlarged body. For this reason, *T. rex* leaves paleontologists with a mouthful of confusion.

In the end, we think it likely that animals such as *Tyrannosaurus* would have been a more than adequate, indeed terrifying, active predator, yet one that wouldn't turn up its nose at a lunch of carrion. Indeed, it is probably better to view all theropods, from the exceptionally large tyrannosaurids down to much smaller troodontids, dromaeosaurids, and coelophysoids, as opportunistic hunters and feeders. They were probably equal-opportunity consumers, taking large herbivores, smaller flesh-eaters, or even the occasional carcass.

What these animals ate has been widely debated. In the case of ornithomimosaurs, these animals must have had a small amount of muscle that closed the jaws, although the muscles were oriented for rather rapid jaw closure. The margins of the jaws were covered with cornified skin or scales that may have been relatively sharp, much like the condition in living ducks. According to M. A. Norell and colleagues, these dinosaurs probably used their beaks to strain food sediment in an

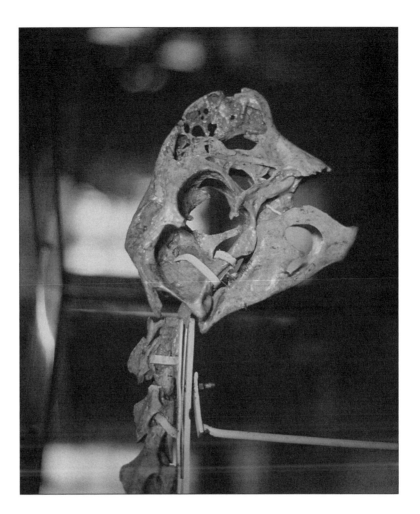

Figure 12.12. The box-like, toothless skull of *Oviraptor*.

aqueous environment, rather than for predation on large animals, and they may have been ecologically tied to food supplies in wetter environments, which would be consistent with their beak morphology. Ornithomimosaurs are also known to have swallowed stones and these gastroliths indicate that at least some of these forms had a muscular gizzard. Therefore, and consistent with the evidence from their cornified beak, these animals were probably herbivorous, using their gastroliths as grit to grind up plant matter like modern plant-eating birds.

The edentulous oviraptorosaurs, on the other hand, had very short, box-like and strong skulls, and the jaw musculature was very well developed (Figure 12.12). In 1977 and again in 1983, R. Barsbold provided a detailed analysis of the mechanics of the oviraptorosaur skull. In these studies, Barsbold was able to reject a diet of eggs that had earlier been ascribed (in name as well as in deed) to these Late Cretaceous theropods, arguing instead that their jaws were designed to feed on hard objects that required crushing. Rather than eating clams, oysters, mussels, and the lot as we humans do, prying or steaming them open with our hands or on the stove, he proposed that these animals cracked them open by the brute force of their jaw muscles acting on the thick horny bill covering the margins of the mouth and the palate, and especially the stout pegs in the center of their palate.

Because theropods were the top predators of their day, how they fitted into their paleocommunities and the processes that shaped these communities have been compared with modern carnivoran mammals, those creatures that superficially share an ecological role similar to that of theropods. B. Van Valkenburgh and R. E. Molnar compared predatory features of theropod dinosaurs and mammalian carnivores, focusing primarily on aspects of their skulls, jaws, and teeth, as well as groupings of sympatric species of predatory theropods and mammals (i.e., trophic guilds). Their conclusions? Even though the theropod jaws were weaker than those of modern predatory mammals, theropods have had a higher species richness than modern carnivores. Overall feeding diversity was lower in theropods and the skulls are built more like those of dogs than those of cats or hyenas. In a similar study that focused on the sympatric species of *Allosaurus* and *Ceratosaurus* from the Morrison Formation, D. Henderson was also able to identify feeding partitioning among theropods, this time represented by three different feeding strategies. Henderson argued that, as these theropods convergently evolved to large body size, they also evolved quite different jaw mechanisms to reduce feeding competition.

Killing prey and ingesting body parts by slicing, dicing, and dismembering are obvious mainstays of predation, but processing bone also forms a major feeding repertoire in hyenas among mammalian carnivores. Were there bone-processing dinosaurs? The answer to this question comes from an unlikely source, a 44 cm long, 13 cm high, and 16 cm wide specimen of fossil feces, otherwise known as a coprolite. That's one huge turd, more than twice as large as any previously reported coprolite. Its great size and stratigraphic and geographical context suggest that a tyrannosaur was the culprit that produced it, most likely *Tyrannosaurus rex*. What is most remarkable is that the specimen contains between 30% and 50% of bone fragments, thought to be the remains of limb bones or parts of a ceratopsian frill. In combination with important information gained from theropod gut contents (see below), this coprolite provides physical evidence that tyrannosaurs crushed, consumed, and incompletely digested large quantities of bone.

Senses To locate and track their prey, theropods of all kinds needed as keen an awareness of their environment as possible. Olfaction – an animal's sense of smell – has been little analyzed for dinosaurs in general. Drawing upon modern lizard behavior, *Jurassic Park* made claims that *T. rex* relied on sight and movement to detect prey rather than olfaction, but information on the size of the brain and the parts used for sense of smell (the olfactory bulbs) in tyrannosaurids appear to tell the opposite tale. Some researchers have suggested that the olfactory bulbs of these animals exceeded the size of the cerebral hemispheres – very large indeed. On the basis of this interpretation, *Tyrannosaurus* was said to have high olfactory acuity. Whether the olfactory bulbs were quite so large is presently under investigation, but certainly theropods depended on their sense of smell to evaluate a variety of environmental cues.

Theropods also needed a sharp sense of vision, so it is not surprising that eye size in these animals was large. Indeed, relatively large eye size is a shared, derived character of coelurosaurs and within this group ornithomimosaurs have exceptionally immense eyes. In dromaeosaurids and troodontids, the eyes have migrated to a more forward-looking position, indicating overlapping fields of vision. Overlapping fields of visions almost certainly means that these animals saw stereoscopically; that is, they merged the two separate independent images from each eye into a single image, much as we, and many modern carnivorous birds do today. What better way to maintain good eye contact with a fleeing meal?

Likewise, hearing is exceptionally important to predatory animals and so it is not surprising that many theropods probably had good sound perception. Indeed, the middle ear cavity of troodontids and ornithomimosaurs was greatly enlarged, suggesting that these theropods were especially able to hear low-frequency sounds. L. M. Witmer has taken analyses of hearing in troodontids even further. He noted that, in these animals, the two ears communicate with each other via a pneumatic channel that passes through the base of the braincase. Because of such connections, this channel might have been used – as it is in modern birds, which also have them – in sound localization. From a purely biomechanical standpoint, airborne sound waves hitting one of the two ear drums would have been transmitted through what Witmer called the interaural ("between the ears") pathway to the internal surface of the other ear drum. In this way, each ear conveys the same vibration to the middle and then the inner ear, but one would have done so out of synch with the other. Out-of-synch perception of the same sound wave provides the cue for perceiving sound directionality. Put another way, the snap of a twig by, say, a baby *Maiasaura* in hiding or the grunt of an incautious solitary *Centrosaurus* would have arrived asynchronously at the outside *and* inside of a troodontid's two ears. It is a relatively simple matter (all hard-wired into the central nervous system) to evaluate the phase differences in sound, turn the head in the correct direction, and bolt off after one's next meal.

Prey Exactly who these "next meals" were is no trivial matter. In our chapters on herbivorous dinosaurs, we spoke about stomach contents and coprolites, which provide a more-or-less direct association of feeders and fodder. For theropods, we similarly have only a rare glimpse of this kind of information. Some of the adult skeletons of *Coelophysis* from Ghost Ranch have the remains of juveniles in their bellies. Are these cases of cannibalism? Most scientists think so. Two additional examples can be found among coelurosaurs. The first, the nearly complete skeleton of

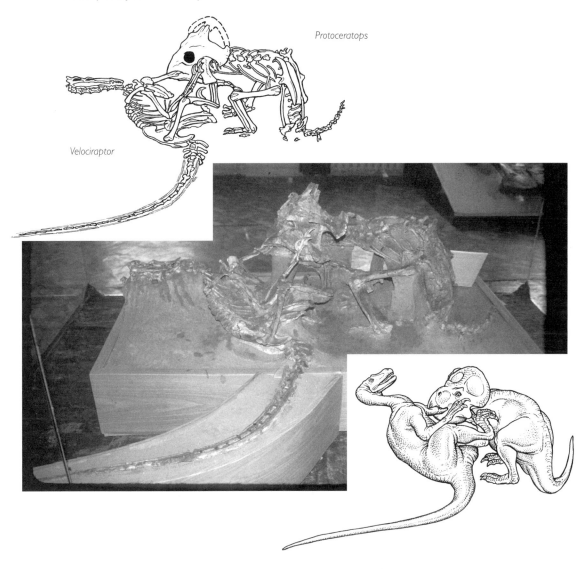

Figure 12.13. The famous fighting dinosaurs, *Velociraptor* wrapped around *Protoceratops*, from the Late Cretaceous of Mongolia.

Compsognathus that led T. H. Huxley to his views on the origin of birds, contains the majority of a skeleton of a fast-running lizard. Not only did *Compsognathus* swallow whole this delectable lacertilian meal, but it must have captured this victim through its own speed and maneuverability. Other evidence of theropod stomach contents and diet come from *Sinosauropteryx* (lizards and mammals), *Baryonyx* (fish remains), and *Daspletosaurus* (hadrosaurid bones). Finally, the prize specimen to have come out of the years of fieldwork in the Gobi Desert also attests to dromaeosaurid diets, a *Velociraptor* with its hind-feet half into the belly of a sub-adult *Protoceratops* and its hands grasping or being held in the jaws of the soon-to-be victim, that provide the most dynamic and irrefutable evidence about the prey of *Velociraptor* (Figure 12.13).

Beyond these few direct, and often astounding, observations of dietary preferences among non-avian theropods, we are left to speculate in a more roundabout way on the issue of who ate whom. These areas of guesswork are often the faunal associations of the predators themselves. For example, it may be that *Marshosaurus* fed on contemporary small- to medium-sized, fleet-footed ornithopods such as *Othnielia* and *Dryosaurus*; after all they are members of the same fossil assemblages. Likewise, *Liliensternus* may have fed on *Plateosaurus*, *Alectrosaurus* may have fed on hadrosaurids, and so forth. Whether true or not, these are the often the best available data that can be used to address the question of theropod diets (see also Chapter 16).

That theropods ate flesh – whether from the backs of contemporary brethren, or from mammals, lizards, snakes, or turtles, or from the inside of mollusc shells – is not a controversial issue. Whether they did it with any great style is. The degree to which dinosaurs did *anything* in a sophisticated fashion may well be reflected in their brain size (Box 15.4). And in this measure non-avian theropods are not at all that badly off. According to J. A. Hopson, all theropods for which there is brain size information have surprising cerebral powers – their brains are every bit as large as one would expect of a crocodilian or lizard "blown up" to the proper body size. And, for some, brain size is fully within the realm of a scaled-up bird. Among this latter group, troodontids had the largest brains for their body size of any of the "conventional" theropods. What this may mean in terms of lifestyle is debatable to a degree, but it is certainly suggestive that these animals probably had more complex perceptual ability and more precise motor-sensory control than some of the smaller-brained dinosaurs. It certainly implies relatively high activity levels and perhaps sophisticated inter- and intraspecific behavior (see Chapter 15).

For predatory animals, complex intraspecific behavior may have taken the form of pack-hunting. In his study of *Deinonychus*, J. H. Ostrom noted that the first remains of this predator to be recovered consisted of three partial skeletons that were found in close association with a skeleton of the iguanodontian ornithopod *Tenontosaurus*. Because of the great disparity in size between these two dinosaurs (*Tenontosaurus* is nearly 7 m long, much larger than a 3.5 m long *Deinonychus*), Ostrom argued that it was unlikely that a rogue *Deinonychus* could bring down a *Tenontosaurus* alone. However, add in a few more *Deinonychus*, each with their "terrible claws" unleashed, and the playing field becomes more than leveled. A coordinated attack by two, three, or more individuals would have easily overpowered even the largest of tenontosaurs, with more than enough spoils going to pack-hunters.

While *Deinonychus* and other small predatory dinosaurs may have pursued their meals in packs, it is likely that the larger forms hunted in a solitary fashion. How this was accomplished has been a matter of some debate. G. S. Paul has suggested that these large predators made shark-like hit-and-run attacks, crippling or killing their victims with a deep bite to their flanks, belly, or neck. They would then wait for weakened prey to succumb from loss of blood and resulting shock, swooping in for the *coup de grâce.* This attack scenario is consistent with the general fleetness of theropods of all kinds and, for animals such as *Tyrannosaurus* and *Carnotaurus*, with their small and ineffective forelimbs.

Alternatively, large and solitary theropods may have dispatched their prey in a more active way, much like large cats do today. According to Farlow and Molnar, it may be that these theropods suffocated their victim by seizing the snout or neck between their jaws, thus holding closed their nostrils and mouth or the trachea. All that remained was to hold on until the prey was weakened or had bled to death. This kind of attack, which is consistent with the large gape – large enough to wrap around the snout or neck of any prey animal – of these predators, has more action and terror than theropods-as-sharks, but in both cases these speculations have defied contact with actual information supplied by the fossil record. We know that non-avian theropods were active predators, but the exact means by which they brought down their prey in the majority of cases is lost in the mists of time.

Social behavior: sex and the *Rex*

As fixated as we might be on how theropods tracked, attacked, and fed, there are other aspects of their biology that are equally fascinating. And here we finally turn to intraspecific social behavior. What can the skeletons of non-avian theropods tell us about how these animals related to each other socially? Like our discussion of crests in hadrosaurids, or frills and horns in neoceratopsians, we turn first to the adornments on the skulls of numerous theropods. Quite a number of these predatory dinosaurs – from *Syntarsus, Dilophosaurus, Proceratosaurus*, and possibly *Ornitholestes*, to *Ceratosaurus, Alioramus*, and *Oviraptor* – sported highly visible cranial crests. Some are made of thin lamellae of bone, while others are hollow, presumably part of the cranial air–sinus system. Beyond these, theropods such as *Yangchuanosaurus, Allosaurus, Acrocanthosaurus*, and the tyrannosaurids bore slightly elevated upper margins on the snout and raised and roughened excrescences over the eyes. These structures are believed to have been cores for hornlets (small horns) made of keratin, which sheathed these roughened bumps (in the way that a cow's horn sheathes a bony base beneath), and which must have given the face a punk-rock spiky look.

The prominent yet quite delicately built crests and the stouter hornlets assuredly functioned in display and – at least for the latter – may also have been used occasionally in head-butting squabbles over territories or mates. But how can we test such a form–function relationship? As with J. A. Hopson's analysis of the function of hadrosaurid crests (see Chapter 10), we can look to a few adjuncts of display in social animals. If crests and hornlets functioned in visual display, particularly in those theropods that lived in large groups (see above), we might expect them to be species specific and probably sexually dimorphic so as to signal a given animal's identity and sex. And likewise we might expect crests to show their greatest development in reproductively mature individuals; youngsters should have small, poorly developed crests and hornlets.

Are these expectations met in those theropods with presumed display structures? To a degree, yes; sexual dimorphism appears to be present in the ceratosaurs *Syntarsus* and *Coelophysis*. In these two theropods, one of the two morphs is characterized by having a relatively long skull and neck, thick limbs, and powerfully developed muscles around the elbow and hip. The other, more gracile form retains a number of juvenile features, including a shorter skull and neck, and slender limbs. Unfortunately, it is unclear whether the prominent cranial crest is sexually dimorphic in size and shape. In one other theropod, however, the case has been made for sexually dimorphic ornamentation: in 1990, Carpenter identified sexual dimorphism in *Tyrannosaurus*, based in part on the development of the bony horn cores on the head. Interestingly, it is the larger, more robust morph that has been identified as female; is the same true for *Syntarsus* and *Coelophysis*?

Regardless of sexual dimorphism in non-avian theropods, what we know about their reproductive biology has been greatly enhanced by the discovery of brooding oviraptorosaurs. Several recently discovered articulated adult oviraptorosaur skeletons are preserved overlying nests of eggs. Laid in a circular pattern of as many as 22, these eggs are identical with another egg containing an articulated oviraptorosaur embryo from the same locality. This nest is of the same species as the adult skeleton overlying it and was parented by the adult.[2] The oviraptorosaur skeleton (Mom? Dad?) is positioned directly above the center of the nest, with its limbs arranged symmetrically on either side and its arms spread out around the perimeter as if protecting the nest. This posture, preserved in a quarter of known oviraptorosaur skeletons, is the same as that seen in birds that brood their nest. In these specimens, not only do we have an amazing window into theropod reproductive biology, but they also indicate that the behavior of sitting on open nests in this posture evolved well before the origin of modern birds.

What we know of the post-hatching ontogeny of non-avian theropods comes mostly from bonebeds (e.g., Ghost Ranch and Cleveland-Lloyd), as well as from some of the Upper Cretaceous localities of the Gobi Desert. For *Coelophysis* and *Syntarsus*, apparently there was a 10- to 15-fold increase in body size from hatchling to adulthood, and this growth is thought to have been quite rapid. Accompanying this rapid growth were proportional changes in the skull (relatively smaller eye socket, enlargement of the jaws and areas for muscles), relative lengthening of the neck, and relative shortening of the hindlimb. Similar changes – when they can be identified – are thought to have occurred in *Tarbosaurus* and *Albertosaurus* as well.

From a vocal point of view, could theropods growl, howl, grunt, sing, and/or groan? The obvious answer is that we don't yet know. Could they hiss? More likely, since any animal that can exhale can (and usually does)

---

2 Earlier studies of these same kinds of egg attributed them to the ceratopsian *Protoceratops*. Their physical association with partially preserved theropod specimens was incorrectly explained by turning the theropod into an egg-predator (*oviraptor* – egg thief), not a parent.

hiss. In any event, it is certainly true that, were theropods capable of making any noises, the great cavernous nature of many of their skull bones, from pneumatic lacrimal bones to sinuses of the maxillae, would have been well placed to act as vocal resonators. Thus we might expect deep and sonorous bellows from the likes of *Albertosaurus* and *Tarbosaurus*, and perhaps higher-frequency screeches from *Troodon* and *Dromiceiomimus*. Not surprising, these kinds of vocal frequency roughly match those thought to have been heard by the theropod ear. Moreover, we know that sound plays an important part in the social interrelationships of living (avian) theropods; is it too radical to propose that non-avian theropods likewise relied upon sound for social interactions?

Before leaving the biological aspects of non-avian theropods, we must address the issue of what sort of integument they had. This most ephemeral of prehistoric anatomy is important, particularly in view of the theropod origin of birds. Were all non-avian theropods covered with scales? If so, what were they like? Or could some or all of them have been covered with feathers, in which case the evolution of feathers precedes avian origin? We will discuss the details of these questions in Chapter 13, but can now present its preamble from the perspective of new research being conducted in Lower Cretaceous strata in Liaoning Province of northeastern China.

Up until the mid-1990s, most researchers thought that all non-avian theropods were covered with scales of some sort, almost certainly based on the evolutionary retention of the condition seen in crocodilians and squamates (lizards and snakes). The only direct information on theropod skin came from the South American ceratosaurian *Carnotaurus*, whose skin was covered with an array of tubercles of modest size surrounded by smaller rounded scales.

When bird material and other exceptionally well-preserved fossils began showing up in what is now known as the Jehol biota of Liaoning, it was perhaps inevitable that non-avian dinosaurs would also be found. Yet who would have imagined the incredible state of preservation of these fossils. Like Solnhofen and *Archaeopteryx* before it, the treasures of the Jehol biota often preserve evidence of soft tissues, including fur on mammals and feathers on birds – and on non-avian dinosaurs! Among the most famous of these are *Caudipteryx*, *Protarchaeopteryx*, *Sinornithosaurus*, and *Microraptor*. Another theropod from the Jehol biota – *Sinosauropteryx* – is extensively covered with filaments that have been interpreted as feather precursors. Thus, on the basis of these Jehol taxa, it is clear that the origin of feathers pre-dates not only the onset of avian flight but also the origin of birds within Theropoda. This, as we will see, has very important consequences for what we might call a bird.

The evolution of Theropoda

In order to understand these mostly predatory dinosaurs, their way of life, and other aspects of the evolution of their lifestyles, we need to know the various kinds of theropod and how they are related to each other. In keeping with our goals of recognizing only monophyletic taxa, we must cast our net wider than just the "conventional" theropod that regularly appears in popular books, television shows, and on

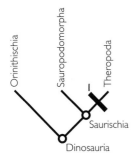

Figure 12.14. Cladogram of Dinosauria emphasizing the monophyly of Theropoda. Derived characters include: at 1 extreme hollowing of vertebrae and long bones, enlarged hand, vestigial fourth and fifth digits, remaining digits capable of extreme extension owing to large pits on the upper surfaces of the ends of the metacarpals.

cereal boxes. In doing so, we ask: What is the largest group that includes not only such theropods as *Tyrannosaurus* and *Coelophysis* but also their common ancestor and *all* the descendants of this common ancestor. According to our current understanding, Theropoda is formed of a diverse array of groups, including ceratosaurs, carnosaurs, therizinosauroids, tyrannosauroids, ornithomimosaurs, troodontids, and dromaeosaurids. Furthermore, it also contains birds. Viewed in this way, Theropoda can be defined as all the descendants of the common ancestor of *Coelophysis* and Aves (Figure 12.14).

There are many derived features that unite the theropod clade. Recent work by P. C. Sereno suggests that Theropoda can be diagnosed at the very least by having extreme hollowing of vertebrae and long bones, an enlarged hand (upward of 50% of the length of the rest of the forelimb), reduction of the bones of the fourth and fifth digits of the hand to the point where we would consider them vestigial, and the

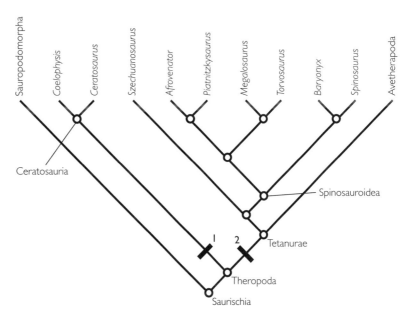

Figure 12.15. Cladogram of Theropoda. Derived characters include: at 1 modification of the neural spines and transverse processes of the vertebrae, fusion of the sacral ribs with the ilium, ventral and lateral flaring of the crest above the acetabulum on the ilium, modification of the knee joint, and fusion between the upper ankle bones; at 2 low ridge demarcating the maxillary antorbital fossa, spine table on axis, reduced, rod-like axial spinous process, prominent acromion on the scapula, loss of digit IV phalanges, metacarpal II nearly twice the length of metacarpal I, reduced femoral trochanteric shelf, prominent extensor groove on femur, fibular condyle on proximal tibia strongly offset from cnemial crest, broadly triangular metatarsal I attached to distal part of metatarsal II.

remaining digits capable of extreme extension owing to large pits on the upper surfaces of the ends of the metacarpals.

Theropods share closest relationship with Sauropodomorpha, and together form a monophyletic Saurischia. Their relationship with Ornithischia is yet more distant (see Chapter 5; introductory text to Part III: Saurischia).

In what follows, we delve into the evolution of Theropoda, considering some of the more and less renowned groups (Figure 12.15). Our synthesis of this rapidly changing field is based upon a combination of Gauthier's original ideas and subsequent analyses and discussion by T. Rowe, T. Holtz, P. C. Sereno, M. A. Norell, and other specialists on such groups as Ceratosauria, Tyrannosauroidea, Therizinosauroidea, Ornithomimosauria, Oviraptorosauria, Troodontidae, and Dromaeosauridae.

Ceratosauria

At its base, Theropoda is the wellspring of the two major descendant groups: Ceratosauria (named after one of its members, *Ceratosaurus*) and Tetanurae (*tetanus* – stiff; *uro* – tail). Ceratosauria was first identified by Gauthier in his 1986 study and subsequently analyzed in detail by Rowe and Gauthier, and R. S. Tykoski and Rowe. Although the monophyly of ceratosaurs has recently been questioned, for the moment it appears to be a reasonably well-supported clade, diagnosed by a number of derived features, among them modification of the neural spines and transverse processes of the vertebrae, fusion of the sacral ribs with the ilium, ventral and lateral flaring of the crest above the acetabulum on the ilium, modification of the knee joint, and fusion between the upper ankle bones.

Within Ceratosauria, there appear to be two major groups, the first – which Holtz has called Coelophysoidea – consists of *Dilophosaurus* and its close relatives, among them *Coelophysis*, *Syntarsus*, and *Liliensternus*, while the second – Neoceratosauria, first recognized and analyzed by J. F. Bonaparte and colleagues in 1990 and 1991 – is formed of *Ceratosaurus* itself, plus *Elaphrosaurus*, *Masiakasaurus*, *Majungatholus*, and members of Abelisauridae, an unusual group of South American theropods.

Tetanurae

The remaining great clade of theropods is Tetanurae, first recognized by Gauthier in 1986. The name refers to the fact that, in this group of theropods, the back half of the tail is stiffened by interlocking zygapophyses, fore-and-aft projections from the neural arches. Members of this group, whose record extends from the Middle Jurassic to the present, share a large number of other derived features, among them modification the antorbital region, a prominent muscular ridge on the scapula, changes in the hand skeleton, and modifications of the femur and knee joint.

Basal among tetanurans are the spinosauroids, a clade of theropods just recently recognized. Consisting principally of *Spinosaurus*, *Irritator*, *Baryonyx*, and *Suchomimus*, the front of the spinosauroid snout is elongate as is the antorbital region, the shoulder region is very muscular, and the elbow is well developed. In keeping with these and other modifications, spinosauroids are generally interpreted as fish-eating predators.

Figure 12.16. Cladogram of Avetheropoda. Derived characters include: at **1** enlarged external naris, participation of lateral surface of nasal in antorbital cavity, presence of nasal recesses, prefrontal excluded from rostral rim of orbit, supraorbital notch between postorbital and prefrontal, paroccipital processes directed strongly ventrolaterally from occiput to below level of foramen magnum, very short basipterygoid processes, mid-cervical centrum length less than twice the diameter of forward articular surface, front margin of spinous processes of proximal mid-caudal vertebrae with distinct kink, spur along front margin of spinous processes of mid-caudals; at **2** presences of a pterygopalatine fenestra, short mandibular process on pterygoid, presence of sternal ribs (three pairs), reduction of coracoid process, semi-lunate carpal, U-shaped ischial obturator notch, loss of transverse groove on astragalar condyle; at **3** crenulated ventral margin of premaxilla, parietal at least as long as frontal, U-shaped mandibular symphysis.

Fitting above spinosauroids are several important lineages that successively lead to *Torvosaurus*, *Eustreptospondylus*, *Megalosaurus*, and *Piatnitzkysaurus*. Only then do we encounter the next great clade of tetanurans known as Avetheropoda (Figure 12.16). First recognized by G. S. Paul in 1988, Avetheropoda (*avis* – bird; a reference to bird-like features of many members of this group) is that group of theropods more closely related to birds than is *Piatnitzkysaurus*. Avetheropods share a number of derived features, including the addition of a fenestra in the maxilla, modification of the back of the skull and palate, forward shift of the zone of stiffening of the tail skeleton, and other features of the pelvis and hindlimb. Avetheropoda consists of two clades, Carnosauria and Coelurosauria.

Carnosauria, a group formerly considered to consist of all large theropods, is now used as the name for a monophyletic group that includes *Allosaurus, Acrocanthosaurus, Carcharodontosaurus, Sinraptor*, and *Monolophosaurus*. Their external nares are enlarged, the sinuses in the facial skeleton become elaborated, the back of the skull and braincase are modified, and there are a host of changes to the vertebral column.

Coelurosauria

The other side of Avetheropoda is Coelurosauria, closer to birds than the taxa that we've so far discussed. Coelurosauria can be diagnosed on the basis of the following derived features: enlargement of the antorbital fossa, rear extent of the jaw joint in front of the most posterior extent of the occipital condyle, flat articulations between the centra of cervical vertebrae, metacarpal III shorter than metacarpal II, pre- and postacetabular blades of ilium subequal in length, very narrow contact between pubis and ischium, ascending process arising out of the complete breadth of the astragalus, and loss of pronounced horizontal groove across cranial face of astragalar condyles.

Basally within Coelurosauria, we encounter forms such as *Marshosaurus, Dryptosaurus, Compsognathus*, possibly *Ornitholestes*, and perhaps Tyrannosauroidea. Tyrannosauroids, those most terrifying of carnivorous dinosaurs have only recently been recognized as coelurosaurs, thanks to Holtz's phylogenetic studies. All of these theropods may have all it takes to be a coelurosaur, but do not to have the wherewithal to join the next-higher clade, Maniraptoriformes (*manus* – hand). Maniraptoriformes consists of Ornithomimosauria and Maniraptora; that is, those theropods with a modified palate, a semi-lunate wrist bone (Figure 12.17), and changes in the ankle that reflect a more in-line stride of the hindlimb. Ornithomimosaurs were a diverse, but distinctive group of highly modified theropods – *Harpymimus, Garudimimus, Ornithomimus, Struthiomimus, Dromiceiomimus, Archaeornithomimus, Gallimimus, Anserimimus* – which all bore a very lightly built skull with a long and low snout, a very large orbit and antorbital opening, no upper teeth and great reduction or loss of the lower teeth in a very long and low mandible, long forelimb, flat claws on the foot, and other modifications of the fore- and hindlimbs. Ornithomimosaurs bear an uncanny resemblance to modern ratites, the non-flying, cursorial group of birds that includes ostriches (*Struthio*) and emus (*Dromiceius*). Although ratites have rather reduced arms and, like all modern birds, a reduced tail, the similarities between ratites and ornithomimosaurs are sufficiently striking to suggest that these creatures may have behaved in similar ways.

Maniraptoran theropods are far from being well understood, but the following relationships are generally agreed upon among recent studies. Two groups – Oviraptorosauria and Therizinosauroidea – together form the basal maniraptoran clade. Oviraptorosaurs share a number of unique features that blend together to make these forms some of the most unusual among dinosaurs. These include: a crenulated ventral margin of the premaxilla; parietal at least as long as the frontal; ascending process of the quadratojugal bordering more than three-quarters of the infratem-

5 cm

Figure 12.17. Semi-lunate carpal (solid wrist bone indicated by arrow) in the left hand of *Ingenia*; an oviraptorosaurian maniraptoriform.

poral fenestra, a [...] dentulous dentary;
and a cranial pro[...] [t]he caudal process.

More unus[...] [r]elatively large and
ponderous, thes[...] [carniv]orous, with an edentulous beak and [...] [m]atter. D. A. and D. E.
Russell analogi[...] [a]ffinities have been a
matter of some [...] [be]tween theropods and
sauropodomor[...] [relation]ship among saurischians, and as a g[...] [th]is last view, now based
on detailed cl[a...] [...]t.

Eumaniraptora

The rest of th[...] [di]nosaurs – that is, those
theropods m[...] [con]sists of Eumaniraptora,
which is ma[...] [Avi]alae (Figure 12.18). Made
famous by [...] [Spi]elberg, deinonychosaurs
evoke more [...] T. *rex*. For these are the
"raptors" of *Jurassic Park* fame. [...] [the]m are the sickle-clawed
troodontids, best exemplified by *Troodon* itself, as well as *Saurornithoides*
and *Byronosaurus* among others. Similar in body form, dromaeosaurids –
among them, *Deinonychus*, *Dromaeosaurus*, *Microraptor*, *Saurornitholestes*, and
Velociraptor – comprise the other half of Deinonychosauria.

With our discussion of Deinonychosauria, we've reached the transition between non-avian and avian theropods. This subject, of some

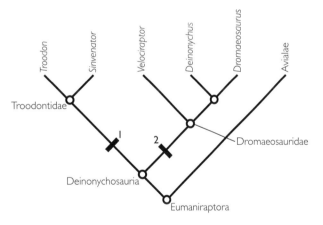

Figure 12.18. Cladogram of Eumaniraptora. Derived characters include: at 1 pneumatic
foramen in quadrate, loss of basisphenoid recess, large number of teeth, close packing of
front dentary teeth, reduced basal tubera, asymmetrical metatarsus, slender metatarsal II
markedly shorter than metatarsals III and IV, and a robust metatarsal IV; at 2 short T-
shaped frontals, a caudolateral overhanging shelf of the squamosal, lateral process of
quadrate that contacts quadratojugal above enlarged quadrate foramen, stalk-like
parapophyses on dorsal vertebrae, modified raptorial digit II, chevrons and
prezygapophyses of caudal vertebrae elongated and spanning several vertebrae,
presence of subglenoid fossa on coracoid.

controversy because of its implications to dinosaur extinction (it seems that theropod dinosaurs didn't quite go extinct after all) and interpretations of the biology of extinct dinosaurs, deserves its own special treatment, which we will take up in Chapters 13 and 18.

Theropods meet history: a short account of their discovery

Theropod discoveries begin auspiciously enough with Oxford University cleric and professor William Buckland's[3] work on *Megalosaurus* (*mega* – great). Based on a tooth-bearing jaw discovered in Oxfordshire, England, in 1819 and later given its name in 1824, *Megalosaurus* was one of the founding members of R. Owen's Dinosauria.

Thereafter, discoveries of new theropods throughout the nineteenth century were dominantly a North American experience. The controversial *Troodon* (*troo* – wound; *odon* – tooth) led the way. Philadelphia's J. Leidy based this theropod on a single tooth from the Upper Cretaceous of Montana in 1856. As we learned in Chapter 8, the affinity – and even the reality – of *Troodon* was a source of confusion until P. J. Currie referred additional, and very significant, skull and postcranial skeletal material to this taxon. Now *Troodon* is known in much greater detail.

In 1861, the small (less than a meter long) and very delicate skeleton of *Compsognathus* (*kompsos* – delicate; *gnathus* – jaw) was described by A. Wagner. This theropod was found in the same year and also in the same lithographic limestone rocks as another famous creature that will figure extensively in our story: the fossil bird *Archaeopteryx* (see Chapter 13). Interestingly, it was *Compsognathus*, rather than its feathered companion, that transfixed Huxley in his quest to understand bird origins. The only subsequent discovery of *Compsognathus* was made in 1972 in Upper Jurassic strata of the Canjuers region of France.

Back in North America, E. D. Cope was enlarging our sense of theropod anatomy and thereby enlivening our image of these dinosaurs with his energetic reconstructions of a new theropod from the Late Cretaceous of New Jersey. Originally called *Laelaps* (named for the mythical hunting dog Lailaps ("storm wind")) when it was described in 1866, this evocative name was dropped (it was unfortunately occupied by an insect), and replaced with *Dryptosaurus* (*drypto* – tearing) by Marsh in 1877.

Still greater theropod discoveries came with the furious efforts by Marsh and Cope to out-do each other in their exploration of the Western Interior of the USA during the 1870s and into the 1890s (Box 6.2). In terms of presently recognized theropods, Marsh got off the first shot with *Allosaurus* (*allo* – other) from the Late Jurassic (Morrison Formation) of the Western Interior of the USA in 1877. Then there were the second, third, and fourth salvos: *Coelurus* (*coel* – hollow; *uro* – tail) again from the Morrison Formation of Utah in 1879, the recognition of Theropoda in 1881, and *Ceratosaurus* (*cera* – horn) from the Morrison Formation of Colorado and Utah in 1884.

---

3 Like many of his Victorian peers and predecessors, Buckland was a naturalist as well as cleric. A most unusual man, one of his life goals was to eat his way through the animal kingdom.

The sole theropod from Cope that comes down to us today is *Coelophysis*. With a name meaning "hollow form," in reference to air spaces within the vertebrae, this dinosaur was recognized in 1889 from rather scrappy material from the Upper Triassic of New Mexico. In recent years, specifically since the discovery of an immense and dense bonebed of *Coelophysis* at Ghost Ranch, New Mexico, in 1947 and additional sites in Arizona, this small (less than 3 m long), agile theropod is one of the best known of all predatory dinosaurs.

Marsh returned to the theropod discovery scene once more before the close of the century with his recognition of *Ornithomimus* (*ornitho* – bird; *mimus* – mimic) in 1890. This theropod – from the Late Cretaceous of the Western Interior of Canada and the USA – heralded a large clade of ostrich-like theropods, the diversity of which we are still in the process of appreciating.

In the early part of the twentieth century, theropods continued to be discovered in North America, either as the result of continued exploration of Morrison Formation outcrops or from fieldwork in Alberta, Canada, that rapidly eclipsed all comers to the dinosaur-collecting arena. Yet the rest of the world was not to be dealt out of the discovery game. In turn, there was Tendaguru and Baharije Oasis in Africa, and the Central Asiatic Expedition in the Gobi Desert.

H. F. Osborn, turn of the nineteenth century vertebrate paleontologist and later director of the American Museum of Natural History in New York, as well as self-styled magnate of evolutionary biology, christened the first theropod in the new century – *Ornitholestes* (*lestes* – robber) from the Upper Jurassic Morrison Formation of Wyoming – in 1903. Because this small (2 m long) meat-eater was exceedingly well preserved and shared many features with *Archaeopteryx*, *Ornitholestes* has long had prominence in debates about the theropod affinities of living birds (see Chapter 13). Two years later, Osborn hit the presses with an additional two new theropods, both from the Late Cretaceous. Unlike *Ornitholestes*, these were among the largest – upward of 13 m long – of all time: *Albertosaurus* (named for Alberta, Canada, where the dinosaur was discovered; it is now known from elsewhere in the Western Interior of North America) and the all-time favorite *Tyrannosaurus* (*tyrannos* – tyrant), now known from Montana, North and South Dakota, Wyoming, and Colorado in the USA, and Alberta and Saskatchewan in Canada.

Then Tendaguru happened. The story of the discovery of Late Jurassic dinosaur bones in this out-of-the-way corner of what is now Tanzania has always focused on the wealth of sauropods found here (see Box 11.1). Nevertheless, between 1909 and 1912, W. Janensch and his co-workers were also collecting important theropod material, the most significant of which – a small (3.5 m long), fleet-looking animal – was named *Elaphrosaurus* (*elaphros* – fleet) by Janensch in 1920.

While the Humboldt Museum für Naturkunde expedition were busy with Janensch's work at Tendaguru, his colleague E. Stromer from Munich was hard at work in the Western Desert of Egypt. By the beginning of the 1930s, Stromer was to describe three new theropod dinosaurs from mid-Cretaceous outcrops at Baharije Oasis. The first, *Spinosaurus* (*spina* – spine), was named in 1915; the second, *Carcharodontosaurus* (named for *Carcharodon*, the great white shark), was recognized in 1931; while the third, *Bahariasaurus* (named for Baharije Oasis, where it was found), had to wait till 1934 for its christening. Although extremely poorly preserved, these theropods verged on the bizarre when their skeletal anatomy was compared with other theropod taxa known at that time. For this and other reasons, it is one of the paleontological tragedies of World War II that the original specimens of *Bahariasaurus* and *Carcharodontosaurus* were destroyed in the allied bombing of Munich at the close of the war. Only recently has work resumed at Baharije Oasis, by J. B. Smith and a field crew from the University of Pennsylvania, and their rewards assure that Stromer's legacy is still yielding important discoveries.

As research in Africa proceeded through the decade, discoveries in Upper Cretaceous strata in Alberta, Canada, heralded the ensuing Canadian Dinosaur Rush. The most important of these early results from the point of view of Theropoda were H. F. Osborn's 1917 description of a new theropod (*Struthiomimus* (*struthio* – ostrich)) in 1917 and the first comprehensive description of the skeleton of any of these animals, by L. M. Lambe for *Gorgosaurus*. This latter study has provided the basis for anatomical comparisons to the present day and was the first to seriously entertain the possibility that theropods (at least the large ones) may have scavenged for a living.

The theropod evolutionary tree began to take shape beginning in 1914 (and later embellished in 1920) through the work of F. von Huene. In these two studies, von Huene divided Theropoda into two major groups, small theropods – which he called Coelurosauria – and the big ones, von Huene's Carnosauria (*carneo* – flesh). These two taxa were to stand in one way or another for all subsequent theropod research and have even been adopted, in modified form, in modern treatments of theropod phylogeny.

With the 1920s, theropod studies exploded. In the course of the decade, seven new genera and species were described, exploration (now by at least three major North American museums) of the badlands of Alberta continued unabated, the American Museum of Natural History immersed itself in exploration of what turned out to be the great fossil fields of the Gobi Desert, and amid all this tumult, in 1927, the first theropod bones were quietly being discovered off the beaten track of central Utah in a place now known as the Cleveland-Lloyd Quarry. It was from this site that one of the largest collections of theropod material, principally of *Allosaurus*, has ever been found.

The 1920s, however, began with the release of Gilmore's important monograph on the anatomy and taxonomy of the theropod dinosaurs known at that time from the USA. In the following year, R. C. Andrews described a small and fragmentary theropod – *Sarcosaurus* (*sarkos* – flesh) – from the Early Jurassic of England; this form was long neglected in discussion of theropod dinosaurs until it was resurrected in the phylogenetic studies of Rowe and Gauthier.

Yet both Gilmore's and Andrew's work was eclipsed by a barrage of new dinosaurs from Canada and Central Asia. By the middle of the decade, the Upper Cretaceous strata of Alberta had yielded two small (2 m long) theropods: *Dromaeosaurus* (*dromaios* – swift), described by W. D. Matthew and B. Brown; and *Chirostenotes* (*chiro* – hand; *steno* – narrow), named by Gilmore. And over the same interval, Osborn had turned out three new and rather exotic theropods, all about 2 m in length, from the Upper Cretaceous of the Gobi: *Oviraptor* (*ovi* – egg; *raptor* – stealer), *Saurornithoides* (*ornithoides* – bird-like), and *Velociraptor* (*velo* – swift). By comparison, von Huene's *Proceratosaurus* (*pro* – before; *cerato* – horn) – known only from a small skull collected from Middle Jurassic strata of England that was described in 1926 – seemed a paltry offering to theropod diversity from elsewhere in the world.

von Huene's extensive 1932 monograph on all saurischian dinosaurs was the first to summarize theropod taxonomy, distribution, and evolution, thereby setting the stage for much of our appreciation of the group until the 1960s. A year later, von Huene described two new, 6 m long, and somewhat enigmatic theropods from the Late Cretaceous of India with C. Matley: *Indosaurus* (*indo* – from India) and *Indosuchus* (*suchus* – crocodile). Around the same time, Gilmore announced another new theropod, *Alectrosaurus* (*alectros* – unmarried; in reference to the enigmatic form of this theropod), collected from the Upper Cretaceous by the Central Asiatic Expeditions; and C. M. Camp described a new, much smaller form, *Segisaurus* (named for Segi Canyon in Arizona), from the Lower Jurassic of Arizona.

World War II exacted an extraordinary toll on humanity and human affairs and, as we have seen, paleontology was not miraculously excluded. In the 1940s, only two new theropods were described: *Caenagnathus* (*caeno* – new) from the Late Cretaceous of Alberta, named in 1940 by R. M. Sternberg and the Late Jurassic *Szechuanosaurus* (named for Szechuan (Sichuan), China), recognized in 1942 by Young. The only other activity of note over this time interval was von Huene's 1942 study of Late Triassic vertebrate faunas from South America, which provides intriguing, yet only fragmentary, evidence of very early theropods from this part of the world. It would take another 21 years for the details and significance of such Late Triassic theropods to be more fully understood.

The 1950s were little better in offering up theropod news. *Acrocanthosaurus* (*akros* – high; *akantha* – spine), described from the Lower Cretaceous of Oklahoma (now also known from Texas) by J. Stovall and W. Langston Jr in 1950, was followed by E. A. Maleev's 1955 announcement of *Tarbosaurus* (*tarbos* – terror), the first of many new Mongolian theropods discovered by workers from the Soviet Academy of Sciences and later by Russian, Polish, and Mongolian paleontologists.

Research during the 1960s not only reinvigorated our general view of dinosaurs, but it was a time when the diversity and complexity of theropods began to be appreciated. The decade began with the recognition of *Metricanthosaurus* (*metrikos* – moderate) from the Late Jurassic of

England, an old dinosaur (an earlier-recognized species of *Megalosaurus*) provided with its new taxonomic clothing by A. D. Walker in 1964. It was 1969, however, that literally blew the lid off theropod dinosaur diversity and paleobiology. That year, Raath described *Syntarsus* (*syn* – fused; *tarsos* – tarsus), a 2 m long, gracile predator from the Early Jurassic of southern Africa and, thanks to the subsequent work of Rowe, from Arizona as well. At the other end of the Mesozoic, E. H. Colbert and D. A. Russell provided an important redescription of *Dromaeosaurus* and Russell began his investigation into brain sizes among various theropod groups.

But the single most important discovery – perhaps in the recent history of all dinosaurs – was the announcement by Yale University's J. H. Ostrom of *Deinonychus* (*deino* – terrible; *onycho* – claw) in 1969. This 2 m long predator comes from the Early Cretaceous of Montana and Wyoming. It is rightly claimed that *Deinonychus* – obviously sophisticated and lethal in its behavior – led Ostrom to reconsider claims about theropod agility, hunting behavior, evolutionary relationships, and, in a broader context, dinosaur metabolism and the affinities of dinosaurs to birds. This watershed effect of discoveries will also be discussed in Chapter 13.

The 1970s were again dominated by the theropod discoveries that came from the Polish–Mongolian Palaeontological Expeditions and the Joint Soviet–Mongolian Paleontological Expeditions in the Upper Cretaceous of the Gobi Desert. From the Poles, we have *Deinocheirus* (*chiro* – hand), known only from a set of immense and intimidating forelimbs described by H. Osmólska and E. Roniewicz in 1970, *Gallimimus* (*gallus* – chicken), the largest yet known ornithomimosaur (at 6 m long), named by Osmólska and her colleagues from Poland and Mongolia in 1972, and Barsbold's 1974 description of important new material of *Saurornithoides*, a small, yet assuredly dangerous predator closely related to *Troodon* from North America.

At the same time, the first results of the Joint Soviet–Mongolian Paleontological Expeditions were becoming available. Beginning with Maleev's description of *Tarbosaurus* published posthumously in 1974, there followed a spate of studies by Barsbold on the taxonomy, evolution, and feeding of oviraptorosaurs (*ovum* – egg), the description of *Alioramus* (*ali* – other; *ramus* – branch, as in another branch of tyrannosaurids) by Kurzanov in 1976. That same year, Kurzanov also described a new and enigmatic theropod from the Upper Cretaceous of Uzbekistan in Central Asia, that he called *Itemirus* (after Itemir, the closest village to where this dinosaur was found).

Also in 1970, S. P. Welles announced *Dilophosaurus* (*di* – two; *lophos* – crest), a 6 m long, lightly built theropod from the Early Jurassic of Arizona, while Russell described a new tyrannosaurid from the Late Cretaceous of Alberta, which he named *Daspletosaurus* (*dasples* – frightful). His work on the anatomy and taxonomy of the Late Cretaceous ornithomimosaurs gave us *Archaeornithomimus* (*archaios* – ancient; *ornis* – bird) from northern China. By the mid-1970s, J. H. Madsen Jr began publishing on the very abundant theropod remains from the Cleveland-Lloyd Quarry (in the Upper Jurassic Morrison Formation) of central Utah.

Some of these remains included brand-new taxa, such as *Stokesosaurus* (named in honor of American paleontologist and geologist W. L. Stokes) and *Marshosaurus* (named in honor of O. C. Marsh). But best of all was Madsen's detailed and thorough description of *Allosaurus*, the most common member of the jumble of bones that is Cleveland-Lloyd.

Elsewhere in North America, the discovery of two other new theropods rounded out the 1970s. One, from the Upper Cretaceous of Alberta, was named *Saurornitholestes* (*lestes* – robber; a lizard-like bird thief) in 1978 by Sues; this small (2 m long) form immediately joined the already rich company of theropods from the Late Cretaceous of southern Alberta and – perhaps more importantly – added to the diversity of small, agile theropods. The other new predatory dinosaur – a 9 m long contemporary of the Cleveland-Lloyd theropods, but from the Colorado side of the Morrison Formation outcrops (Upper Jurassic) – was described as *Torvosaurus* (*torvus* – savage) in 1979 by Galton and J. A. Jensen.

To the south, new theropods turned up again in 1979 with the announcement of *Piatnitzkysaurus* (named for Argentinean geologist A. Piatnitzky). This medium-sized (4.5 m long) dinosaur was described by Bonaparte from Middle Jurassic beds of southern Argentina. At the same time across the Pacific, renewed collecting and museum research efforts were beginning to turn out their share of new theropods. In 1977, Dong described a quite complete skeleton of a nearly 2 m long enigmatic theropod from the Late Cretaceous of Xinjiang Province that he named *Shanshanosaurus* (for Shanshan zhan, the name of the locality from which this dinosaur was collected). *Yangchuanosaurus* (named for Yangchuan County) was his next conquest. Described in 1978, this now well known, 8 m long theropod comes from the Late Jurassic of Sichuan Province, China.

The rewards from the joint Polish and Mongolian efforts also continued through the 1980s. Through Osmólska's studies, we have *Elmisaurus* (in Mongolian, *elmyi* means hindfoot) – a small, but as yet poorly known theropod – in 1981. *Hulsanpes* (Khulsan is the locality in Mongolia where this dinosaur was discovered; *pes* – foot) followed in 1982. And in 1987, she gave us a new troodontid that she named *Borogovia* (named for the borogoves in Lewis Carroll's poem "Jabberwocky"[4]). Barsbold named a new, primitive ornithomimosaur (*Garudimimus*; Garuda is a monstrous bird in Asian mythology) and a delicate new oviraptorosaur named *Ingenia* (named for Ingeni, Mongolia, near the locality that yielded this dinosaur) in 1981. Thereafter, he presented us with *Adasaurus* (named for Ada, an evil spirit in Mongolian mythology), a new and as yet poorly known dromaeosaurid described in his comprehensive 1983 monograph on these small predatory dinosaurs. The following year, it was *Harpymimus* (*harpyiai* – harpy), a primitive ornithomimosaur described jointly with A. Perle from the Early Cretaceous of Mongolia. During the latter half of the 1980s, Barsbold described a new oviraptorosaur, *Conchoraptor* (*conch* – shell), in 1986 and a new ornithomimosaur,

4 "Twas brillig and the slithy toves Did gyre and gimble in the wabe: All mimsy were the borogoves And the mome raths outgrabe." (Lewis Carroll, 1872. *Through the Looking Glass*.)

Anserimimus (*anser* – duck), in 1988. Finally, one of the most peculiar of these new Mongolian discoveries by the Joint Soviet–Mongolian Paleontological Expeditions is that of *Avimimus* (*avis* – bird), an extremely gracile, long-legged, and perhaps toothless 1.5 m long theropod originally described by Kurzanov in 1981. Originally thought to be close to the origin of birds, this theropod was subsequently placed above *Archaeopteryx* in avian phylogeny, but has most recently come to roost as a member of more "conventional" Theropoda.

By comparison with Mongolia, new theropods from the rest of the world just dribbled in over the course of the 1980s. *Liliensternus* (named in honor of German paleontologist H. R. von Lilienstern), a 5 m long theropod from the Late Triassic of Germany originally thought to be a species of the poorly known theropod called *Halticosaurus* (*haltic* – leaping), was described by Welles in 1984. In 1985, Dong and Z. Tang described a new and much better preserved Middle Jurassic theropod from Sichuan Province, China, which they named *Gasosaurus* (in reference to the Dashanpu petroleum-mining company).

At the same time, Bonaparte and colleagues were busy making discoveries in Argentina. The first, a small Late Cretaceous theropod from Salta Province of northeastern Argentina named *Noasaurus* (in Spanish, NOA stands for norte-oeste Argentina) in 1980 by Bonaparte and J. Powell, was followed by *Abelisaurus* (named for R. Abel, Director of the Museo de Cipolletti in Argentina), a 6 m long theropod from the Late Cretaceous described in 1985 by Bonaparte and F. E. Novas. *Abelisaurus* was mere adumbration to what followed later the same year: a third new theropod, the mid-Cretaceous *Carnotaurus* (*taurus* – bull), was a short-faced, horned, 8 m long brute with almost hideously diminutive arms. Between *Abelisaurus* and *Carnotaurus*, these new theropods provided the first clear suggestion that at least some theropods – Bonaparte's Abelisauridae – evolved in isolation within the southern landmasses.

Finally, the first new English theropod to be named in nearly a quarter of a century was described in 1986 by A. J. Charig and A. Milner under the name *Baryonyx* (*bary* – heavy). This possibly piscivorous (fish-eating) theropod comes from the famous Lower Cretaceous Wealden beds, the same strata that provided England with its greatest collection of dinosaurs.

Other important studies of the 1980s include Currie's research on the anatomy and relationships of *Troodon* in 1987, Colbert's important work on the anatomy, growth, and evolutionary significance of *Coelophysis* from Ghost Ranch in 1989, Rowe's cladistic analysis of a new species of *Syntarsus* from the southwest USA, and Paul's comprehensive work on the taxonomy, anatomy, reconstruction, and restoration of all predatory dinosaurs. In fact, the only new theropod discovery was the "rediscovery" of a small skull from the Late Cretaceous of Montana that had earlier been thought to be *Albertosaurus* from the Late Cretaceous of Montana. This well-preserved skull of a dwarfed tyrannosaurid was given the moniker *Nanotyrannus* (*nano* – small) by Bakker and his colleagues in 1988.

Theropods continued to tumble out of the rock across the globe during the 1990s. Numerous discoveries were again made in the Upper

Cretaceous rocks of Mongolia: *Tochisaurus* (in Mongolian, *tochi* means ostrich) described in 1991 by Kurzanov and Osmólska, *Bagaraatan* ("little predator" in Mongolian) described in 1996 by Osmólska, *Rinchenia* (named for Bjambyn Rinchen, Mongolian ethnographer, historian, philologist, linguist, translator of European *belles-lettres*, writer, novelist, and father of Rinchen Barsbold) named by Barsbold in 1997, and *Achillobator* ("Achilles (tendon) hero") described by Perle and colleagues in 1999. China also yielded an important theropod stash. From Liaoning, home of the Early Cretaceous feathered dinosaurs, came *Sinosauropteryx* ("Chinese dinosaur wing") named in 1996 by Q. Ji and S. Ji, *Caudipteryx* ("tail feather") described in 1998 by Q. Ji and co-workers, and, in 1999, *Sinornithosaurus* ("Chinese bird lizard") and *Beipiaosaurus* (named for the Liaoning city of Beipiao) both named by X. Xu and co-workers in separate papers. In addition, China produced *Monolophosaurus* ("single-crested lizard") and *Sinraptor* ("Chinese hunter"), two carnosaurs from the Middle Jurassic of China named by X. Zhao and Currie in 1994, and the therizinosauroid *Alxasaurus* (named for the Alxa Desert of Inner Mongolia, China) and the troodontid *Sinornithoides* ("Chinese bird-like"), both named by Russell and Dong in 1994 from the Late Cretaceous of northern China. The remaining Asian discovery made during the decade is *Caenagnathasia* ("Asian *Caenagnathus*"), an oviraptoro-saur named by Currie and co-workers in 1994 from the Late Cretaceous of Uzbekistan.

Not quite as productive as Asia during the 1990s, Europe did see the discovery of six new theropods. *Pelecanimimus* ("Pelican mimic"), from the Early Cretaceous of Spain, was described B. Pérez-Moreno and colleagues in 1994. From Upper Jurassic rocks in neighboring Portugal, *Lourinhanosaurus* (named for Lourinha, Portugal) was named by O. Mateus in 1998. France produced two theropods: the Early Cretaceous *Genusaurus* (*genu* – knee) described by M. Accarie and co-authors in 1995, and the Late Cretaceous *Variraptor* ("Var thief" – Var is the area of southern France in which this theropod was discovered) named by J. Le Loeuff and E. Buffetaut in 1998. Finally, *Neovenator* ("new hunter") was named by S. Hutt and colleagues in 1996 from the Early Cretaceous of England, and *Scipionyx* ("Scipio's claw" – named in honor of Scipione Breisla, an Italian geologist) was described by C. Dal Sasso and M. Signore in 1998 from the Early Cretaceous of Italy.

Only two theropods were discovered in North America during the 1990s, both from the Late Triassic of the desert southwest of the USA: *Gojirasaurus* (Gojira is the Japanese name for the movie monster called "Godzilla" in English) named by Carpenter in 1997 and *Camposaurus* (named for C. L. Camp, American vertebrate paleontologist) described by A. Hunt and co-authors in 1999.

Gondwana saw its share of theropods in the 1990s. From South America, there were four from Argentina and one from Brazil. This latter form, the Early Cretaceous *Irritator* ("the one that irritates") was named by D. Martill and colleagues in 1996. Argentina, on the other hand, produced the truely immense *Giganotosaurus* ("giant southern lizard") from the mid-Cretaceous named by R. A. Coria and L. Salgado in 1995, the

Early Cretaceous *Ligabueino* ("Ligabue's little one," named in honor of Giancarlo Ligabue) described by Bonaparte in 1996, and finally the Late Cretaceous *Megaraptor* ("large thief") and *Ilokelesia* ("flesh-eating reptile") separately named by F. Novas, and Coria and Salgado in 1998.

Three theropod discoveries were made in Africa during the 1990s. *Afrovenator* ("African hunter") and *Suchomimus* ("crocodile mimic"), from the Early Cretaceous of Niger, were successively described by Sereno and co-workers in 1994 and 1998, while in Madagascar new material of the Late Cretaceous *Majungatholus* ("Majunga dome") described in 1998 by S. Sampson and co-authors provided ample evidence that this dinosaur is a theropod and not a pachycephalosaur.

Finally, from what must be the most isolated and frigid of dinosaur localities, we have the Early Jurassic *Cryolophosaurus* ("frozen-crested lizard"), named by W. Hammer and W. Hickerson in 1994, from the Transantarctic Mountains of Antarctica.

Theropod discoveries continue to be made from 2000 onward. So far, six come from Asia. *Fukuiraptor* ("Fukui robber") named by Y. Azuma and Currie in 2000 comes from the Early Cretaceous of Japan. Another four are known from the Late Cretaceous of Mongolia, two in 2000 and another two in 2001. The former consist of *Nomingia* (for the Nomingiin Gobi, that part of the Gobi Desert where this theropod was discovered) named by Barsbold and colleagues and *Byronosaurus* (for Byron Jaffe and his support of the Mongolian Academy Sciences–American Museum of Natural History Paleontological Expeditions) described by Norell and co-workers. The announcement of two new oviraptorosaurs, *Citipati* ("lord of the cemetery") and *Khaan* ("ruler" in Mongolian), by J. Clark, Norell, and Barsbold, came the next year. Another feathered Liaoning theropod, *Microraptor* ("small robber"), was named by Xu and colleagues in 2000.

North America, Europe, and Africa have produced the rest of the burgeoning crowd of this century's theropods. *Bambiraptor* ("Bambi robber" – Bambi was the nickname for this dinosaur when it was collected), a new dromaeosaurid named by D. A. Burnham and co-workers in 2000, is known from the Late Cretaceous of Montana, and *Nothronychus* ("sloth-like claw"), described by J. Kirkland and D. Wolfe in 2001, is a new therizinosauroid from the Late Cretaceous of New Mexico. In Europe, we have *Pyroraptor* ("fire thief") from the Late Cretaceous of France, described by R. Allain and P. Taquet in 2000, and from Africa comes *Masiakasaurus* ("vicious lizard") from the Late Cretaceous of Madagascar, named by Sampson and collaborators in 2001.

Clearly the depth of theropod discoveries have not been fully plumbed since their original discovery of *Megalosaurus* in the early 1800s. Nor is it likely to be even close. As long as there are dinosaur paleontologists working in the sundry badlands throughout the world, the richness of the theropods world should continue to grow, with no end in sight.

Important readings

Clark, J. M., Maryańska, T. and Barsbold, R. 2004. Therizinosauroidea. In Weishampel, D. B., Dodson, P., and Osmólska, H. (eds.), *The Dinosauria*, 2nd edn. University of California Press, Berkeley, pp 151–164.

Currie, P. J. Koppelhus, E. B., Shugar, M. and Wright, J. (eds.) 2004. *Feathered Dragons – Studies on the Transition from Dinosaurs to Birds.* Indiana University Press, Bloomington, IN, 361pp.

Holtz, T. R. 2004. Tyrannosauroidea. In Weishampel, D. B., Dodson, P. and Osmólska, H. (eds.), *The Dinosauria*, 2nd edn. University of California Press, Berkeley, pp. 111–136.

Holtz, T. R., Molnar, R. E. and Currie, P. J. 2004. Basal Tetanurae. In Weishampel, D. B., Dodson, P., and Osmólska, H. (eds.), *The Dinosauria*, 2nd edn. University of California Press, Berkeley, pp. 71–110.

Makovicky, P. J., Kobayashi, Y. and Currie, P. J. 2004. Ornithomimosauria. In Weishampel, D. B., Dodson, P. and Osmólska, H. (eds.), *The Dinosauria*, 2nd edn. University of California Press, Berkeley, pp. 137–150.

Makovicky, P. J. and Norell, M. A. 2004. Troodontidae. In Weishampel, D. B., Dodson, P. and Osmólska, H. (eds.). *The Dinosauria*, 2nd edn. University of California Press, Berkeley, pp. 184–195.

Norell, M. A. and Makovicky, P. J. 2004. Dromaeosauridae. In Weishampel, D. B., Dodson, P. and Osmólska, H. (eds.), *The Dinosauria*, 2nd edn. University of California Press, Berkeley, pp. 196–209.

Osmólska, H., Currie, P. J. and Barsbold, R. 2004. Oviraptorosauria. In Weishampel, D. B., Dodson, P. and Osmólska, H. (eds.), *The Dinosauria*, 2nd edn. University of California Press, Berkeley, pp. 165–183.

Tykoski, R. S. and Rowe, T. 2004. Ceratosauria. In Weishampel, D. B., Dodson, P. and Osmólska, H. (eds.), *The Dinosauria*, 2nd edn. University of California Press, Berkeley, pp. 47–70.

CHAPTER 13

Theropoda II:
The origin of birds

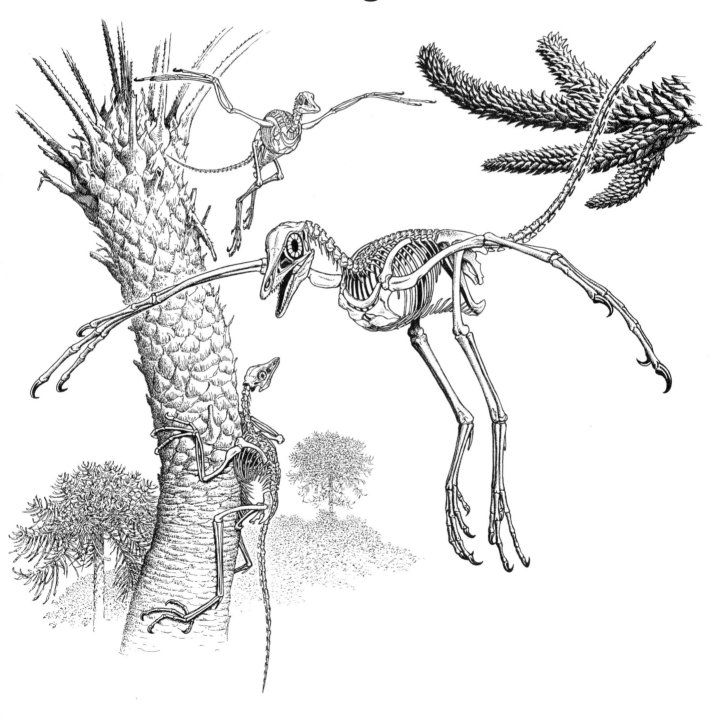

Q: Which came first, the chicken or the egg?
A: The egg. But it wasn't from a chicken.

Birds are dinosaurs. We do not mean that they are somehow *related* to dinosaurs (although if they are dinosaurs, they must be related to particular members of this group). We do not mean that they *come from* dinosaurs (although they obviously evolved from something that was itself a dinosaur). We mean that birds *are* dinosaurs. Claiming that birds are dinosaurs is no more radical than saying that humans are mammals. The issue boils down to a basic cladistic point: humans are mammals because humans have the diagnostic features of mammals. Birds are dinosaurs because they possess the diagnostic features of dinosaurs.

But what is a bird? Even as recently as nine years ago (when the first edition of this book was published), what was and what was not a bird seemed obvious: you looked for feathers, and if it had feathers, it was a bird. After all, are there any birds without feathers? Yet as the fossil record now shows, this view is an illusion caused by looking exclusively at living faunas. Recent startling discoveries from the Cretaceous of China have turned this equation (feathers = birds) on its head. The important question is *not*, "Are there any birds without feathers?"; the important question is, "Are there any feathered creatures that are not birds?" And if they are not birds, then what exactly are they?

The key to the identity and relationships of organisms – as we have so often stressed – is phylogenetic systematics, and so we will apply cladistic methods to this branch of vertebrates. In doing so, we will consider a number of thorny questions about birds.

1 What is a bird? (Which, as we have seen elsewhere, is the same question as "Where do birds come from?")
2 Where do feathers come from?
3 What are the fingers in the bird hand?
4 How did bird flight evolve?

We start by looking at *living* birds, because these are most familiar to us, and because, as we shall see, these form a monophyletic group based upon a hefty suite of characters.

Living birds

Living birds are obviously vertebrates; they possess a nerve chord and a backbone. Moreover, they are clearly tetrapods (they bear four limbs) and amniotes (their eggs have amniotic membranes). Living bird skulls belong to the great diapsid clade (they have upper and lower temporal openings), and within the diapsids, clearly show they are archosaurs by the possession of an antorbital opening. But birds don't superficially resemble crocodiles (the other living archosaurs), because they possess a unique suite of highly evolved features superimposed upon the archosaur body plan.

Feathers
All living birds have feathers. Feathers are complex, both in terms of their structure and development (Figure 13.1a). All feathers are

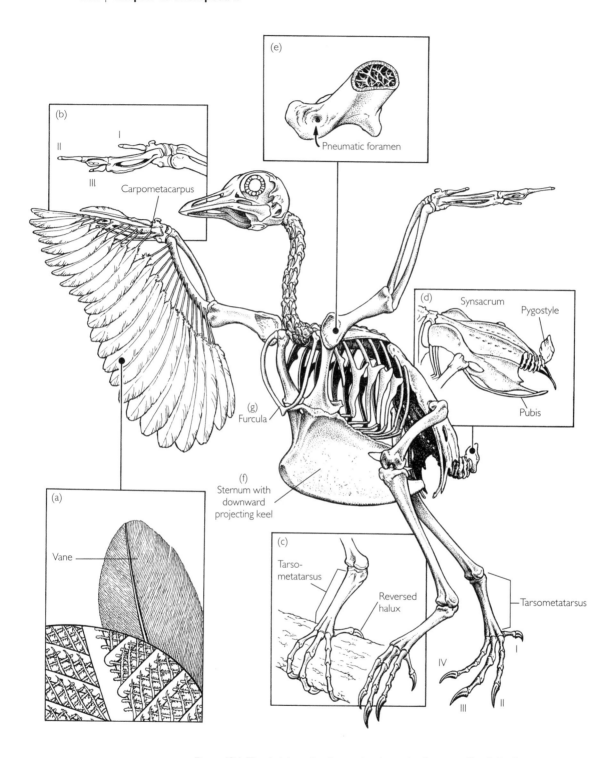

Figure 13.1. The skeleton of a pigeon, showing major features of its skeletal anatomy.
(a) Detail of feather structure; (b) carpometacarpus with digits labeled; (c) tarsometatarsus;
(d) synsacrum (fused pelvic bones) with pygostyle; (e) hollow bone with pneumatic
foramina; and (f) sternum with large downward-projecting keel; and (g) furcula.

composed of keratin, the material from which fingernails and claws are made. They consist of a hollow, central shaft that decreases in diameter toward the tip. Radiating from the shaft are barbs, processes of feather material that, when linked together along the length of the shaft by small hooks called barbules, form the sheet of feather material called the vane. Feathers with well developed, asymmetrical vanes are usually used for flight and are therefore called flight feathers. Feathers in which the barbules are not well developed and tend to be puffy with poorly developed vanes are called downy feathers or down and, as we know from sleeping bags, comforters, and ski parkas, are superb insulation. Birds that have lost the ability to fly, such as ostriches and emus, tend to have reduced numbers of barbules; their feathers tend to be more hair-like and serve exclusively as insulation.

Absence of teeth
No living bird has teeth. The jaws of birds are covered with a beak (rhamphotheca) made of keratin. One should not confuse the absence of teeth with jaw strength or dietary preference, but the absence of teeth, implies that no bird chews its food in the way that mammals do; processing of food is left to organs further down the gut.

Large brains and advanced sight
Mammal chauvinists that we are, we speak derogatorily of "bird brains." In fact living birds have well-developed brains, certainly by comparison to their living diapsid brethren (lizards, snakes, and crocodiles), and even by comparison to many mammals. Along with large brains, birds generally have well-developed vision. This, along with large brain size, is thought to have been associated with sophisticated motor activities, such as flight.

Carpometacarpus
The wrist and hand bones in the hand of modern birds are fused into a unique structure called the carpometacarpus (Figure 13.1b). Most of us are familiar with the carpometacarpus; it forms the flattened tips of the "buffalo wings" that we savor as appetizers. Everybody agrees that the carpometacarpus is composed of three fingers; the outstanding question, however, is "Which three?" Paleontologists argue on the basis of phylogeny that the three fingers are I (the thumb), II, and III. Embryologists, looking at evidence from the growth of modern birds, argue that the fingers are actually II, III, and IV. As this is one of the important questions of avian phylogeny, we'll revisit this subject later in the chapter.

Foot
The feet of all living birds are distinctive. There are three toes in front (digits II, III, and IV), and a smaller a toe (digit I) at the back. All birds have well-developed, strong claws on their feet. The three central metatarsals (foot bones, to which the toes attach; in this case II, III, and

IV)) are fused together, and with some of the ankle bones, to form a structure called a tarsometatarsus (Figure 13.1c).

Pygostyle

No living bird has a long tail. Instead, the bones are fused into a compact, vestigial structure called a pygostyle (*pygo* – rump; *stylus* – stake; Figure 13.1d). The flesh surrounding this tail remnant rejoices under the nickname of "Pope's (or Parson's) nose" at Thanksgiving time. At least two groups of birds do not possess a pygostyle, but neither do they possess a well-developed tail.

Pneumatic bones

Living birds have an extremely complex and sophisticated system of air sacs throughout their bodies. The bones are extremely light, with thin walls and a minimal series of splint-like buttresses bracing them internally. This type of bone structure is called pneumatic, reflecting that the bulk of the volume in bird bones is taken up by air spaces. The key features of pneumatic bones are pneumatic foramina – openings in the wall of the bone for the air sacs to enter the internal bone cavities (Figure 13.1e). Traditionally, pneumatic bones have been interpreted by ornithologists as an adaptation for flight: tremendous strength is achieved with minimal weight. Indeed, soaring birds such as albatrosses and vultures have pneumatic bones throughout the body, while diving birds, such as the loon (or diver), have much less pneumaticity in their bones. While there is no doubt that the lightness associated with pneumatic bones is important in flight, we shall see that hollow bones, at least, have a long and distinguished ancient history entirely unrelated to flight.

Rigid skeleton

Bird skeletons have undergone a series of bone reductions and fusions to produce a light, rigid platform to which the wings and the muscles that power them can attach.[1] The chest region of birds is a kind of semi-flexible basket, in which fused vertebrae in the back are connected to a well-developed breastbone, or sternum, by ribs with upper and lower segments. The pelvic region is fused together into a synsacrum, a single, locked structure consisting of many sacral vertebrae. The sternum itself is commonly quite large and, particularly in flapping flyers, has developed a broad, deep keel, or downward-protruding bony sheet, for the attachment of flight muscles (Figure 13.1f).

Rigidity is maximized in the shoulder region, where the powerful flight muscles attach. Pillar-like coracoid bones buttress against the front of the sternum, the scapulae (shoulder blades), and paired, fused

---

1 This works just as a racing bicycle. Racing bicycles are designed to maximize the power output of the rider. For this reason, they are light and the frames are stiff. Stiff frames maximize the power translated from the rider's legs to the back wheel. A frame that flexes absorbs energy that could otherwise be delivered to the rear wheel.

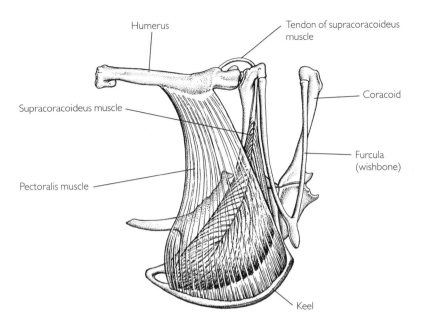

Figure 13.2. The two major muscles for flight: the pectoralis and the supracoracoideus. The pectoralis is the muscle used in the downward (power) stroke, while the supracoracoideus is used in the recovery stroke.

clavicles (furcula; Figure13.1g). At the dinner table we know the furcula as the "wishbone." No living organism except birds has a furcula.

Finally, modern flying birds have a unique arrangement of muscles to effect the wing recovery stroke. The downward stroke is obtained by the pectoralis muscle that attaches to the front of the coracoid and sternum, and to the furcula and humerus. The recovery stroke, however, is carried out by the supracoracoideus muscle. This muscle inserts at the keel of the sternum, runs up along the side of the coracoid bone, and inserts tendinously at the top of the humerus through a hole formed by the coracoid, furculae, and scapula (Figure 13.2). Modern birds can fly if that tendon is cut; however, as noted by ornithologist R. Raikow, take-off is greatly impaired. The hole through which the supracoracoideus tendon reaches the top of the humerus is an important, unique adaptation of modern birds to a particular type of flight stroke.

Archaeopteryx lithographica and the ancestry of living birds

We now turn to the fossil record, and to *Archaeopteryx lithographica*. *Archaeopteryx* (*archaeo* – old; *pteryx* – wing) was first known as a feather (Figure 13.3). Recall that names are applied to complete fossils if possible, but any part of a fossil organism can be named if the material that is found is sufficiently different from existing taxa to warrant a new name. So it was with *Archaeopteryx*. The name was first applied to a feather impression, found in 1860 in fine-grained deposits of carbonate mud located near the town of Solnhofen, Bavaria (southern Germany). These deposits were the remnants of a Late Jurassic, stagnant, poorly oxygenated coastal lagoon. The fine-grained, smooth, milk-colored carbonate

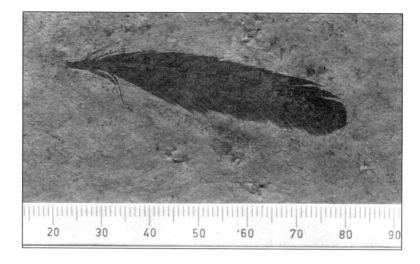

20   30   40   50   ·60   70   80   90

Figure 13.3. The first evidence for Jurassic-aged birds: the feather *Archaeopteryx lithographica*, described in 1861, from the Solnhofen quarry (scale in centimeters). (Photograph courtesy of J. H. Ostrom; images copyright Yale Peabody Museum.)

rock splits beautifully into thin slabs, which are then used for floor tiles (it is still possible to obtain Solnhofen tiles) or, before the advent of metal engraving, as printing plates (hence the species name *lithographica*). Over hundreds of years of quarrying, remarkably well-preserved, delicate fossils had been found at this locality; ancient crabs, lobsters, fish, insects, and even fragile pterosaurs. The low energy of the lagoon allowed things to settle undisturbed, and the lack of oxygen hindered microorganisms that would otherwise have destroyed the carcasses.

The excitement caused by the isolated feather was nothing compared to that generated a year later, when a specimen – feather impressions and bones – was quarried out of Solnhofen. The half-meter long fossil was chimeric, because it had bird feathers peacefully co-existing with obviously "reptilian" features, such as a tail and hands with claws. Nobody had every seen anything like it, and it caused a furor.[2]

This second specimen of *Archaeopteryx* quickly got into the entepreneurial hands of a local doctor, who made it his business (literally) to sell the thing to the highest bidder. So the first *Archaeopteryx* with bones was sold to an astute Sir Richard Owen (and the Trustees of the Natural History Museum (London); hence it is now called the London specimen), who made it his mission to obtain the fossil; it was ironic that, as late nineteeth century nationalism underwent the nascent rumblings that eventually led to World War I (and beyond), the prized fossil of Germany ended up in Britain.

---

2 Darwin had just published the *On the Origin of Species* in 1859, proposing that species evolved into other species. Here, a mere two years later, was discovered an apparent "missing link" that had both "reptilian" and avian features. Interestingly – and bafflingly – Darwin never referred to *Archaeopteryx* in the five subsequent editions of *On the Origin of Species*.

Figure 13.4. The beautifully preserved, complete Berlin specimen of *Archaeopteryx*. (a) Main slab preserving most of specimen; (b) counter-slab, preserving opposite side of specimen, primarily impressions. Note the exquisite feather impressions radiating out from the wings and tail. (Photograph courtesy of J. H. Ostrom; images copyright Yale Peabody Museum.)

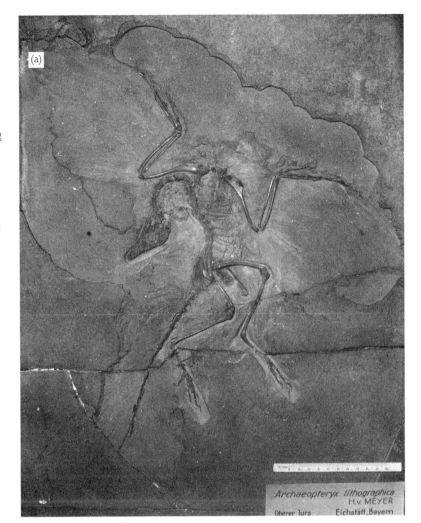

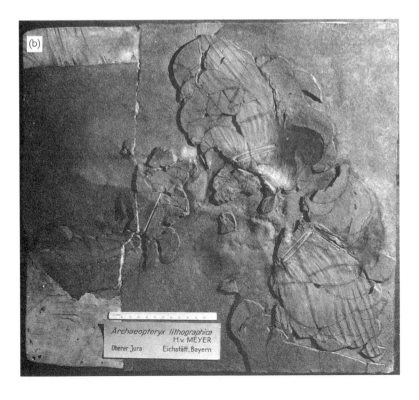

The next *Archaeopteryx* specimen, retrieved from Solnhofen in 1877, finished up (after no small amount of negotiating, bickering, and money) securely in German hands (this is called the Berlin specimen, which is where it presently resides)![3] This second specimen is a gorgeous thing; complete feathers spread out in natural position, and with a skull clearly showing teeth (Figure 13.4). Since then, five other specimens have been found, all from Solnhofen: one in 1951 (the Eichstatt specimen), one in 1956 (the Maxberg specimen), one that was collected in 1955, but not identified as *Archaeopteryx* until 1970 (the Teyler specimen), one in 1989 (the Solnhofen specimen), and yet another reported as recently as 1993.[4] Thus, to date, *Archaeopteryx* is known from seven specimens and one feather.

Anatomy of *Archaeopteryx*

Skull

The skull of *Archaeopteryx* (Figure 13.5a) is approximately triangular in shape, with nasal, antorbital, and eye openings present. The eye opening is rather large and round, and possesses a sclerotic ring, a series of plates that supported the eyeball. The temporal region is not well preserved, although hints of lower and upper temporal openings remain. *Archaeopteryx* has blade-like, unserrated, recurved teeth.

Arms and hands

The arms are quite long ($\geq$ 70% of the length of the legs), and by comparison with modern birds, relatively unspecialized. The hands on *Archaeopteryx* are about as large as the feet, and each hand bears three fully moveable, separate fingers. Each finger is tipped with a well-developed, recurved claw. The wrist of *Archaeopteryx* bears a semi-lunate carpal (see Chapter 12).

Legs and feet

The foot of *Archaeopteryx* has three toes in front, and a fourth toe apparently behind (Figure 13.5d).[5] The three in front are more or less symmetrical around digit III, and all the toes all have well-developed claws.

The ankle of *Archaeopteryx* is a modified mesotarsal joint (see Chapter 5). Although there has been some disagreement, it is now generally agreed that a small splint of bone rises up from the center of the astragalus, the ankle bone, to form a tall ascending process. The three foot bones (metatarsals II, III, and IV) are elongate, narrow, and

---

3 The thoroughly unwholesome bartering that distributed the various specimens of *Archaeopteryx* around Europe is recounted in A. Desmond's book, the *Hot-Blooded Dinosaurs*, as well as in A. Feduccia's book *The Age of Birds*.

4 This last specimen, described by P. Wellenhofer, was deemed on the basis of its long legs, sternum, and other features sufficiently different from all previous *A. lithographica* specimens to merit a new specific designation: *A. bavarica*.

5 It has generally been assumed that digit I of the foot of *Archaeopteryx* points back as in modern birds; for such is how it appears on several specimens. Indeed, analysis of the strong curvature of the rear claw suggested to University of North Carolina ornithologist A. Feduccia that *Archaeopteryx* was a tree-dwelling animal. Nonetheless, recent reanalysis by Brown University paleontologists K. Middleton and S. Gatesy suggests that, as in all other theropods, digit I of *Archaeopteryx* actually lies to the side of the foot. The apparent rearward orientation of this digit in the specimen is considered by Middleton and Gatesy to be an artifact of preservation.

pressed closely together, but clearly unfused. The legs are slightly longer than the arms, and the thigh bone has a gentle S-shape, with the ridge-like head turned at 90° to the shaft. The thighs are considerably shorter than the shins (tibia and fibula), although the fibula itself is attenuate; that is, it becomes smaller and smaller as it approaches the ankle.

Long bones
Archaeopteryx has thin-walled long bones with large hollow spaces within.

Trunk and tail
The axial skeleton of *Archaeopteryx* lacks many of the highly evolved features that characterize modern birds. The body is relatively long, and shows none of the foreshortening or fusion that one sees in the vertebrae of birds. The sternum is relatively small, with a small keel. A large, strong furcula is present (Figure 13.5f). Also present are gastralia, or belly ribs, which primitively line the belly in vertebrates (Figure 13.5c).

Archaeopteryx lacks a synsacrum and instead has a more typical (i.e., generalized) "reptilian" pelvis. In the Berlin specimen, the pubis seems to be directed rearward; however, the bone has apparently been broken where it connects with the ilium. In other specimens it points downward. The footplate on the pubis is well developed, although the front part is absent.

Interestingly (for a feathered creature), *Archaeopteryx* has a long, straight, well-developed tail. Projections from the neural arches (zygapophyses) are elongate, meaning that the tail is inflexible and has little potential for movement along its length.

Feathers
Archaeopteryx has well-preserved feather impressions.[6] The best-preserved feathers are clearly flight feathers (Figure 13.5b). Those on the arms are in number and arrangement very much like those seen in

---

6 The feathers of *Archaeopteryx* have caused no small amount of comment, much of it frustratingly incorrect. The most egregious example of this was a series of publications in the *British Journal of Photography* in the middle 1980s on the possiblity that the feathers of *Archaeopteryx* were forged. The authors were two distinguished astronomers, F. Hoyle and N. C. Wickramasinghe, and some collaborators. Observation by low-angle photographic techniques led them to the conclusion that the feather impressions were actually made by carving and by the addition of a gypsum paste. There were some obvious misidentifications (such as claiming the tail to be a single feather), and the charges were thoroughly refuted by a variety of scientists. The gist of one of the most important refutations goes as follows: *Archaeopteryx* specimens come from slabs of Solnhofen limestone that have been split, revealing the specimen. The fossil is found on the so-called "main slab," but impressions and bone fragments are also found on its opposite, the "counterslab." It would have to be one extraordinary forger to produce the precise, but opposite, feather pattern on the counterslab. P. Wellehofer, in describing the 1988 Solnhofen specimen, banged the final nails in the coffin by noting that the 1988 specimen also has feather impressions; it would have to be a spectacular forger, indeed, who could carry on a case of superhumanly skillful forging for 130 years!

Figure 13.5. A reconstruction of *Archaeopteryx*, surrounded by photographs taken from the actual specimens. (a) Skull, seen from right side, note teeth; (b) feather impressions showing vanes and shaft superbly preserved; (c) trunk region seen from left side, note gastralia; (d) foot (four-toed and clawed, with symmetry

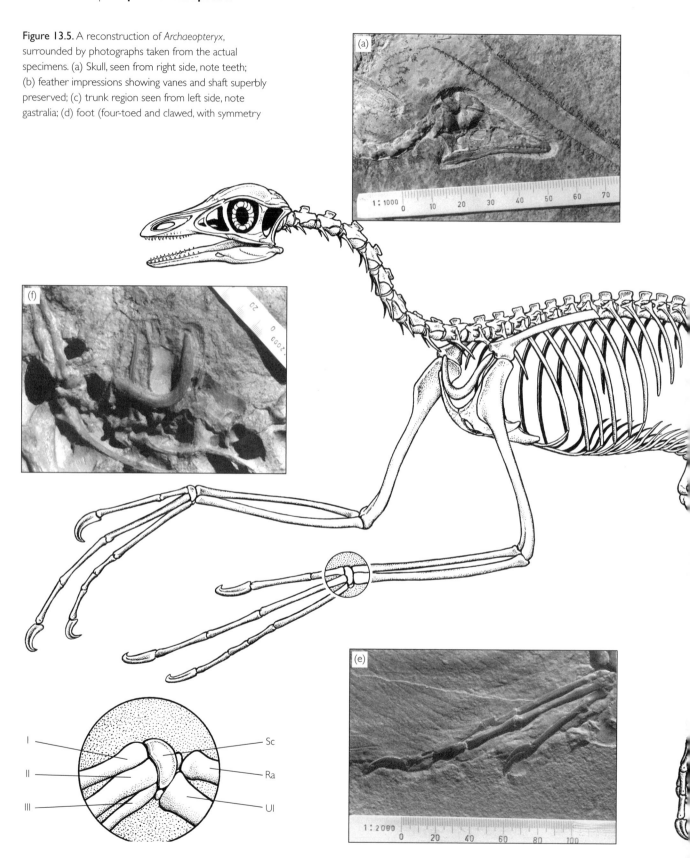

Figure 13.5. (*cont.*)
around digit III; digit I opposite digits II, III, and IV);
(e) right hand and wrist with clawed digits (in ascending
order, I, II, and III), *inset* drawing of left wrist, showing
semi-lunate carpel (Ra, radius; Ul, ulna; Sc, semi-lunate
carpel); (f) robust theropod furcula. (Photographs
courtesy of J. H. Ostrom; images copyright Yale Peabody
Museum.)

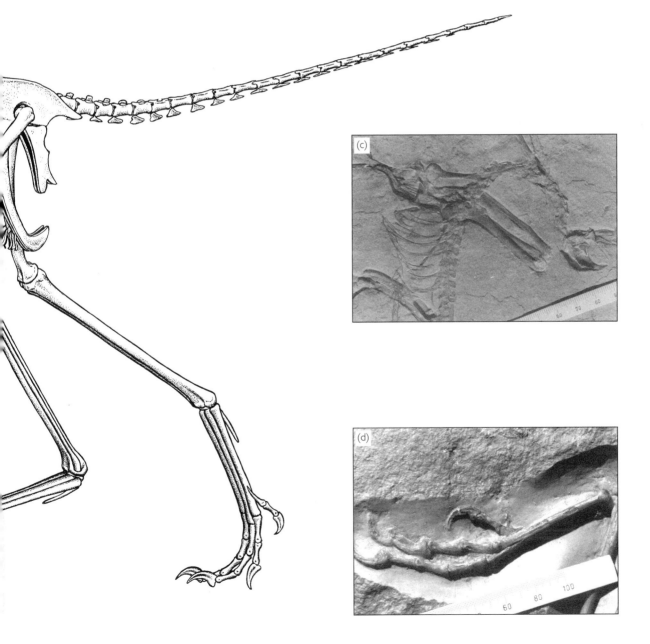

living birds, and as in living birds the vanes are asymmetrical. Moreover, as in modern birds, the asymmetrical vanes produce an airfoil cross-section in the wing. Unlike living birds, however, there are feathers also lining the well-developed tail. These radiate out from the vertebrae, and form an impressive tail plume.

The ancestry of living birds

With *Archaeopteryx* firmly in hand (and possibly in the bush as well; see the arboreal hypothesis below), the ancestry of Aves seemed to be a done deed. All birds have feathers; *Archaeopteryx* has feathers; *Archaeopteryx* must be a bird. The problem simply became one of finding out about the ancestry of *Archaeopteryx*. But as it turned out, things were not so simple as that. The ancestry of *Archaeopteryx* now appears to be pretty well understood, but as we shall see, the problem lies not with the identity (and ancestry) of *Archaeopteryx* but rather what we mean by "bird." The clade *Archaeopteryx* + living birds, known as Avialae, is well supported (Figure 13.6), but ultimately the fossil record will require us to entertain a concept of birds that is considerably expanded over one developed from just living representatives of the clade.

Living birds as dinosaurs

Before we construct the relationships of living birds to other dinosaurs, a subtle point must precede the analysis. We have already established that living birds are a monophyletic group, based upon the large suite of characters detailed at the beginning of this chapter. And, we know that there is a robust clade that consists of *Archaeopteryx* + living birds (Avialae; Figure 13.6). If this is true, then the primitive characters that *Archaeopteryx* bears must have something to do with the ancestry of Aves. Because Avialae is a monophyletic group, we do not need to go to the highly evolved living taxa to learn about their origins; we can go through the most primitive member of the group known – *Archaeopteryx* – to investigate the ancestry of living birds.

Higher taxonomic relationships of Archaeopteryx

We can now identify some of the higher taxa to which *Archaeopteryx* belongs. *Archaeopteryx* has an antorbital opening, and indeed, all living birds have an antorbital opening. Therefore, birds in general and *Archaeopteryx* in particular are archosaurs. In addition, a glance at your Thanksgiving turkey (or any living bird) will convince you that all living birds (as well as *Archaeopteryx*, but you won't see *that* on the table!) have a fully erect stance, in which the shaft of the femur is at 90° to the head. No living bird has a femur with a ball at the head (as in Mammalia); instead

Figure 13.6. Cladogram of Avialae (all living birds + *Archaeopteryx*). Avialae is a well-supported clade with a suite of diverse characters including very long arms, narrowing of the face and reduction of the size and number of the teeth, enlargement of the braincase, reduction of the fibula toward the ankle, and, significantly, the presence of feathers. As we shall see, although this last character had been valid since the discovery of *Archaeopteryx*, it has become clear in the last seven years that it diagnoses a group significantly more inclusive than Avialae.

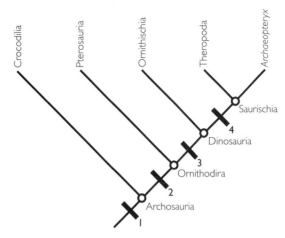

Figure 13.7. Cladogram depicting the position of *Archaeopteryx* within Archosauria. Diagnostic characters for each node include: at **1** antorbital opening (Archosauria); at **2** four-toed clawed foot, with symmetry around digit III; digit I reduced, lying closely appressed to and alongside of digit II (Ornithodira); at **3** semi-perforate acetabulum (Dinosauria); and at **4** ascending process on the astragalus (Saurischia). The cladogram shows that *Archaeopteryx*, and therefore birds, share the diagnostic characters of Dinosauria. Birds are dinosaurs. Within Dinosauria, the character at node **4** among others (see introductory text to Part III: Saurischia) clearly indicates that *Archaeopteryx*, while a bird, is also a saurischian dinosaur.

the head is elongate. Moreover, the ankle of *Archaeopteryx* (and all birds) is a modified mesotarsal joint.

Consider the foot of ornithodirans. Ornithodiran feet are four-toed. Three toes point forward (digits II, III, and IV), and the fourth (digit I) is reduced. This last-mentioned toe lies part way down the bone adjacent to it, along the side. Ornithodiran feet are symmetrical around digit III, and all toes bear large, arcuate claws (see Chapter 4). This condition is found in all living birds although in these, digit I is rotated to the back of the foot (rather than alongside). Birds are ornithodirans, albeit modified ones (Figure 13.7).

Within Ornithodira, Dinosauria is diagnosed by the possession of three or more vertebrae in the pelvis, a reorientation of the shoulder, reduction of the fourth finger in the hand, and a semi-perforate acetabulum. *Archaeopteryx* clearly has five pelvic (or sacral) vertebrae, the reorientation of the shoulder (from the primitive archosaurian condition), and a perforate acetabulum (see Chapter 5).

In summary, then, the characteristics that distinguish Dinosauria within Ornithodira clearly apply to *Archaeopteryx*. *Archaeopteryx*, because it bears the diagnostic characters of Dinosauria, is itself a dinosaur. This is not so revolutionary; look at the illustrations within this chapter and those in Chapter 12 and you will see that *Archaeopteryx* looks like a dinosaur. What is more revolutionary is the conclusion that *Archaeopteryx* is a dinosaur, and so therefore must be within Avialae *as well as* Dinosauria (Figure 13.7). That being the case, living birds must be a subset of Dinosauria, and both of them should be subsumed within an expanded Reptilia.

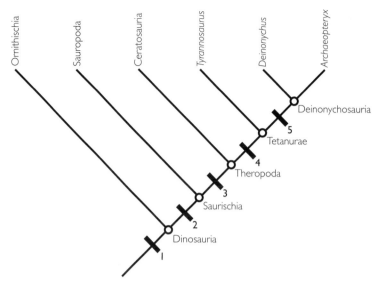

Figure 13.8. Cladogram depicting the position of *Archaeopteryx* within Dinosauria. Some of the characters defining each node are as follows: at **1** is the same as **3** in Figure 13.7; at **2** is the same as **4** in Figure 13.7. At node **3**, one obvious diagnostic character (of many) is the hollow bones possessed by all theropods. Node **4** is united by the shared elongation of the hemal and neural arches, and by possession of a furcula. Node **5** is indicated by an opisthopubic pubis.

Archaeopteryx as a theropod

Archaeopteryx likewise bears the diagnostic characters of theropods. As we have seen, since the late 1800s, it has been recognized that the vertebrae, leg, and arm bones in all theropods are thin walled, and hollow. This distinctive characteristic makes them easily recognizable in the field, even when very little fossil material is preserved. *Archaeopteryx* (and all modern birds, for that matter) have hollow bones; this is an inheritance from the original condition found in theropod leg and arm bones. Moreover, *Archaeopteryx* bears an enlarged three-fingered hand with the deep pits at the end of the metacarpals so diagnostic of Theropoda (Figure 13.8; see Chapter 12). If we conclude that *Archaeopteryx* is a theropod, it must also be concluded that the living birds, too, are theropods, albeit highly derived ones.

Archaeopteryx as a tetanuran

As we saw in Chapter 12, theropods are divided into two groups: Ceratosauria and Tetanurae. *Archaeopteryx* has tetanuran affinities, with elongate zygapophyses (leading to a stiffened tail), shortened tooth row, and astragalar groove. With a high ascending process on its astragalus, *Archaeopteryx* is clearly not at the base of Tetanurae (Figure 13.8).

Archaeopteryx as a coelurosaur

Tetanurans, as we have seen, come in a fantastic variety of shapes and sizes. Within Tetanurae, however, *Archaeopteryx* possesses diagnostic characters of the coelurosaur clade, placing it squarely within the group. Coelurosaurs, it will be remembered, all have the distinctive semi-lunate carpal in their wrist. This half-moon-shaped bone is unique in Theropoda. Also seen in *Archaeopteryx* are the shortened ischium (far shorter than the pubis) and large, circular orbits. *Archaeopteryx* is a coelurosaur.

Archaeopteryx as a maniraptoran

As befits their name, all maniraptoran coelurosaurs have a grasping, three-fingered hand that is a modification of the ancestral theropod condition. The distinctive maniraptoran addition to that hand is an elongation of the middle digit (II). *Archaeopteryx* has this feature. Maniraptorans have a fusion of the clavicles to form a furcula. The furcula of non-avian maniraptorans is a large, robust bone and very unlike the splinter-like wishbone with which we are familiar. *Archaeopteryx* has a thick, U-shaped furcula, rather like that found in *Oviraptor* and *Ingenia*. Other maniraptoran features found in *Archaeopteryx* include a highly flexed neck, elongate forelimbs, and the distinctive bowed ulna. *Archaeopteryx*, and thus birds, are clearly members of Maniraptora.

Archaeopteryx as a deinonychosaur

Within Maniraptora, an unnamed[7] monophyletic group consisting of deinonychosaurs and *Archaeopteryx* is united by a variety of features. The most obvious of these is to be found in the pelvis, which is opisthopubic and missing the front portion of the pubic footplate. This pubic condition in maniraptorans compares nicely with that in *Archaeopteryx*. Moreover, the tail vertebrae of this group all show an extensive elongation of the hemal and neural arches. This character, as we have seen, leads to a reduction of the movement of which the the tail is capable. *Archaeopteryx* also has lost the fourth trochanter on the femur (Figure 13.8).

What can we conclude from all this? That *Archaeopteryx* is also the theropod dinosaur. Within Theropoda, *Archaeopteryx* is diagnosable as a maniraptoran coelurosaur. Finally, within Maniraptora, *Archaeopteryx* is most closely allied with deinonychosaurs. This means that the closest relatives of *Archaeopteryx* (and thus, the closest relatives of birds) were highly predatory and active creatures, such as *Deinonychus*, *Troodon*, and *Velociraptor*. This leaves us with the conclusion that living birds must be a type of deinonychosaur.

Pneumatic bones and feathers: adaptation versus inheritance

The hypothesis that living birds are dinosaurs has some important implications with regard to the questions posed at the outset of this chapter. We can now answer the first question "Where do living birds come from?" by stating that they evolved from the common ancestor of deinonychosaurs and *Archaeopteryx*, an animal that bore the diagnostic features of that clade. This maniraptoran would have been a highly

---

7 Obviously, not every branching point on a cladogram needs a name. In this case, the lack of a name simply means relief from the already too complex nomenclature of dinosaur paleontology.

predaceous, gracile, active, biped of small to medium size, that extensively used its grasping, dextrous hands in a variety of prey-catching and manipulation functions.

Birds are commonly cited as supreme examples of adaptation. Indeed, pneumatic bones and feathers are singled out as marvelous adaptations to maintain lightness and permit flight. There is no doubt that pneumatic bones maintain avian lightness, and that feathers are a superb adaptation for flight. The question is, however, "Did pneumatic bones and feathers evolve for lightness and flight, respectively?" In both cases, the paleontological answer is "No" – although our "No" in the first case is qualified. As we have seen, hollow bones are a theropod character (recall that even the name Coelurosauria contains a reference to the hollow bones in these dinosaurs). However, true pneumatic bones (with pneumatic foramina) are not found in any non-avian dinosaur; moreover, no Cretaceous bird has pneumatic bones, yet many are believed to have been excellent flyers. Truly pneumatic bones developed within Cretaceous avialians. But in the case of feathers, the story is very different.

The revolution of the 1990s: feathered dinosaurs

It would be hard to underestimate the revolution in our understanding of the evolution of birds that has occurred *within* the last 10 or so years. As recently as the first edition of this book we hypothesized that, because the ancestors of living birds were small and highly active, feathers might easily have developed as insulation for a warm-blooded (i.e., endothermic; see Chapter 15) creature, rather than for flight. The insulatory properties of down feathers, at least, are well known. For all of these reasons, we and many others speculated that feathers originated not as a flight adaptation, but as an insulatory mechanism in a small, non-flying, cursorial, endothermic dinosaur. This idea becomes less far fetched when it is remembered that all living birds are endothermic; it simply became a matter of how far back endothermy occurred in the bird lineage.

This flew in the face of what ornithologists had heretofore reasonably assumed, namely that feathers evolved for flight. How could it be otherwise? Feathers, with their light weight and contoured form, are so ideally suited for flight.

But the paleontological idea could be tested. If feathers were actually an adaptation for high levels of cursorial activity, then fossils of non-flying theropods ought to be found with feathers. When the evidence finally came – and it came from both paleontology and embryology, it was spectacular. We consider here first the evidence from embryology and then the evidence from paleontology.

The origin of feathers

Embryology

In the early 1900s, G. Heilmann postulated, almost by default (because he had demonstrated that birds were more closely related to "reptiles" than to mammals), that feathers are an outgrowth of "reptilian" scales. Somehow the scales, he supposed, grew longer and divided into barbs and barbules. And for almost 100 years, Heilmann's ideas guided research, although in fact very little real progress was made in under-

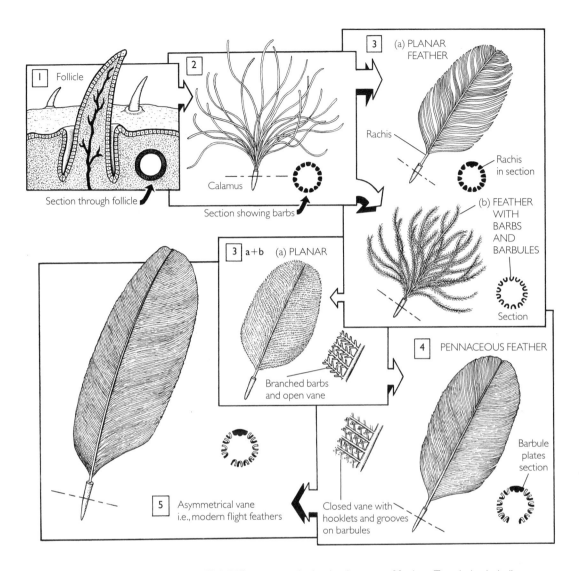

Figure 13.9. Different stages in the development of feathers. Type 1: simple, hollow, cylindrical, hair-like filaments. Type 2: tufts of multiple, elongate filaments, attached at one end; Type 3: filament tufts align in a single plane (Type 3a) while also developing barbs and barbules (Type 3b). Eventually, a new planar (vaned), barbed form evolves (Type 3a+b). Type 4: Vane becomes "closed"; that is, tiny hooks on the barbule attach to grooves on adjacent barbules, producing an integrated semi-rigid vane that does not allow much air to pass through. Finally Type 5: Vane becomes asymmetrical, the diagnostic feature of flight feathers. (Figure redrawn from Prum and Brush, 2003.)

standing how feathers evolved. Work in the past six years, however, suggests that the development of feathers occurs by the interaction of specialized follicles and a series of regulatory genes controlling the onset and termination of growth. Even more tantalizing for us, however, evolutionary biologists P. O. Prum and A. H. Brush recognize five stages of feather evolution, each stage represented in feather types on living birds as well as marking a developmental modification of the previous stage. These stages are: (1) formation of a hollow cylinder (the shaft); (2–3) loosely associated, unconnected, unhooked barbs (these stages correspond to down feathers); (4) hooked barbs on a symmetrical vane (corresponding to contour feathers, such as wrap around the body); and finally (5) hooked barbs on an asymmetrical vane (flight feathers; see Figure 13.9). These developmental modifications, they suggest, can be linked to phylogeny, in some striking ways (see below and Figure 13.13).

Fossils

Until the 1990s, the fossil record likewise had little to offer on the subject of feathers. *Archaeopteryx* bears feathers effectively indistinguishable from those of living birds, and thus it was assumed that considerable evolution had to have taken place prior to the Late Jurassic. A Triassic amniote, *Longisquama*, bears elongate feather-like structures that were once interpreted as potential feathers; although these enigmatic structures are not yet fully understood, they are no longer widely believed to have any relationship to the origin of feathers (Figure 13.10). In 1978, Russian paleontologist A. S. Rautian described a Jurassic "feather" that he interpreted as an early stage in the evolution of feathers. He called the fossil *Praeornis sharovi*. Other workers, examining the specimen, did not conclude that an early stage in feather evolution was represented by the fossil. The oldest unambiguous feather known is still the Late Jurassic *Archaeopteryx* feather specimen, discovered in 1860. This specimen, a flight feather, represents the most advanced stage in Prum and Brush's scheme.

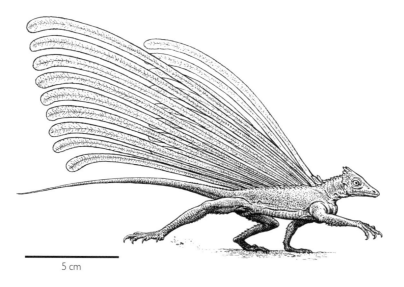

5 cm

Figure 13.10 A reconstruction of the Triassic tetrapod *Longisquama*, claimed by some to possess a row of feather-like projections along its back.

The Liaoning fossils   It seemed as if the fossil record ought to produce a feathered, non-flying theropod, particularly if our evolutionary scenario regarding the origin of feathers had any validity. For this reason, the hunt for a feathered, non-flying theropod was aggressively joined by a variety of workers interested in the origin of birds. Although not without a misstep or two along the way,[8] by the middle 1990s, spectacular feathered theropod dinosaurs from the Lower Cretaceous Yixian Formation, Liaoning Province, China, began to be recovered.

The rocks of Yixian look superfically like those of Solnhofen; thinly bedded (laminated) siltstones and shales interbedded with volcanic rocks. The volcanic units have proven to yield reliable isotopic dates of about 124 Ma, making the fossils late Early Cretaceous in age. The laminated rocks that preserve the fossils can be split, and a slab and counter-slab recovered for each fossil. Preservation is spectacular; the specimens are generally complete and completely articulated, and the fine mudstones in which they are preserved show not only the impressions of the integumentary coverings but also some darkened staining, possibly reflecting the preservation of some of the original organic matter that composed the integument and internal organs (Figure 13.11). With such superb preservation, there is little room for doubt about either the osteology (bone structure) of the organism or the type and distribution of its covering. As noted by Chinese paleontologist Ji Qiang and his co-authors:

> Because feathers are the only integumental covering in vertebrates that have a tufted or branched structure, the occurrence of (such structures in these theropods), coupled with (their) phylogenetic position near the base of birds, is strong evidence that these structures are feather homologues. The myriad findings of flightless dinosaurs from Liaoning with similar integumentary structures … provide imporant evidence that the origin of feathers is unrelated to the origin of flight [in birds].
> (Ji *et al.*, 2001)

First there was the 1997 discovery of *Sinosauropteryx*, a small coelurosaur with curious – and initially controversial – impressions of barb-like filaments. The animal had proportionally much shorter forelimbs than those found on avialans, and obviously had no pretensions to any type of flight. The filaments were eventually recognized as a feathers of Types 1–2; in short, they formed a very primitive downy coat insulating a clearly non-flying theropod. Next came the somewhat larger, toothless *Caudipteryx*, once thought to be a flightless bird but then clearly revealed to be an oviraptorisaurian. *Caudipteryx* bears Type 4 feathers – those with well-developed barbs, barbules, and symmetrical vanes. Even more startling was *Beipiasaurus*; a very large (ostrich-sized) therizinosauroid (see Theropoda I) also with no obvious ability to fly.

---

8 The legacy of a forged feathered dinosaur that once caused a considerable scientific "flap" is forthrightly related by investigative reporter Lewis Simons in the pages of the October, 2000, issue of *National Geographic* (pp. 128–132.)

Figure 13.11. Feathered non-flying dinosaurs from Liaoning Province, China. Clockwise from upper left-hand corner: (a) *Protarchaeopteryx*; (b) *Sinornithosaurus*; (c) *Caudipteryx*; (d) juvenile *Sinosauropteryx*; and center (e) an unnamed feathered dromaeosaur.

Beipiasaurus has relatively primitive feathers (Types 2–3) with barbules. Even this large dinosaur was insulated, suggesting to some that a feathered integument was present in most, if not all, theropods. A non-flying deinonychosaur was also found: *Sinornithosaurus*. This organism bears Type 5, barbed, barbuled, and hooked vaned feathers, comparable in every way to those of living birds. As of this writing, over a dozen specimens of non-flying theropods with feather coverings have been recovered from the Liaoning deposits.

And we may be sure that the riches of the Yixian Formation have not yet been completely plundered. In January, 2003, X. Xu and collaborators unleashed *Microraptor gui* on a very unsuspecting world. This was not the first specimen of the animal known: *Microraptor* is moderately well known as a diminuitive (crow-sized) deinonychosaur. But this specimen was covered with Type 5 (asymmetrically vaned) flight feathers, not only on its arms (wings) but – incredibly – on its legs. They fan out from the back legs of the specimen like a hand of cards (Figure 13.12). Flight feathers were heretofore unknown on the back leg of any organism, nor were they known in organisms other than avialans, and the authors interpreted this creature as a primitive *flying* deinonychosaur. Xu and his co-authors suggested that *Microraptor* used all four limbs in flight, much as a modern flying squirrel does. The ramifications of this interpretation for the origin of flight are discussed below.

With the discovery of the Liaoning fossils, the development of feathers can be linked with the fossil record. Prum and Brush note a tantalizing correlation between the tetanuran cladogram and the distribution of

Figure 13.12. *Microraptor gui*, with broad feathers on all four limbs. (Specimen courtesy of X. Xu.)

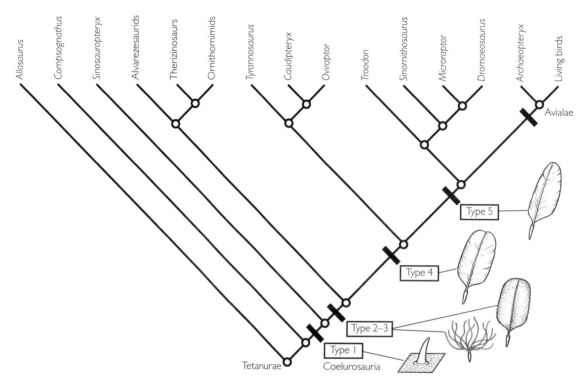

Figure 13.13. Cladogram with selected tetanurans (Theropoda) with the different stages of feather development superimposed (see text and Figure 13.9). If Prum and Brush are correct, primitive feather coats may have graced most derived tetanuran theropods. (From Prum and Brush, 2003; courtesy of Patricia J. Wynne.)

feather types in theropods (Figure 13.13). It appears that more basal tetanurans (e.g., coelurosaurs, therizinosauroids) bear more basal types of feathers, and more derived tetanurans (e.g., oviraptorosaurs and deinonychosaurs) bear more derived feathers. Thus, the development of feathers appears to track the development of tetanurans. This, in turn, provides real insights into what the origin of feathers was all about. The prediction that feathered non-flying dinosaurs would eventually be discovered was correct: it is likely that, long before flight was the evolutionary ambition (as it were) of any theropod, feathers provided the insulation that is a prerequisite for endothermy, allowing theropods to maintain high levels of activity for extended periods of time (see Chapter 15).

What, if anything, is a bird?

The discoveries from China require us to completely rethink our views on what is and is not a bird. Clearly, the old equation (bird = feathers) won't fly; there are now many examples of feathered, non-flying dinosaurs below Avialae on the cladogram. Likewise, the equation (warm-blooded = bird) also doesn't work; these feathered dinosaurs were surely warm-blooded (endothermic). Should Aves – traditionally, birds – be restricted to all those organisms bearing the distinctive suite of characters of living birds? That would, of course, exclude

Archaeopteryx, which certainly has a plausible claim on the designation "bird." As we shall see in Chapter 14, the fossil record also gives a good account of the evolutionary changes that take us from the skeleton of *Archaeopteryx* to that seen in living birds, and provides some guidance about what a bird is and which organisms the traditional bird taxon – Aves – should encompass.

Dissent

The concept of living birds as dinosaurs has historically provoked controversy because of the position of the pubis and, more recently and problematically, the identities of the fingers in the hand, as well as the time disparity between the Chinese finds and *Archaeopteryx* (see Box 13.1).

Pubis

As we have seen, the pubis points backward in both modern birds and ornithischian dinosaurs (a similarity reflected in the name "Ornithischia"). A number of workers, as early as the late 1800s, observed this shared character, but by the early part of the twentieth century, thanks to the exhaustive studies of G. Heilmann, the similarities between ornithischians and birds were recognized as convergent. This leaves the pubis pointing forward in saurischians and backward in birds. Although this might appear to be a stumbling block in the hypotheses presented here, several facts strongly support our analysis: (1) recall that the pubis pointing forward is a primitive character within Tetrapoda and it should not be used, therefore, to falsify the monophyly of dinosaurs and *Archaeopteryx*; (2) the pubis in deinonychosaurs does not point forward, but points downward and, as we have seen, this is now believed to be the position of the pubis in *Archaeopteryx*. This suggests that in the history of dinosaurs, the pubis rotated backward three times: once in ornithischians, once in therizinosauroids, and once in avialans. It is significant that, in modern bird embryos, the pubis begins directed forward (the primitive condition) and rotates backward as the embryo develops.

Fingers

Recall from your last brush with "buffalo wings" that the carpometacarpus of living birds is a fused structure composed of three fingers. Which three, however, is not so easy to determine, because in their fused state, they are hardly recognizable. Still, paleontology suggests an obvious answer: if Avialae is real, then the fingers must in fact be I, II, and III. This is what we would expect given the condition that the fingers of the deinonychosaur hand are well known and unequivocably I, II, and III. Indeed, even in *Archaeopteryx*, the hand is unfused and the fingers are clearly identifiable as (not unexpectedly) I, II, and III.

Suppose for a moment that you weren't convinced of the monophyly of dinosaurs and modern birds. You could attack this problem in a slightly different fashion. Because the fingers fuse into the carpometacarpus as the hand forms during development, you could study the hand of modern

BOX 13.1

Plus ça change …

The relationship of birds to dinosaurs as outlined here is not new. The famous early Darwinian advocate T. H. Huxley, as well as a variety of European natural scientists from the middle and late 1800s recognized the connection between the two groups. Indeed, one did not have to be a Darwinian to recognize the important shared similarities, and Huxley's opinions were widely accepted at the time. As noted in 1986 by J. A. Gauthier, Huxley outlined 35 characters that he considered to be "evidence of the affinity between dinosaurian reptiles and birds," of which 17 are still considered to be valid today.

So what happened? Why is it news that birds are dinosaurs? During the very early part of the twentieth century, Huxley's ideas fell into some disfavor, as it was proposed that many of the features shared between birds and dinosaurs were due to convergent evolution.

What evidence was there to argue for convergence in the case of dinosaurs and birds? Really, not very much. But in light of limited knowledge of dinosaurs at the time, the group simply seemed too specialized to have given rise to birds. Moreover, clavicles were not known from theropods (then, as now, the leading contender as the most likely dinosaurian ancestor of birds). Thus fused clavicles (furcula) in birds had to originate outside Dinosauria. What was needed was a more primitive group of archosaurs that did not seem to be as specialized as the dinosaurs.

In the early part of the twentieth century, such a group of archosaurs, the ill-defined "Thecodontia," was established by Danish anatomist G. Heilmann as the group from which all other archosaurs evolved. Since this was, by definition, the group that gave rise to all archosaurs and since birds are clearly archosaurs, it was concluded that birds must have come from "thecodonts." Heilmann had in mind an ancestor such as *Ornithosuchus* (note the name: *ornitho* – bird; *suchus* – crocodile), a 1.5 m carnivorous bipedal archosaur that, among living archosaurs, looks a bit like a long-legged crocodile. For over 50 years, Heilmann's detailed and well-argued analysis held sway over ideas about the origin of birds.

Several events caused the thecodont ancestry hypothesis to fall into general disfavor. The first was that clavicles were found in coelurosaurians among theropods. Moreover, it later came to be recognized that "Thecodontia" is not monophyletic; that is, it is defined by no unique, diagnostic characters pertaining to all its members and no others. How could one derive birds (or anything else) from a group that had no diagnostic characters?

The renaissance of the dinosaur–bird connection must be credited to J. H. Ostrom, of Yale University. In the early 1970s, through a series of painstakingly researched studies, he spectacularly documented the relationship between *Archaeopteryx* and dinosaurs, in particular coelurosaurian theropods. His ideas inspired R. T. Bakker and P. M. Galton, who in 1974 published a paper suggesting that birds should be included within a new class, Dinosauria. The idea didn't catch on, in part because it involved controversial assumptions about dinosaur physiology and because the anatomical arguments on which it was constructed were not completely convincing. In 1986, however, J. A. Gauthier applied cladistic methods to the origin of birds, and, with well over 100 characters, demonstrated that *Archaeopteryx* (and hence, birds) are indeed coelurosaurian dinosaurs.

birds as it develops into a carpometacarpus. In 2002, ornithologist A. Feduccia and embryologist J. Nowicki did just that on ostrich embryos, and reaffirmed conclusions of embryologists dating back to the 1870s that the fingers of the bird hand seem to be II, III, and IV. If this were the case developmentally, then the homology of the fingers of living birds departs from that of theropods, and avian origins must lie within some other group.

The "wrong" fingers in the hands of birds has been a serious problem for the dinosaur relationship to birds, so much so that the title of a 1997 report in the journal *Science* was "The forward march of the bird-dinosaurs halted?"[9] Just two years later, however, J. A. Gauthier and embryologist G. P. Wagner proposed a solution to this apparent conundrum. Embryologists, they noted, identify the sequence of "condensations"; that is, primordial buds of material that later develop into fingers. Hence, condensation I begets digit I, condensation II begets digit II, and so on. Gauthier and Wagner instead proposed that as development of the avian hand progresses, a "frameshift in the developmental identities" of the fingers takes place. According to them, the bud considered embryological condensation II becomes adult digit I, condensation III becomes digit II, and condensation IV becomes digit III. In this way, the digital homologies for living birds and for extinct theropods are rendered the same.

Wagner and Gauthier's suggestion was a relief for many paleontologists, who had been unable to resolve the discrepancies between embryology and paleontology. Nonetheless, for Feduccia and Nowicki, nothing had been resolved, since they did not believe that a frameshift is possible in the development of the avian hand, because events are too "constrained." They claimed that digit reduction occurs in the bird hand as it does in the hands and feet of many other vertebrates: symmetrically, from the outside digits inward.

While this leaves us with conflicting proposals, we note, as did Wagner and Gauthier, that the identities of the digits – even when disarticulated – are easy to determine, and those of *Archaeopteryx* (and other theropods) are obviously I, II, and III. Is it probable that those of living birds could really be II, III, and IV? It would make the number of convergences between living birds and dinosaurs in all their other characters improbably large. As we have seen, other neontologists (including embryologists) are far more comfortable with the dinosaur–living bird relationship, and have amassed other neontological evidence to sustain it.

Time

A few scientists – Feduccia, for one – note that *Archaeopteryx* is a Late Jurassic fossil; yet the fossils from China, which represent the morphological condition from which *Archaeopteryx* is supposed to be derived, are Early Cretaceous. Indeed, these researchers point out that no deinonychosaur is known from pre-Late Jurassic sedimentary rocks. Because the Chinese specimens post-date *Archaeopteryx*, they argue that the Chinese fossils cannot bear upon the origin of birds (which clearly preceded *Archaeopteryx*). This approach is essentially non-cladistic and, given that, it is not surprising that many scientists that hold this view still adhere to a modified form of the "thecodont" ancestry hypothesis. Because thinking in terms of "thecodonts" is not phylogenetic, it involves non-parsimonious character distributions and convergences, and it is ultimately untestable. Moreover, C. R. Brochu and M. Norell

9 Hinchliffe, R. 1997. The forward march of the bird-dinosaurs halted? *Science*, **278**, 596–597.

have demonstrated quantitatively that although the Chinese fossils occur in Lower Cretaceous deposits, their ancestry must considerably pre-date their appearance in the Liaoning rocks. Brochu and Norell found that the inferred time of ancestry of the Chinese feathered dinosaurs (not their actual appearance in the rocks) is concordant with the Late Jurassic age of the more highly evolved *Archaeopteryx* (see also Box 16.1 for a discussion of this type of approach). Cladistic methods – upon which our analysis is based – use chararcter distributions because these are testable; stratigraphic position, conversely, is not (see Chapter 3).

Other

Another tiny minority of specialists derives birds from crocodiles, basing the hypothesis of crocodile–bird affinities on aspects of the braincase. While a few of these characters may appear valid, the crocodile–bird relationship is generally rejected, again on grounds of parsimony. If the crocodile–bird hypothesis were true, then the remarkable number of characters shared between birds and coelurosaurians would have to be ascribed to convergent evolution, a prospect that is unlikely.

Flight

The origin of bird flight itself is shrouded in mystery. Feathers, whenever they first occurred, were clearly a prerequisite to flight in theropods at least (even if flight may not have been a prerequisite to feathers).

Two opposing endpoints on a spectrum exist as regards the origin of bird flight (Figure 13.14). The first is the so-called arboreal (or trees down) hypothesis: that bird flight originated by birds gliding down from trees (Figure 13.14a). In this hypothesis, gliding is a precursor to flapping (powered) flight; as birds became more and more skillful gliders, they extended their range and capability by developing powered flight. Perhaps flapping developed as a modification of the motions used in controlling flight paths. In this scenario, a quadrupedal ancestor is usually postulated, and a "Protoavis," or early-stage gliding quadruped, is also implicated. *Archaeopteryx* is viewed as a later stage in the process; according to at least one interpretation (that of Feduccia), it had a perching foot, and the unfused fingers of the hand were used for grasping tree limbs. It would have flown primarily as a glider from tree to tree, using flapping flight for short bursts, and only when necessary. The arboreal hypothesis has had many adherents, the longest-term proponent of whom is probably W. J. Bock of Columbia University.

Antithetical to the arboreal hypothesis is the cursorial (or ground up) hypothesis for the origin of flight (Figure 13.14b). The cursorial hypothesis states that bird flight originated by an ancestral bird running along the ground. In this scenario, perhaps as obstacles were avoided, the animal became briefly airborne. Flapping (powered) flight appeared early on, as the animal strove to overcome more fully the force of gravity. This idea obviously requires a highly cursorial ancestor, in which

Figure 13.14. The arboreal and cursorial hypotheses for the origin of bird flight. (a) The arboreal hypothesis, which suggests that bird flight evolved by gliding down from trees. (b) The cursorial hypothesis, which suggests that bird flight evolved by running along the ground until the animals became airborne.

feathers were already present. In this hypothesis, the legs, feet, and hands of *Archaeopteryx* are viewed as an inheritance from a cursorial maniraptoran ancestor. The cursorial hypothesis has been strongly advocated by J. H. Ostrom,[10] J. A. Gauthier, and K. Padian.

Which to choose?

Both the arboreal and cursorial hypotheses have much to recommend them. The arboreal hypothesis is intuitively appealing, for it requires little imagination to produce a flying creature from an arboreal-dwelling one. The advantages to be gained are obvious, and potentially difficult evolutionary challenges such as getting airborne are easy. On the other hand, the cursorial hypothesis is strongly supported because, based upon the detailed anatomy reviewed in this chapter, ultimately the ancestor of birds had to have been a cursorial creature.

A problem with the cursorial hypothesis is that it has so far proven nearly insurmountable to "design" a cursorial theropod that developed flight by running along the ground. Until very recently, functional morphologists – that is, scientists who study the function of particular

---

10 Ostrom originally suggested that flapping flight evolved as a prey-catching mechanism. He imagined a small, insectivorous theropod "harvesting" the air by using its wings to generate air currents toward its mouth as it ran. The idea was heavily criticized as creating a power stroke exactly opposite that necessary for flight.

anatomical structures – have been unable to satisfactorily model the running-to-flight transition in early birds: the speed and power necessary for flight are hard to achieve. For this reason, an arboreal stage intermediate in the development of flight has been attractive to many scientists.

A potential insight bearing on this has recently come from the work of functional morphologist K. Dial of the University of Montana. Dial's 15-year-old son observed partridges running up bales of hay. As they climbed the bales, they augmented their bipedal running by flapping. Investigation by Dial into the phenomenon showed that, as birds mature, they attempt ever-steeper slopes, augmenting their developing leg musculature and coordination with wing movements. In an adult bird, Dial likened the motion to the spoiler on a racer car: just as the spoiler keeps the driving wheels (rear) on the ground to maximize traction, so the flapping wings help the animals get a purchase on steep slopes, and even overhangs. It is not too far-fetched an idea, Dial argues, to extend this approach into the phylogeny of deinonychosaurs, and to argue that flapping flight evolved through just this process of enhancing running – a behavior immanent to deinonychosaurs – and climbing, potentially as a prerequisite to flight.

In a sense, both the arboreal and cursorial hypotheses got a shot in the arm from the discovery of the feathered specimen of *Microraptor*. *Microraptor* is a deinonychosaur, and thus ultimately cursorial. On the other hand, the feathered legs of *Microraptor* are very reminiscent of those of tree-dwelling gliders. Xu and colleagues imagined *Microraptor* as a tree-dweller that would have glided with limited powered-flight capability. Did *Microraptor* augment its cursoriality by wing flapping, in the manner suggested by Dial? In short, did it both run and glide? In this view, ultimately, the ancestor of living birds was cursorial; however proximally, that ancestor may have had arboreal specializations not generally attributed to theropods. After all these years *Microraptor* might indeed prove to be the hypothetical creature once called "Protoavis."[11]

Indications of cursorial ancestry are present in all living birds. One has but to observe the characteristic bipedality in living birds – including the thigh length-to-shin length ratios and the digitigrade stance – to recognize the hallmarks of a running ancestor. Indeed, it is not coincidence that bird limbs are little changed from the non-flying coelurosaurian condition. This, on the other hand, leaves us with a problem for *Microraptor*. As reconstructed by Xu and colleagues, the rear

11 The "Protoavis" model for *Microraptor* is not without its problems. "Protoavis" has historically been imagined as having been somewhat like a flying squirrel: a quadruped that developed a membrane between its splayed arms and legs, and glided from tree to tree. But all theropods (including *Microraptor*) are bipeds and are designed differently from mammals. For one, they do not develop the membranes connecting the front and back legs. For another, the head of the thigh bone (femur) in theropods is elongate (see Chapter 12) and not a ball such as that seen in mammals. Because of this, a theropod splaying its legs like a mammal is hard to imagine (e.g., think of a chicken in that posture). Yet the four-winged model of Xu and colleagues requires this of the rear legs of *Microraptor*.

legs of *Microraptor* would have been splayed outward as a second set of wings, such as one sees in gliding mammals. But the hip joint in dinosaurs (and living birds) is very different from that seen in mammals. Because the head of the mammalian thigh bone (femur) is a ball, the legs can rotate outward. On the other hand, the head of the dinosaurian femur is an elongate ridge (see Figure 4.15), which, we argued in Chapter 4, was part of the parasagittal, erect stance of all dinosaurs. The hip joint of dinosaurs does not generally allow that kind of rotation; indeed, imagine a chicken or turkey with its legs splayed outward. The image would not be comfortable: the animal's legs would have to be disarticulated to attain the splayed position.

Flight in Archaeopteryx

Flight in *Archaeopteryx* – once the sole insight that we had into the origin of avian flight and thus a hotly contested battleground between the two opposing camps – has become more or less relegated to a footnote by the insights afforded from the Chinese specimens and from work such as Dial's. Still, the question remains, how did well *Archaeopteryx* fly? In short, mediocre by the standards of living birds. As we have seen, it lacks many of the skeletal specializations of modern birds. Instead, the creature has a primitively elongate trunk, no synsacrum, gastralia, no carpometacarpus, weakly developed coracoids, a small sternum without much of a keel, none of the supracoracoideus adaptations of living birds, and no pneumaticity in its bones.

J. M. V. Rayner of Leeds University has attempted to analyze flight in *Archaeopteryx*. His conclusion is that *Archaeopteryx* could flap its wings, attaining moderately high speeds, but could not perform the kind of slow flight that a running take-off might require. For this reason, Rayner reasoned that *Archaeopteryx* had to have been primarily a tree-dweller. Feduccia's conclusion that the position of digit I in the foot and the curvature of the claw suggests perching adaptations, fit well with this viewpoint (see note 5, p. 308) Ultimately we are left with a possibly arboreal animal capable of powered flight, but not of the kind available to living birds.

Conclusions   We began this chapter by promising to answer some fundamental questions about birds. To the question, "Where do birds come from?" our answer is, of course, dinosaurs. With so much of this book behind you, however, you can now see that a better way to phrase this question would be "What are birds?" because what they are dictates their ancestry. The transition from primitive theropod dinosaurs through non-flying feathered deinonychosaurs, through *Archaeopteryx*, to modern birds is, despite a very imperfect fossil record, remarkably complete. In their mosaics of primitive theropod and advanced characters both the Chinese fossils and *Archaeopteryx* are just about as satisfactory intermediates as one could envision. Yet the very presence of the Chinese forms requires us to redefine what we think of as birds. The old idea that feathers are diagnostic of birds is outdated, unless you care to call a truckload of non-flying dromaeosaurs birds as well. Cladistically, feathers carry back at least to coelurosaurs, which means that, as Prum and Brush so trenchantly put it, "many of the most charismatic and culturally iconic dinosaurs, such as *Tyrannosaurus* and *Velociraptor*, are likely to have had feathered skin."

Our second question, "Where do feathers come from?" has begun to be answered by evolutionary biologists in both a mechanistic and an evolutionary way, and now a correlation appears to exist between how derived is the theropod and how derived are the feathers it bears. Moreover, the evidence from the fossil record is abundant and clear that the origin of feathers was initially an adaptation for an endothermic metabolic state and not for flight.

Our third question involved which fingers make up the bird hand. Here the paleontological and phylogenetic evidence calls strongly for I, II, and III. Yet, embryological evidence suggests II, III, and IV. Wagner and Gauthier's work, however, proposes frameshifts in the development of the digits, and thereby suggests a compromise between these two apparently unreconcilable positions.

Finally, we asked, "How did flight evolve?" Here, the answer is not fully laid out in the fossil record; however, it is clear now that the ultimate ancestors of birds were highly adapted for cursorial behavior. The feathered specimen of *Microraptor* could be some version of the long-sought "Protoavis" creature. But the remarkable thing about deinonychosaurs is that many of the adaptations necessary for flight were already present in that clade. For example, dromaeosaurids – even those that didn't fly – were animals that already had long arms and large hands, in nearly the proportions found in avialans. Deinonychosaurs didn't use those arms and hands for flight, but rather for manipulation of prey. Likewise, feathers were apparently developed for insulation; avialans and some deinonychosaurs simply co-opted the limb proportions and integument for their own purposes.

This reminds us of a fundamental property of evolution. Structures are not commonly invented wholesale in evolution. Evolution modifies existing structures. Here, the feathers and grasping arms of non-avialan deinonychosaurs were modified – remarkably little – to permit flight. It was a breathtaking evolutionary sleight-of-hand.

Important readings

Bock, W. J. 1986. The Arboreal origin of avian flight. In Padian, K. (ed.), *The Origin of Birds and the Evolution of Flight*. California Academy of Sciences Memoir no. 8, pp. 57–82.

Brochu, C. A. and Norell, M. A. 2000. Temporal congruence and the origin of birds. *Journal of Vertebrate Paleontology*, **20**, 197–200.

Chiappe, L. M. and Witmer, L. M. 2002. *Mesozoic Birds: Above the Heads of Dinosaurs*. University of California Press, Berkeley, 520pp.

Desmond, A. 1975. *Hot-Blooded Dinosaurs: A Revolution in Paleontology*. Blond and Briggs, London, 238pp.

Dial, K. 2003. Wing-assisted incline running and the evolution of flight. *Science*, **299**, 402–404.

Dingus, L. and Rowe, T. 1997. *The Mistaken Extinction*. W. H. Freeman and Company, New York, 332pp.

Feduccia, A. 1980. *The Age of Birds*. Harvard University Press, Cambridge, MA, 196pp.

Feduccia, A. and Nowicki, J. 2002. The hand of birds revealed by early ostrich embryos. *Naturwissenschaften*, **89**, 391–393.

Gauthier, J. A. 1986. Saurischian monophyly and the origin of birds. In Padian, K. (ed.), *The Origin of Birds and the Evolution of Flight*. California Academy of Sciences Memoir no. 8, pp. 1–56.

Hecht, M. K., Ostrom, J. H., Viohl, G. and Wellenhofer, P. (eds.) 1984. *The Beginnings of Birds*. Proceedings of the International *Archaeopteryx* Conference Eichstatt: Freunde des Jura-Museums Eichstatt, Willibaldsburg, 382pp.

Ji, Q., Currie, P. J., Ji, S. and Norell, M. A. 1998. Two feathered dinosaurs from northeastern China. *Nature*, **399**, 350–354.

Ji, Q., Norell, M. A., Gao, K.-Q., Ji, S.-A. and Dong, R. 2001. The distribution of integumentary structures in a feathered dinosaur. *Nature*, **410**, 1084–1088.

Martin, L. D. 1983. The origin and early radiation of birds. In Brush, A. H. and Clark, G. A. Jr (eds.), *Perspectives in Ornithology*. Cambridge University Press, New York, pp. 291–353.

Norell, M., Chiappe, L. and Clark, J. M. 1993. New limb on the avian family tree. *Natural History*, **9**(93), 38–43.

Olson, S. L. 1985. The fossil record of birds. *Avian Biology*, **8**, 79–239.

Ostrom, J. H. 1974. *Archaeopteryx* and the origin of flight. *Quarterly Review of Biology*, **49**, 27–47.

Ostrom, J. H. 1976. *Archaeopteryx* and the origin of birds. *Biological Journal of the Linnean Society*, **8**, 91–182.

Ostrom, J. H., Gall, L. F. and Gauthier, J. (eds.) 2002. *New Perspectives on the Origin and Early Evolution of Birds*. Proceedings of the International Symposium in Honor of John H. Ostrom, February 13–14, 1998. Yale University Peabody Museum, Yale University Press, New Haven, CT, 613pp.

Padian, K. 2004. Basal Aviale. In Weishampel, D. B., Dodson, P. and Osmólska, H. (eds.), *The Dinosauria*, 2nd edn. University of California Press, Berkeley, pp. 210–231.

Paul, G. 2002. *Dinosaurs of the Air: The Evolution and Loss of Flight in Dinosaurs and Birds*. Johns Hopkins University Press, Baltimore, 436pp.

Prum, R. O. and Brush, A. H. 2003. Which came first, the feather or the bird? *Scientific American*, **288**, 84–93.

Rayner, J. M. V. 1988. The evolution of vertebrate flight. *Biological Journal of the Linnean Society*, **34**, 269–287.

Schultze, H.-P. and Trueb, L. (eds.) 1991. Birds. In Schultze, H.-P. and Trueb, L. (eds.), *Origins of the Higher Groups of Tetrapods – Controversy and Consensus*. Cornell University Press, Ithaca, NY, pp. 427–576.

Sereno, P. and Rao, C. 1992. Early evolution of avian flight and perching: new evidence from the Lower Cretaceous of China. *Science*, **255**, 845–848.

Shipman, P. 1998. *Taking Wing*. Simon and Schuster, New York, 336pp.

Wagner, G. P. and Gauthier, J. A. 1999. 1,2,3 = 2,3,4: A solution to the problem of the homology of the digits in the avian hand. *Proceedings of the National Academy of Sciences USA*, **96**, 5111–5116.

Xu, X., Tang Z. and Wang X. 1999. A therizinosaurid dinosaur with integumentary structures from China. *Nature*, **399**, 350–354.

Xu, X., Wang X. and Wu X. 2000. A dromaeosaurid dinosaur with a filamentous integument from the Yixian Formation of China. *Nature*, **401**, 262–266.

Xu, X, Zhou, Z., Wang, X., Kuang, X., Zhang, F. and Du, X. 2003. Four-winged dinosaur from China. *Nature*, **421**, 335–340.

Zhou, Z.-H., Wang X.-L., Zhang F.-C. and Xu, X. 2000. Important features of *Caudipteryx* – evidence from two nearly complete new specimens. *Vertebrata Palasiatica*, **38**, 241–254.

CHAPTER 14

Theropoda III: The early evolution of birds

A*rchaeopteryx* is obviously not, in a taxonomic sense (or any other sense, for that matter), a living bird. It has many features that are far from the condition found in living birds, including teeth, an unfused hand, a tail, no synsacrum, and gastralia. How and when did the changes take place that distinguish living birds from *Archaeopteryx*?

For many years, it appeared as if the fossil record was not going to be very informative when it came to birds, largely because their bones are fragile and preservation required extraordinary conditions – such as the quiet, oxygen-poor lagoon that once was Solnhofen. Still, the past decade has revealed much that was previously unknown, allowing us to reshape our understanding of Mesozoic avialan evolution.

The bird is a word

We saw in Chapter 13 that our view of what is or is not a "bird" has been enriched (and complicated) by new discoveries, largely from China. But because we now know that all feathered creatures are not really birds, we are left with some uncertainty about what organisms should be called birds and where to put the venerable taxon Aves.

Carolus Linnaeus (see Chapter 3) established Aves in 1758 to include birds – but in this case, living birds; the thought of extinct birds never crossed his mind. Since then the word "bird" has always been synonymous with members of Aves.[1] However, with feathered theropods that are obviously neither Aves nor birds now appearing, it is necessary to unlink the term Aves from "birds." We'll call anything avialan a bird for it is a fact that if you saw an avialan perched in a tree or scooting down a garden path, you'd surely think it was a bird. As for Aves, we choose to retain Linnaeus' original definition (if not his meaning, for he was not thinking in evolutionary terms) and make Aves synonymous with living birds. And, we'll continue working within Avialae, the clade that includes *Archaeopteryx*, Aves, and everything in between.

The Mesozoic avialary Within Avialae, the next step up the cladogram from *Archaeopteryx* is *Rahonavis* ("menace-from-the-clouds bird") from the Late Cretaceous of Madagascar described by C. Forster and collaborators in 1998. It was slightly larger than *Archaeopteryx* (the size of a crow; Figure 14.1) and possessed an enlarged sickle claw on its feet (exactly like the condition in dromaeosaurids and troodontids), and a long, *Archaeopteryx*-like tail. Although removed from *Archaeopteryx* by 25 million years, it nevertheless was clearly more like living birds than *Archaeopteryx* – it had

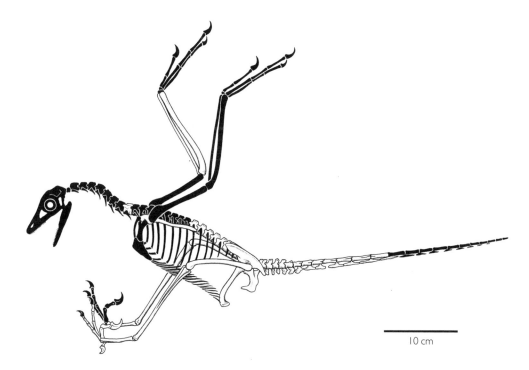

10 cm

Figure 14.1. *Rahonavis*, from the Late Cretaceous of Madagascar.

1 Hence, there are English words relating to birds such as "avian" and "aviary."

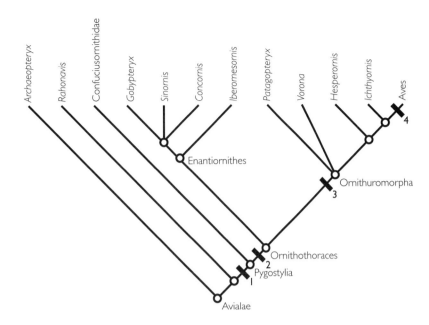

Figure 14.2. Cladogram of Mesozoic birds, depicting some of the steps in early avialan evolution. Derived characters include: at **1** pygostyle; at **2** reduction in number of trunk vertebrae, flexible furcula, strut-like coracoid, alula, carpometacarpus, fully folding wings; at **3** further reduction in the number of trunk vertebrae, loss of gastralia, final rotation of the pubis to lie parallel with the ilium and ischium; at **4** loss of teeth.

pneumatic foramina leading into internal cavities (pleurocoels) in its thoracic vertebrae (implying a metabolism closer to that found in living birds; see Chapter 15), fusion of six sacral vertebrae (one more than *Archaeopteryx* and more basal theropods) into a synasacrum, apparent quill knobs on its long forelimbs suggesting that the animal had feathered wings and could fly, and a fibula that doesn't reach to the ankle. Birds, as we now know them, are beginning to be assembled (Figure 14.2).

Flight proficiency seems to have been a driving force in avialan evolution. Subsequent events appear to have included shortening of the tail to form a pygostyle, an increase to seven vertebrae in the synsacrum, and structural reinforcing of the shoulder by elongation of the coracoid bones. Consisting of Confuciusornithidae, a group containing *Confuciusornis* (named for the Chinese philosopher Confucius; Figure 14.3) and *Changchengornis* (Chang Cheng is the Chinese name for the Great Wall) from the Early Cretaceous of China, and all remaining birds, this overarching group is quite naturally known as Pygostylia.

Thereafter, birds altered the anatomy of their antorbital cavity, reduced the number of trunk vertebrae, altered the shoulder joint, began fusing the digits of the hand into a carpometacarpus, and evolved an alula (the so-called "bastard-wing," a set of three or four quill-like feathers placed on the small, highly modified first digit of the hand, providing modern birds with low-speed flight capabilities and maneuver-

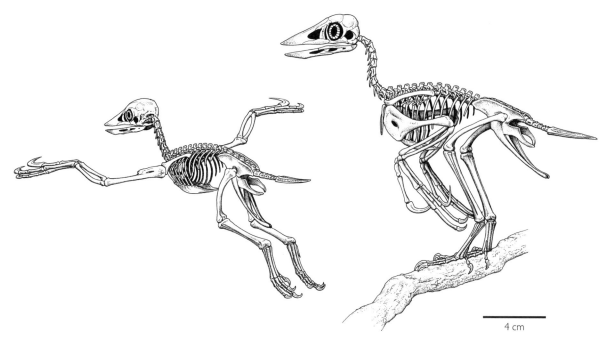

4 cm

Figure 14.3. *Confuciusornis*, from the Early Cretaceous of China.

ability). These transformations are found in Ornithothoraces ("bird chest"), consisting on the one hand of the most diverse clade of Mesozoic birds (which went extinct before the close of the Era) called Enantiornithes, and on the other Ornithuromorpha, the lineage leading to Aves (see Figure 14.2).

Enantiornithes

First recognized by paleornithologist Cyril Walker at the Natural History Museum in London in 1981, Enantiornithes ("opposite bird") is known from an abundance of beautiful and often complete material. All sparrow-sized, these small, arboreal birds (Figure 14.4) had well-developed flight capabilities, a robust, keeled sternum, a relatively slim furcula similar to that found in modern birds, a synsacrum composed of eight or more co-ossified vertebrae, a strut-like coracoid, powerful, well-developed arms (including humerus, radius, and ulna), and a partially fused carpometacarpus. The wrist joint is also rather more like that of modern birds than the deinonychosaur-like semi-lunate carpal found in *Archaeopteryx*. Modern birds have a modified wrist joint to allow the wing to fold tightly against the body; Enantiornithes had this, but more primitive birds did not. Enantiornithes also has adaptations indicative of perching: the first toe is positioned opposite to the others. The perching foot is yet another clue that these Mesozoic birds lived in trees, reinforcing other evidence that flight was an integral part of their life habits.

Still, one would hardly call Enantiornithes modern birds – they had gastralia across their belly (a carry-over from the primitive archosaurian

Figure 14.4. *Iberomesornis,* a member of Enantiornithes from the Early Cretaceous of Spain.

3 cm

condition), relatively numerous back vertebrae (a number intermediate between the 13–14 found in *Archaeopteryx* and the 4–6 found in living birds), an unfused tarsometatarsus, and an unfused pelvis.

Enantiornithes apparently had a worldwide distribution. *Nanantius* (*nan* – dwarf) is from Australia, *Iberomesornis* ("Iberian intermediate bird") hails from Spain, and *Sinornis* ("Chinese bird") comes from China. Others include *Kizylkumavis* (named for the Kizylkum Desert, where it was found) and *Sazavis* ("clay bird") from Uzbekistan (Asia), *Alexornis* (named for American paleornithologist Alexander Wetmore) from Mexico, *Enantiornis* itself and *Avisaurus* ("bird lizard") from Argentina and the USA. Most are rather fragmentary; however, distinctive aspects of the unfused tarsometatarsus allow them to be recognized as enantiornithiform birds.

Ornithuromorpha

Returning to the line leading to Aves, Ornithomorpha ("in the form of a bird") is represented by *Vorona* ("bird" in Malagasy) from the Late Cretaceous of Madagascar, *Patagoptyeryx* ("Patagonian wing") from the Late Cretaceous of Argentina, and all remaining birds, the clade known as Ornithurae ("bird tail"). These birds have modified the articulations between vertebrae, as well as the scapula, and hand, including fusions of the metacarpus.

Ornithurae is one of the most robust of all avialan clades, united by at least 15 unambiguous features, including reorientation of the pubis to lie more or less parallel to the ilium and ischium, further reduction of the number of trunk vertebrae, decrease in the size of the acetabulum, and a patellar groove on the femur. Who were these ornithurans? Not surprisingly, they include not only the closest relatives to living birds (Hesperorniformes and Ichthyorniformes), but also the crown of avialan evolution – Aves – as well.

Hesperornithiform birds were a monophyletic clade of large, long-necked, flightless diving birds that used their feet to propel themselves, much like modern loons (divers) (Figure 14.5). They had highly reduced arms and developed powerful hindlimbs for propulsion (as opposed to swimming by "flapping" of the arms, such as that seen in modern penguins). The hindlimbs were oriented to the side of the creature, and could not be brought under the body. For this reason, locomotion on land was clumsy at best; a kind of seal-like waddling is implied. In the water, on the other hand, hesperornithiform birds were clearly well adapted. The long, flexible neck would have been useful in catching fish, a behavior indicated from fossilized feces preserved with their skeletons. In many respects, the group is quite close to modern birds, in that all members have the shortened, fused trunk, a carpometacarpus, pygostyle, a completely fused tarsometatarsus, and a synsacrum composed of at least 10 vertebrae. Moreover, no member of this group bears gastralia. All differ from modern birds, however, in that they retain teeth in their jaws. Like modern diving birds, some of the pneumaticity in the bones has been lost. Presumably because of the loss

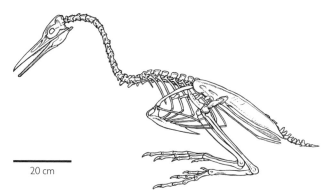

Figure 14.5. *Hesperornis*, the diving bird from the Late Cretaceous of the USA.

of the flight adaptation, the furculum, coracoid, and forelimb bones are highly reduced. All of these adaptions indicate that this group had a long evolutionary history prior to the Late Cretaceous. But who knows exactly what that history was, or when it took place?

Hesperornithiforms have been known for some time. *Enaliornis* ("sea bird") from the Early Cretaceous of England has been known since the late 1800s from well-preserved skull material and from an ankle with a shin bone attached (tibiotarsus). *Hesperornis* (*hesper* – western bird) and its smaller relative *Baptornis* (*bapt* – dipped or submerged) were described by O. C. Marsh in the late 1800s. Within the last decade, L. D. Martin has extensively studied hesperornithiform birds, and described *Parahesperornis* (*para* – near). These last-mentioned hesperornithiform birds are from North America.

Closer related yet to Aves is Ichthyornithiformes. Unlike hesperornithiforms, ichthyornithiforms were long-necked, toothed, gull-like birds, probably with equivalent flying skills (Figure 14.6). *Ichthyornis* (*ichthy* – fish bird), from the Late Cretaceous of North America, had a massive keeled sternum and an extremely large process toward the head of the humerus (deltoid crest) that was probably an adaptation for powerful flight musculature. In other respects, it shared many of the adaptations of modern birds, including a shortened, fused trunk, a carpometacarpus, a pygostyle, a completely fused tarsometatarsus, and a synsacrum formed of 10 or more fused vertebrae.

The Mesozoic fossil record of Aves – all from the Late Cretaceous – is very fragmentary and scattered, although equally tantalizing. These remains have been assigned to a variety of clades within Aves. Aves itself is a well-supported clade, involving as many as 11 characters of the skull, pelvis, and ankle.[2] The group includes screamers and waterfowl (Anseriformes), loons (Gaviiformes), and possibly shorebirds such as sandpipers, gulls, and auks (Charadriiformes), landfowl (Galliformes), wing-propelled divers such as modern petrels (Procellariiformes), and parrots (Psittaciformes). Clearly, these early records of modern birds speak,

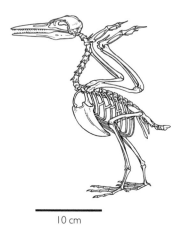

10 cm

Figure 14.6. *Ichthyornis*, a gull-like bird from the Late Cretaceous of the USA.

---

2 These characters, and indeed those of all the basal clades of modern birds, are discussed by ornithologist J. Cracraft and paleoornithologist J. Clarke (Cracraft and Clarke, 2001).

however incompletely, to the origin, intial radiation, and establishment of Aves by the closing moments of the Cretaceous.

Enigmata

One small evolutionary radiation, known as Alvarezesauridae, from the Late Cretaceous, has been difficult to place in the phylogenetic scheme we've outlined here. This is due largely to the very dramatic and unusual specializations of their skeletons. Take *Mononykus* (*mono* – one; *onyx* – claw; Figure 14.7) for example: this Late Cretaceous form, described as a running, flightless bird in 1993 by A. Perle and collaborators, apparently lived in a desert. The fossil came from a Late Cretaceous erg (sand sea). *Mononykus* is a strange beast; from the center of its back, toward the tail, it looks like a typical non-avian theropod dinosaur; strong, elongate, well-developed hindlimbs, and a long straight tail. The arms, however, are clearly those of a bird, albeit it an aberrant one; the digits are fused into a short, stout carpometacarpus, and the arms are stout and short, with a large process (the olecranon process) for developing power at the elbow joint. Perhaps it used its shortened, yet strong arms for burrowing to obtain food. Among its avian characters is a keeled sternum. Unfortunately, no feathers are preserved with the specimen. Subsequent years saw the discovery of other alvarezesaurids – *Parvicursor* ("small runner") and *Shuvuuia* ("bird" in Mongolian) from the Gobi Desert, and *Alvarezesaurus* (named for Argentinean historian Don Gregorio Alvarez) and *Avisaurus* ("bird lizard") from Argentina and the USA, but, despite this increase in our knowledge of the members of the group, the position of this clade within Theropoda is still hotly debated.

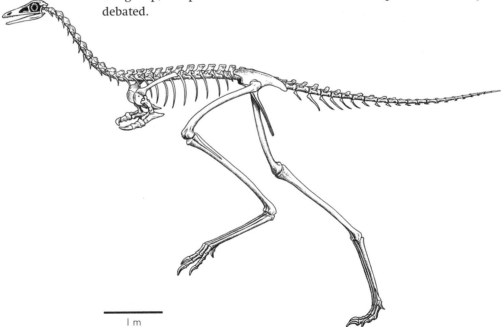

I m

Figure 14.7. *Mononykus*, an alvarezesaurid from the Late Cretaceous of Mongolia.

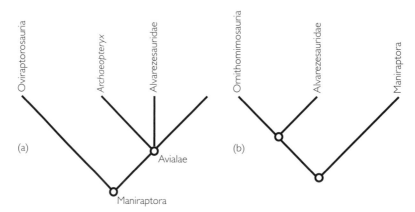

Figure 14.8. Two interpretations of the position of Alvarezesauridae. (a) Alvarezesaurids as birds; (b) alvarezsaurids as relatives of Ornithomimosauria.

The unique mosaic of primitive and advanced characters of *Mononykus* presented its discoverers with a conundrum. With all its primitive characters, it appeared to be far back in the evolution of birds; indeed, further back than Pygostylia. On the other hand, in this position it was bearing what appeared to be a well-developed carpometacarpus, a relatively advanced character that does not appear until Ornithurae.

The early studies of *Mononykus* placed it closer to modern birds than *Archaeopteryx* (Figure 14.8a), but this conclusion was severely criticized, mostly from a non-phylogenetic perspective. If the caropmetacarpus had developed twice independently, the argument went, then perhaps flight occurred twice in birds: once in the ancestor of *Mononykus*, and once in some ancestor of Avialae. That was unlikely, and so the conclusion was that *Mononykus* must be an aberrant non-avian theropod of some kind. Since then, cladistic analyses of all the more recently discovered taxa have provided two decidedly different interpretations of the affinities of alvarezsaurids. P. C. Sereno's phylogenetic work placed alvarezsaurids in a more basal, non-avian position, as the sister-group to ornithomimosaurs (Figure 14.8b). However, the latest word, by L. M. Chiappe, is that they fall out either directly above or below *Archaeopteryx* within the context of avian origins. Considering these two views, there is no doubt that the affinity of Alvarezesauridae will play a critical role in our understanding of theropod relationships, avian or otherwise.

Getting to be a modern bird

Each of these steps in the evolution of birds from *Archaeopteryx* to the likes of modern vultures, sparrows, ostriches, and hummingbirds shows a sequence of the evolution of derived from primitive characters. Using *Archaeopteryx* as a representative of the primitive condition and Aves as representative of the derived condition, we can begin to establish the sequence though which birds acquired a modern aspect (see Figure 14.2). Among the primitive characters that are retained in the earliest evolution of birds are teeth, an unfused hand, an unfused pelvis, and gastralia. Derived characters include a pygostyle, a decrease in the number of trunk vertebrae, a synsacrum that progressively increases in number of vertebrae (from less than 7 to 15 or more in living birds) a well-developed bony sternum, and pillar-like coracoids (implying a supracoracoideus musculature like that found in modern birds). These features make up the bulk of the adaptations necessary for flight at a level equivalent to that found in modern birds. Later to evolve was the carpometacarpus, which would be *de rigeur* in birds by Late Cretaceous time. Moreover, it implies that, although the bird ancestor may have been cursorial, the habit of arboreal perching was indubitably established by the Early Cretaceous. The fact that virtually all the necessary ingredients for modern bird flight were present by just 15 million years after *Archaeopteryx* lived (i.e., by the Early Cretaceous) has sparked some observers to speculate that either the rate of evolution in birds was greatly accelerated during the Late Jurassic–Early Cretaceous interval, or that *Archaeopteryx* itself was kind of atavistic creature, a throwback that, although it lived in the Late Jurassic, was rather primitive even for its time.

In fact, both viewpoints may be true. Because the oldest bird known is *Archaeopteryx*, only finds of equivalent age or older can resolve without question whether or not *Archaeopteryx* was an anachronism. On the other hand, the presence of a specialized diving bird, *Enaliornis* (a hesperornithiform from England), suggests that even as far back as the Early Cretaceous, birds were capable of exploiting a variety of environments (in this case, arboreal and marine environments), and had evolved into some truly specialized forms.

For all of its limitations, the Mesozoic record of birds still provides us with insights into the transition from the condition in *Archaeopteryx* (and even below this nexus, including dromaeosaurids and troodontids; see Chapter 13), through the remarkable avian discoveries recently made in China, to that found in modern birds. Figure 14.2 demonstrates the character sequence: **1** development of the perching adaptation in the foot, in combination with limited flapping flight capabilities (the *Archaeopteryx* condition); **2** development of advanced flapping flight adaptations, including a pygostyle, a fully folding wing, strut-like coracoids and a flexible furcula; **3** fusion of the hand into a carpometacarpus and the foot into a tarsometatarsus, and a loss of gastralia; and **4** loss of teeth (modern birds).

A number of important questions still remain with regard to the origin and diversification of modern birds. Most significantly, the modern group of flightless birds such as emus and ostriches (Ratitae) appears to retain in the cranium primitive features whose origin may be found in Mesozoic birds. Unfortunately, the homologies between the rare skull bones of Mesozoic birds and those of modern birds have proven both complex and controversial; this is a story that must wait for another day.

Important readings

Chiappe, L. M. 1995. The first 85 million years of avian evolution. *Nature*, **378**, 349–355.

Chiappe, L. M. and Dyke, G. J. 2002. The Mesozoic radiation of birds. *Annual Review of Ecology and Systematics*, **33**, 91–124.

Chiappe, L. M. and Witmer, L. M. (eds.) 2002. *Mesozoic Birds: Above the Heads of Dinosaurs*. University of California Press, Berkeley, 520 pp.

Cracraft, J. and Clarke, J. 2001. Phylogenetic relationships among basal birds. In Gauthier, J. and Gall, L. F. (eds.), *New Perspectives on the Origin and Early Evolution of Birds*. Special Publication of Yale Peabody Museum, pp. 143–156.

Gauthier, J. A. and Gall, L. F. (eds.) *New Perspectives on the Origin and Early Evolution of Birds*. Peabody Museum of Natural History, Yale University, New Haven, CT, 613pp.

Padian, K. 2004. Basal Avialae. In Weishampel, D. B., Dodson, P. and Osmólska, H. (eds.), *The Dinosauria*, 2nd edn. University of California Press, Berkeley, pp. 210–231.

PART IV

Endothermy, environments, and extinction

CHAPTER 15

Dinosaur thermoregulation: some like it hot

How shall we imagine dinosaurs? As cold-blooded and cast-iron, the very epitome of obsolescence (much as they were seen during the first two-thirds of the twentieth century) or as vibrant – even hot-blooded[1] – beasts capable of behaving like birds and mammals (see Box 15.1)? Our understanding of dinosaurs really hinges on this question, and some of the greatest intellectual battles associated with dinosaur paleontology have been fought over it.

An enigma explored

By Early to Middle Triassic, therapsids, the clade of synapsids to which mammals and their ancestors belong, gave every appearance of being poised to dominate terrestrial vertebrate faunas throughout the rest of the Mesozoic. But that didn't happen for 160 million years. Instead, dinosaurs, which appeared at approximately the same time as the earliest mammals (Late Triassic), radiated and dominated terrestrial vertebrate ecosystems (see Figure 5.8). Indeed, it was only after the extinction of the dinosaurs that mammals radiated and assumed the importance among vertebrate faunas that they presently occupy.

Dinosaur paleontologist R. T. Bakker, looking at this transition, asked: "Why did this happen?" "Why were therapids overtaken by

---

1 To use Adrian Desmond's (1975) gushing characterization.

BOX 15.1

Warm-bloodedness: to have and have not

Although "warm-bloodedness" is characteristic of birds and mammals, it is by no means restricted to these groups. Physiologists have known for some time of plants (!) that can regulate heat in a variety of ways, the most common being to decouple the metabolic cycle (see Box 15.2) so that energy, instead of being stored or used in growth, is released as heat. Several snakes are known to generate heat while brooding eggs, although this is accomplished by muscle exertion. Certain sharks and tunas can retain heat from their core muscles by counter-current circulation, and a variety of insects, including moths, beetles, dragonflies and bees, are known to regulate their body temperatures. "Warm-bloodedness" is not characteristic of these groups of organisms. Simply, it is known that some of them do maintain temperatures warmer than those of the medium (air or water) in which they are living. Indeed, it has been estimated that "warm-bloodedness" has evolved independently at least 13 different times.

This, of course, differs from the idea that "warm-bloodedness" is diagnostic of a particular group. Indeed, "warm-bloodedness" is characteristic of but two groups: birds and mammals.

---

dinosaurs and shunted off to the sidelines?" His answer, expressed in an influential 1975 article, was, at the time, truly startling:

> One is forced to conclude that dinosaurs were competitively superior to mammals as large land vertebrates. And that would be baffling if dinosaurs were "cold-blooded." Perhaps they were not.
> (Bakker, 1975.)

Implicit in his conclusion that dinosaurs were warm-blooded were the ideas that (1) dinosaurs directly competed with mammals (and/or their therapsid ancestors) and (2) being "warm-blooded" is better than being "cold-blooded," particularly in the case of dinosaur-sized vertebrates. The first assumption – that the earliest dinosaurs out-competed Middle and Late Triassic therapsids (including mammals) is, as we have seen from Chapter 5, problematical. Mammals and dinosaurs probably never really directly competed.

The second assumption inherent in Bakker's statement is that warm-bloodedness in large land vertebrates is clearly superior to cold-bloodedness. If it were otherwise, the success of the dinosaurs would not be so "baffling." A look at the global distribution and sizes of cold-blooded tetrapods makes a case for the advantages of warm blood. Virtually all cold-blooded tetrapods live within 45° north or south of the equator; most of the diversity of cold-blooded tetrapods is concentrated within 20° of the equator. Temperature can play a central role in directly controlling behavior: until a lizard can, by basking or other warming strategies, bring its core temperature to that necessary for optimum physiological functioning (see "Temperature regulation among vertebrates," below), it cannot function up to speed.

Along with having restricted global distributions, modern cold-blooded tetrapods tend to be relatively small creatures. At least in part,

Figure 15.1. Surface area and volume. As size increases, surface area increases as a square function whereas volume increases as a cubic function. In the case of these fishbowls, the radius of the larger bowl is twice that of the small bowl, whereas the volume of the large bowl is three times that of the smaller bowl. This has important consequences for the fish living in each bowl.

this derives from the relationship between the *surface area* of the body relative to its *volume*. Elementary arithmetic tells us that surface area is a square function and volume is a cubic function. For organisms, this has important implications: as an organism grows bigger and bigger, its volume becomes proportionally greater than its surface area (Figure 15.1). Thus, to maximize its surface-to-volume ratio, the organism must be very small. "Cold-blooded" animals, however, depend upon their surface area to elevate their body temperatures: they bask in the sun, or lie on warm substrates to obtain heat. The larger a "cold-blooded" creature is, the harder it is for the surface areas to warm the proportionally much vaster interiors. The result is like a baked Alaska dessert: the outside gets burned, but the inside remains frozen.

So "cold-bloodedness" confirms some limitations on the bearer. On the other hand, it is a fact that most living vertebrates are "cold-blooded." If the vast majority of vertebrates were inferior (as implied by Bakker's statement), why are they so numerous (that is to say, successful)? Why aren't warm-blooded creatures the most numerous?

Is warm-bloodedness superior to cold-bloodedness?

Temperature regulation among vertebrates

R. Silverberg, writing in *Harpers* in 1981, termed "warm-blooded" and "cold-blooded" "convenient ideational shorthand." In fact, many "cold-blooded" animals develop "warm" blood when they are operating at peak efficiency. Two terms that describe far more meaningfully vertebrate temperature regulation (and should be used instead of "warm-blooded" and "cold-blooded") are ectotherm (*ecto* – outside; *therm* – heat) and endotherm (*endo* – inside).

All vertebrates regulate their temperature; it's just a matter of how. Ectotherms do so using external sources of heat, while endotherms regulate temperature internally. We can add a second dimension to our thinking by noting that in some organisms, called poikilotherms (*poikilo* – changing), temperature fluctuates, but in others, called homeotherms (*homeo* – same), the temperature remains constant. Humans are endothermic homeotherms: when we are unable to maintain our body temperature, we get sick. Most lizards – ectothermic poikilotherms – are best transported from place to place by keeping them cool in food coolers: the animals simply slow down. A lizard or a snake is functionally homeothermic: it functions optimally at constant, relatively elevated temperatures. The difference between it and an endotherm is that it can survive in a cooler environment by simply allowing its metabolism to slow down, in a way that an endotherm cannot. Ectotherms can tolerate core decreases in core temperature, while endotherms must internally regulate their core temperatures.

On the other hand, perhaps more foreign to us are endothermic poikilotherms such as bats and hummingbirds, which have different levels of activity associated with different core (body) temperatures. For example, a bat goes into a trance-like state called torpor, in which the temperature of the animal is considerably lower than when the bat is flying. The same kind of thing occurs during hibernation in bears, in which the core temperature is somewhat lowered relative to the active state. The hyrax, a small mammal, sunbathes to raise its core temperature.

It is generally thought that the terms ectotherm and endotherm represent two biochemically and biophysical different methods of obtaining heat. The terms poikilotherm and homeotherm, however, are endpoints in a spectrum that runs from maintaining a constant temperature to having a fluctuating temperature. While many animals do cluster at the familiar metabolic endpoints, some do not.

Endothermic and ectothermic metabolism

Temperature control is not the only issue in endothermic and ectothermic tetrapods. Indeed, it is the very nature of the metabolism itself – that is, the sum of the chemical reactions in the cells of the organism – and the effect that the differences in endothermic and ectothermic metabolisms have on activity that are significant (Box 15.2).

Metabolism in endotherms and ectotherms

With regard to the onset of anaerobic metabolism, endotherms and ectotherms are very different. At body temperatures equal to that of an endotherm, ectotherms can increase their oxygen consumption between 5- and 10-fold. It's very much like the throttle on an engine: when the throttle is opened, the speed of the engine increases, its power output increases, and its fuel consumption increases. Ectotherms increase the amount of oxygen they consume 5–10 times above their warmed resting rate, before lactic acid begins to accumulate in their muscles (causing the familiar ache of tired muscles) .

An endotherm, on the other hand, is also capable of increasing its oxygen consumption 5–10 times above its resting rate. Its resting rate,

BOX 15.2

A primer on metabolism

In all organisms, energy is obtained from the breakdown of a molecule called ATP (adenosine triphosphate) to ADP (adenosine diphosphate) in the following way:

$$ATP \rightarrow ADP + energy$$

The goal, therefore, is to store energy in the form of ATP, so that when energy is required – for example, for muscle movement – it can be released as a high-energy phosphate bond if ATP is broken down to produce ADP.

The storage of energy in the form of ATP occurs through cellular respiration, the breakdown of carbohydrates through a series of oxidizing reactions. Carbohydrates are a family of molecules, including sugar, whose chemical bonds, if broken, release energy. Oxidation means that oxygen has bonded with whatever is being oxidized. In respiration, the chemical bonds in the carbohydrates are broken as oxidation takes place, and energy is released as heat, as well as stored in ATP. Cellular respiration takes place as a result of a series of rather complex reaction pathways, each of which involves the oxidation of the carbohydrate and a number of reaction intermediates. These pathways produce several molecules of ATP. In fact, the oxidative breakdown of a single molecule of glucose (a simple carbohydrate) can produce 32 new molecules of ATP. The system, however, is not nearly 100% efficient: ATP production captures about 40% of the energy of the bonds of the carbohydrates. The remaining 60% is released as heat. This is shown diagrammatically in Figure B15.2.1.

Organisms respire oxygen to drive these complex oxidation reactions, so that energy may be stored in the form of ATP. As the energy output of the organism is increased, the amount of ATP needed is increased, and hence more oxygen is consumed and more heat is produced. This is why breathing and heart rates increase when we exercise: we are using more energy, requiring more ATP to be generated, and thus we need more

oxygen. Likewise, we get hot when we exercise because of the heat released as ATP is produced.

There is a point, however, at which the organism reaches its maximum rate of oxygen consumption. When this point is exceeded, lactic acid forms within the cells as a product of the oxidative breakdown of the carbohydrates. The presence of lactic acid in too great a concentration is detrimental to the organism, and thus when the respiration becomes anaerobic (without oxygen; that is, the need for oxygen exceeds the ability of the cells to respire it), the organism is fast approaching its maximum energy output.

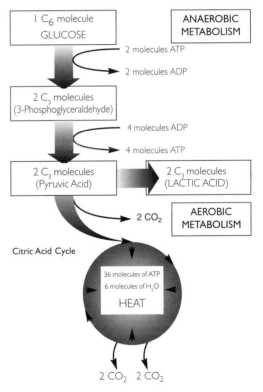

Figure B15.2.1. Respiration consists of the breakdown of 6-carbon (C_6) molecules of glucose into 3-carbon (C_3) products. The energy from the destruction of the glucose is stored in the form of the molecule ATP for eventual use by muscles.

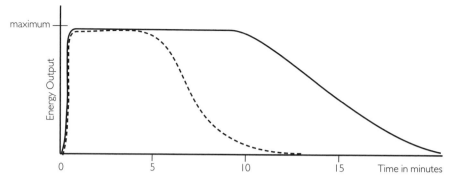

Figure 15.2. Energy output versus time in ectotherms and endotherms. The curves show that the muscles of both endotherms and ectotherms achieve their maximum energy output virtually instantaneously. In general, however, an endotherm sustains maximum energy output more than twice as long as an ectotherm.

however, is approximately where an ectotherm's point of lactic acid formation occurs. Thus the oxygen consumption *and* energy output of an endotherm is considerably greater than that of an ectotherm.

This has major consequences for activity. The *amount* of energy required to move equivalently sized endotherms and ectotherms is similar; that is, it takes about as much energy for a 1 kg lizard to move as for a 1 kg mammal. However, an endotherm can produce far more energy than can an ectotherm before it (the endotherm) begins anaerobic metabolism (lactic acid production). This means that generally, *endothermic tetrapods are capable of higher levels of activity sustained over longer periods of time than are ectothermic tetrapods* (Figure 15.2).

Ectothermic tetrapods have compensated for this difference in a variety of significant ways. Most importantly, many have developed ways to sustain short bursts of anaerobic respiration. Such short bursts allow for about 2–5 minutes of relatively high activity, and allow ectotherms a range of active behaviors, including guarding territory and young, hunting using speed and endurance, and active defense. Indeed, there is reason to believe that the anaerobic production of adenosine triphosphate (ATP) in ectotherms leads to greater initial power output than is possible in endotherms. The differences between endotherms and ectotherms, therefore, are even greater than simply the source of heat necessary to optimize the metabolic reactions carried on in the organism. These metabolic differences translate into some very real differences in physical capabilities. One final factor, however, must be figured into our understanding of these organisms: cost in energy.

By "cost," we are simply asking the question "How much does it cost to maintain an endotherm as opposed to an ectotherm?" Here the answer is clear: endotherms are much more costly in terms of energy use than ectotherms. Stated another way, it has been estimated that it costs 10–30 times as much energy to maintain an endothermic metabolism as to maintain an ectothermic metabolism. Much of the energy is expended on maintaining a constant body temperature. An analogy can again be made between endotherms and an engine with no idling speed. The machine is always revving at high speeds (because it has no idle setting, a lower adjustment) and using up large quantities of fuel (energy).

In a world of finite resources (ours), this means that a limited source of energy can support far fewer endotherms than ectotherms. Suppose, for example, that an organism had available to it all the energy resources on a hectare of land. Because each endotherm requires more energy than each ectotherm, that hectare of land – all other things being equal – is capable of supporting fewer endotherms than ectotherms. Thus, if the number of organisms is any measure of success, ectotherms will be more numerous than endotherms, because ectotherms are more economical: they simply do not cost as much to maintain.

We return, therefore, to the question we asked earlier: "If endothermy is superior to ectothermy, why are ectotherms still so numerous and diverse?" A possible answer to that question is to propose that endothermy is not superior to ectothermy (or vice versa), but rather that endothermy and ectothermy are simply two different metabolic strategies for optimizing use of resources. One depends upon costly sustained levels of activity, while the other substitutes for sustained activity short bursts of activity and greater economy.

Dinosaur endothermy: the evidence

Understanding the metabolism of ancient vertebrates requires circumstantial evidence drawn from a wide variety of sources. Unfortunately, no indicator is unambiguous, and, as in the courtroom, circumstantial evidence may not be enough to convict. Still, a consensus about the physiology of dinosaurs is slowly developing among paleontologists. Dinosaurs apparently were neither endotherms in the mammalian sense nor ectotherms in the crocodilian sense. They were something else, and possibly different strategies were adoped by different dinosaurs. Here, we review the evidence about fossil metabolisms into "Anatomy," "Histology," "Ecology," "Zoogeography," "Phylogeny," and "Geochemistry," following the categories used by J. H. Ostrom, a pioneer in the field of dinosaur metabolism.

Anatomy

Stance

As we observed in Chapter 4, all dinosaurs are thought to have maintained a fully erect stance. Dinosaur trackways reinforce this, showing that dinosaurs moved with a fully erect stance in which the feet were placed – like our own – directly beneath the body (Figure 15.3).[2] Among living vertebrates, a fully erect stance occurs only in birds and mammals, both of which are endothermic. The fully erect limb position in dinosaurs is therefore suggestive of endothermy.

Is there a causal relationship between stance and endothermy? The original idea was that the fine neuromuscular control necessary to maintain a fully erect stance would be possible only within the temperature-controlled environment afforded by an endothermic metabolism.

---

2 In some cases, for example birds and other theropods, the feet are actually placed toward a midline between the two legs.

Figure 15.3. A theropod trackway from the Middle Jurassic Entrada Formation, Utah, USA. Note how closely the right and left print are placed, suggesting a fully erect stance. When spectacular trackways like this are found, it is not hard to imagine the ghostly image of the trackmaker leaving a row of footprints in the soft mud.

Work by D. R. Carrier (see also Box 4.4) suggests another relationship between endothermy and stance. Carrier observed that when an animal with sprawling stance moves, the trunk of the organism flexes from side to side as the animal walks. Such flexion reduces the amount of air that can fill the lungs on the scrunched side (Figure 15.4). In the case of tetrapods with sprawling or semi-erect stances, Carrier noted an irony: because the lungs are alternately being compressed during locomotion (through side-to-side flexion of the trunk), when the animal needs the most air, it gets the least. Carrier suggested, therefore, that the evolution of a fully erect stance was actually a means by which lung volume could be maximized during high-speed locomotion.

Carrier's hypothesis suggests a causal relationship between posture and endothermy. Enhanced stamina in locomotion – such as that associated with endotherms – would require rates of respiration only allowable with a fully erect posture. Considered in this way, a fully erect posture could be a prerequisite for an endothermic metabolism.

Two other simple correlations have been noted between anatomy and endothermy. The first, that long-leggedness is characteristic of living endotherms while living ectotherms possess relatively stubby limbs, was noted by Bakker. Certainly dinosaurs possess rather long limbs which, as we shall see below, have important ramifications for stride length. The

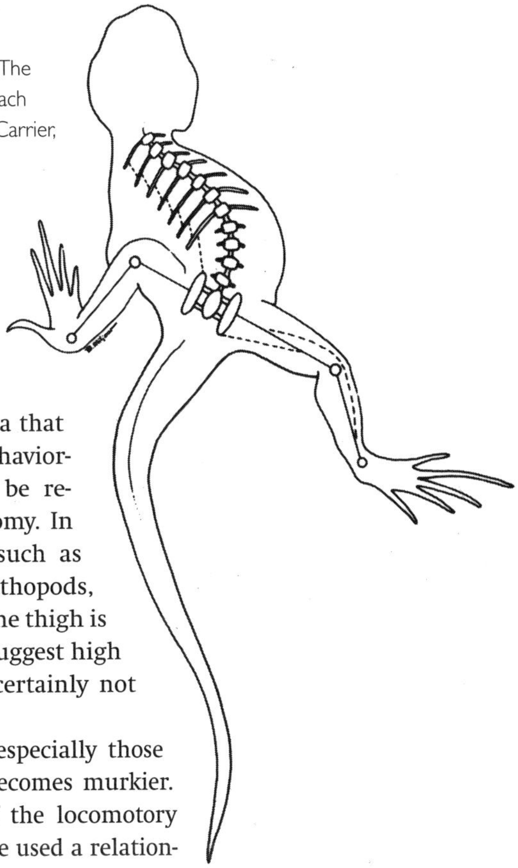

Figure 15.4. A sprawling vertebrate running quickly. The trunk alternately compresses the lung capacity on each side as the animal runs. (Drawing courtesy of D. R. Carrier, *Paleobiology*, **13**, 327.)

second correlation between anatomy and endothermy is the observation that among living tetrapods, the only bipeds are endotherms.

Limb anatomy and inferred activity levels

A number of scientists have had the idea that metabolism could be reflected in such behaviorial differences, which in turn might be reflected in aspects of the skeletal anatomy. In a variety of small bipedal dinosaurs such as dromaeosaurids and the smaller ornithopods, thin, elongate (gracile) bones (in which the thigh is short relative to the length of the calf) suggest high levels of running activity – behaviors certainly not characteristic of modern ectotherms.

And what of the larger dinosaurs, especially those that were not bipedal? Here the issue becomes murkier. In 1987, Bakker published estimates of the locomotory speeds of dinosaurs. To calculate them, he used a relationship pioneered by British functional morphologist R. McN. Alexander. Alexander in 1976 determined that the walking speeds of all tetrapods can be calculated from a combination of footprint spacing (stride length) and the length of the hindlimb (see Box 15.3). He reasoned that, if the relationship applies to living tetrapods, it could also apply to dinosaurs, where the hindlimb length and stride length (from trackways) were known. He concluded that those dinosaurs studied moved relatively slowly, at least while making tracks. Bakker continued Alexander's work, using studies of modern animals and applying these to a variety of fossil tetrapods, many of which were dinosaurs. Bakker imagined a 5- or 10-tonne *Triceratops* achieving a thundering "gallop" (Figure 15.5).

The idea of galloping dinosaurs, however exciting, is rather unlikely. The primitive gait in amniotes involves a hindlimb–forelimb mixture (i.e., right hindlimb → right forelimb → left hindlimb → left forelimb). Galloping, on the other hand, is a distinctive means of locomotion that involves an arcuate bound from the hindlimbs to the forelimbs. It is today unique to Mammalia. As mammals became larger, the backbone stiffened considerably (presumably to support the increased weight), but the bounding motion was retained in the form of the gallop, which still involved the hindlimb-to-forelimb motion. Dinosaurs, as we have seen, have no history of a small bounding quadrupedal ancestor; they evolved from bipedal forms. Thus, for large quadrupedal varieties to evolve a true

BOX 15.3

In the tracks of the dinosaurs

Trackways, the most tangible record of locomotor behavior, provide one aspect of an animal's walking and running capabilities, and the only independent test of anatomical reconstructions. Footprints arranged into alternating left–right–left–right patterns demonstrate that all dinosaurs walked with a fully erect posture. But how can trackways can also give us an indication of locomotor speed?

We begin with stride length; that is, the distance from the planting of a foot on the ground to its being planted again. When animals walk slowly, they take short strides, and when animals are walking quickly or running they take considerably longer strides. This much is intuitive for anyone trying to catch a bus about to pull away from the curb. Now, consider the situation when you are being chased by something smaller than you. The creature chasing you must take long strides for its size, and more of them too, just to keep up. So there is clearly a size effect during walking and running, and these will probably be different for the different kinds of animal under consideration.

How, then, to relate stride, body size, and locomotor speed? British biomechanician R. McN. Alexander provided an elegant solution to this problem by considering dynamic similarity. Dynamic similarity is a kind of conversion factor: it "pretends" that all animals are the same size and that they are moving their limbs at the same rate. With these adjustments for size and footfall, it doesn't matter if you're a small or large human, a dog, or a dinosaur. All will be traveling with "dynamic similarity"; only speed will vary. That variable Alexander terms "dimensionless speed."[1] It is dimensionless speed that has a direct relationship with relative stride length. Stride length, of course, can be measured from trackways, which in turn allows us for the first time to calculate locomotor speed in dinosaurs.

To see how all this works, let's use Alexander's example of the trackway of a large theropod from the Late Cretaceous of Queensland, Australia. The tracks are 64 cm long, which Alexander, from other equally sized theropods, estimated must have come from a theropod with a leg length of about 2.56 m. The stride length of these tracks is 3.31 m, so the relative stride length (stride length : leg length) is 1.3. The dimensionless speed for a relative stride length of 1.3 is 0.4. And from all these measures, this Australian theropod must have been traveling reasonably quickly, at about 2.0 m/s, or 7.2 km/h.

As complicated as this approach appears, it represents the best method for estimating the actual speeds implied by trackways. But what about the fastest speeds of which a dinosaur might have been capable? In 1982, R. A. Thulborn (University of Queensland) developed a method by which absolute locomotor abilities could be calculated. Thulborn's work relied heavily upon Alexander's slightly earlier studies on speed estimates from footprints and the relationship between body size, stride length, and locomotor speed among living animals. For both approaches, Alexander determined that relative stride length has a direct relationship with locomotor speeds at different kinds of gait (e.g., walking, running, trotting, galloping). Explicitly (for the quantitatively oriented):

$$\text{locomotor velocity} = 0.25(\text{gravitational acceleration})^{0.5}$$
$$\times (\text{estimated stride length})^{1.67}$$
$$\times (\text{hindlimb height})^{-1.17}$$

Thulborn used this equation to estimate a variety of running speeds for more than 60 dinosaur species. The first group of estimates were for the walk/run transition, where stride length is approximately two to three times the length of the hindlimb. A potentially more important estimate – especially for dinosaurs fleeing certain death or pursuing that all-important meal – is maximum speed, which

1 Dimensionless speed may appear oxymoronic, but is in fact equivalent to real speed divided by the square root of the product of leg length and gravitational acceleration.

BOX 15.3 (cont.)

Thulborn calculated using maximum relative stride lengths (which range from 3.0 to 4.0) and the rate of striding, called limb cadences (estimated at 3.0 × hindlimb length$^{-0.63}$).

Although we report some of these speeds elsewhere in this book, it is of some value to summarize the overall disposition of Thulborn's estimates. Small bipedal dinosaurs – which would include certain theropods and ornithopods – appear capable of having run at speeds of up to 40 km/h. Ornithomimids, the fastest of the fast, may have sprinted up to 60 km/h. The large ornithopods and theropods were most commonly walkers or slow trotters, probably averaging no more than 20 km/h. Thus the galloping, sprinting *Tyrannosaurus*, however attractive the image, does not impress Thulborn (or us) as likely.

Then there are the quadrupeds. Stegosaurs and ankylosaurs walked at a no more than a pokey 6 to 8 km/h. Sauropods probably moved at 12 to 17 km/h. And ceratopsians – galloping along full throttle like enraged rhinos? Thulborn estimated that they were capable of trotting at up to 25 km/h.

Are these estimates accepted uncritically by all? P. Dodson has argued that these calculations would suggest that humans can run as quickly as 23 km/h. So it is possible that these calculations overestimate the speeds at which dinosaurs could run. On the other hand, anatomically, humans are not dinosaurs and these calculations could be valid for dinosaurs but be unapplicable to humans. At a minimum, they give some indication of the relative speeds of dinosaurs; for example, how quickly *T. rex* might have run by comparison with *Triceratops*.

gallop would require a complex rewiring of their neuromuscular systems. For this reason, the idea of galloping dinosaurs appears unlikely.

Could large dinosaurs have moved at the speeds Bakker proposed? To do so, they would need to be built like the fastest running mammals. In many, the forelimbs of quadrupedal dinosaurs are treated like those of mammals: they are fully erect, and move in a predominantly fore–aft plane. This is a point of anatomy by no means accepted by all workers. In the nineteenth and first two-thirds of the twentieth centuries, most quadrupedal dinosaurs were reconstructed as having semi-erect forelimbs. Today, most paleontologists recognize some flexure at the elbow; so the question is: "Did the elbow bow out or did the flexure facing rearward (e.g., see Figure 15.5)?" While proponents of fully erect forelimbs have marshalled a variety of arguments to support their contention, the issue remains very contentious.

Again, ancestry may play a telling role. In mammals, a fully erect posture evolved in a quadrupedal ancestor; however, in dinosaurs the fully erect posture evolved in a biped. Quadrupedal dinosaurs are thought to have evolved their four-legged stance secondarily (see Chapter 5) and may therefore be very unlike mammals in design.

Assorted adaptations for processing high volumes of food
Remember that endotherms require more energy than ectotherms. Hence, if it could be shown that all dinosaurs required large amounts of food so that they could function, an endothermic metabolism for Dinosauria might be implied. Considered individually, a number of dinosaur groups show features that suggest efficient processing of high volumes of food, such as might be expected in endotherms. Yet no consistent pattern can be applied to all dinosaurs.

Figure 15.5. Gregory Paul's dynamic restoration of several galloping *Triceratops*. (Illustration courtesy of G. S. Paul.)

All ornithischians (except, perhaps, *Lesothosaurus*; see introductory text to Part II: Ornithischia) have deeply inset tooth rows, suggesting that the animals had well-developed cheeks. Within Ornithischia, the sophisticated chewing adaptations of ceratopsians and ornithopods, particularly hadrosaurids, are noteworthy. In hadrosaurids and ceratopsians, at least, where there was a coincidence of cheeks, dental batteries, and the evidence of tremendous jaw musculature, food was being processed in the mouth to a far greater extent than is found in living ectotherms such as snakes, lizards, crocodilians, and turtles.

Secondary palates, because they allow breathing and chewing to take place simultaneously, are commonly associated with more efficient feeding. Indeed, all mammals possess a secondary palate. Ankylosaurs and hadrosaurids both bear well-developed secondary palates, suggesting that these animals had the benefit of being able to breathe and chew simultaneously.

The relationship of these diverse specializations to metabolism is by no means clear. Hadrosaurids and ceratopsians clearly had developed chewing mechanisms at least as efficient as those found in modern herbivorous mammals. Birds, however, which are fully endothermic, do not have secondary palates. Their theropod forebears likewise show none of these specialized feeding features (many of whom, on the basis of their skeletal design, were good candidates for endothermy). Ankylosaurs, though, possessed a secondary palate, had very small teeth and none of the dental batteries characteristic of the ornithopods (see Chapters 9 and 10). Indeed, secondary palates are known in modern turtles and crocodiles (as well as mammals), so their significance in terms of endothermy is not clear. Sophisticated feeding mechanisms can't provide an absolute guide to who is endothermic and who is not.

Hearts

While in this speculative vein, it is interesting to note – as did R. S. Seymour in 1976 – that all living endotherms possess four-chambered hearts. Seymour argued that the four-chambered heart system, in which the oxygenated blood is completely separated from the deoxygenated blood, is probably a prerequisite for endothermy. The argument is based upon the idea that endothermy requires relatively high blood pressures in order to constantly perfuse complex, delicate organs such as the brain with a constant supply of oxygenated blood. Such high blood pressures, however, would "blow out" the alveoli in the lungs. For this reason, mammals and birds separate their blood into two distinct circulatory systems: the blood for the lungs (pulmonary circuit) and the blood for the body (systemic circuit). The two separate circuits require a four-chambered heart – a pump that can completely separate the circuits. Part of Seymour's argument was based upon the inference that, in sauropods such as *Brachiosaurus*, tremendous blood pressures would be required to bring the oxygenated blood to the brain (see Chapter 11). Thus the heart would *have* to be a double pump (one for oxygenated and one for unoxygenated blood), four-chambered organ.

Blood from a stone

Would such a heart be possible in dinosaurs? The nearest living relatives of dinosaurs – crocodiles and birds – both possess four-chambered hearts and the distribution of birds and crocodiles on the cladogram suggests that it is quite conceivable that a heart with a double-pumping system was present as early as basal Archosauria. This idea was strongly reinforced by the discovery in 2000 of what was inferred to be a four-chambered heart with an aorta. The "heart" was preserved as an ironstone mass within the thoracic cavity of the basal ornithopod *Thescelosaurus*, and recognized using computed tomography (i.e., a CT scan). The author leading the research, North Carolina State University's P. K. Fisher, interpreted the find as evidence for an "intermediate-to-high metabolic rate" in *Thescelosaurus*. The discovery, while tantalizing, left some paleontologists unimpressed; they could not convince themselves that the ironstone was originally soft tissue within the animal.

Brains

Most dinosaurs appear to have been considerably smarter than was supposed in the first half of the last century. In the late 1970s, the EQ concept (Box 15.4) was applied to dinosaurs by University of Chicago

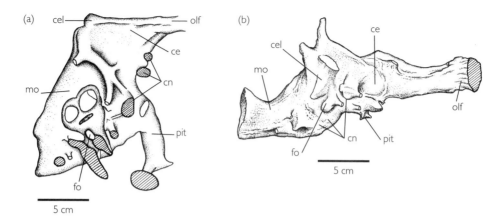

Figure 15.6. Brain endocasts of (a) *Plateosaurus* (see also Figure 4.8); (b) *Tyrannosaurus.* fo, fenestra ovalis (where vibrations from the ear drum are sensed by the brain); ce, cerebrum; cel, cerebellum; cn, cranial nerves (not all shown); mo, medulla oblongata; pit, pituitary (only part shown in *Tyrannosaurus*); olf, olfactory region (sense of smell).

paleontologist J. A. Hopson. Hopson used brain endocasts, internal casts of the braincases of dinosaurs. This was done by painting latex on the inside of a fossilized braincase. When the latex dried and was peeled off, it produced a cast of the inside of the braincase (Figure 15.6).

How good an indicator of the brain is the internal shape of the braincase? The braincases of living birds and mammals closely conform to the shape of their much-enlarged brains. However, in living "reptiles" this is not the case: the brains of crocodilians, lizards, snakes, and turtles are smaller than their bony braincases. Hopson, however, corrected for this situation in dinosaurs, producing minimum brain size estimates. He concluded:

> [I]f the brain required by a vertebrate is primarily a function of its total level of activity and therefore reflects its total energy budget, then coelurosaurs appear to have been metabolically as active as living birds and mammals, carnosaurs and large ornithopods appear to have been less active but nevertheless significantly more so than typical living reptiles, and other dinosaurs appear to have been comparable to living reptiles in their rates of metabolic activity.   (Hopson, 1980, p. 309.)

The nose knows

Studies of modern endotherms by W. Hillenius (College of Charleston, South Carolina) and J. Ruben (Oregon State University) suggested that endothermy requires the lungs to replenish their air (ventilate) at a high rate. And high rates of ventilation lead to the loss of body water, unless something is done to prevent it. What modern mammals and birds do is to grow convolute sheets of delicate bone, called respiratory turbinates, in the nasal cavities. The surface areas of the mucus-covered turbinates pulls moisture out of the air before it leaves the nose, thus conserving water (Figure 15.7).

BOX 15.4

Dinosaur smarts

How can we measure the intelligence of dinosaurs?[1] The short answer is "Not easily." However, it is clear that, at a very crude level, there is a correlation between intelligence and brain:body weight ratios. Brain:body weight ratios are used because they allow the comparison of two differently sized animals (i.e., brain:body weight ratios allow one to compare a chihuahua with a St Bernard dog). The correlation suggests that, in a general way, the larger the brain:body weight ratio, the smarter the organism. Indeed, mammals have higher brain:body weight ratios than fish and are generally considered more intelligent. But how smart is a very large dinosaur with a laughably miniscule brain (e.g., Box 6.1)?

It is a well-known biological fact that proportions of organisms change as they increase in size; this is known as allometry. And it turns out that brain:body weight ratios follow allometric principles as well: brains need not increase in size proportionally in the same amount as the rest of the animal. For example, the brain of a 0.5 m rattlesnake is proportionally larger than the brain of a 3 m anaconda. Does this mean that the anaconda is significantly stupider than the rattler? Obviously not. So, when considering how big or small a brain is in an animal, there has to be a way to compensate meaningfully for size. A quantitative method of doing this was first proposed by psychologist H. J. Jerison who, in the early 1970s, developed a measure called the "encephalization quotient" (EQ). Jerison constructed an "expected" brain:body weight ratio for various groups of living vertebrates (reptiles, mammals, birds) by measuring many brain:body weight ratios among these animals. Jerison noted that living vertebrates cluster into two groups, endotherms and ectotherms. The endotherms show greater encephalization (larger brains) and the ectotherms showed lower encephalization (smaller brains). Thus, for Jerison, living endotherms and ectotherms could be distinguished by brain size.

Having constructed a range of expected brain:body weight ratios, he could account for size in different organisms (and accommodate what might at first seem like an extraordinarily large or small brain). Noting that some organisms still didn't exactly fit into his ectotherm or endotherm group (by virtue of having brains either larger or smaller than expected), he measured the amount of deviation, and then termed this EQ.

Paleontologist J. A. Hopson, now knowing what he could expect for living vertebrates, measured how much the estimated brain:body weight ratio EQ of extinct vertebrates deviate from the expected brain:body weight ratios of their living counterparts. Figure B15.4.1 shows the EQs for several major groups of dinosaurs as reconstructed by Hopson. Because dinosaurs are "reptiles," he measured the deviation of various groups relative to a "reptilian" norm, in this case a crocodile. Significantly, many ornithopods and theropods show a brain:body weight ratio that is significantly larger than would be expected if a modern reptilian level of intelligence is being considered.

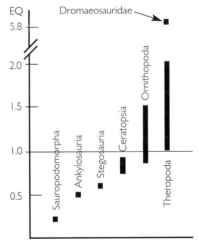

Figure B15.4.1 EQs of dinosaurs compared. The line at 1.0 represents the crocodile "norm," and suggests that many groups of dinosaurs had larger brains than would be predicted from a conventional reptilian model (the crocodile). Note also the break between 2.00 and 5.80; if these measures mean anything, apparently coelurosaurs significantly outdistanced other dinosaurs in brain power. (Data from Hopson, 1980.)

---

1 A more fundamental question is: what is intelligence? As applied here, intelligence refers to the ability to learn, and perhaps the capability for abstract reasoning. The measurement of "intelligence" in humans is freighted with a notoriously racist history and consequently much emotional baggage (see Gould, S. J. 1981. *The Mismeasure of Man*. W. W. Norton, New York, 352pp.) Here, we are discussing intelligence in far cruder terms; that is, at the level of the comparison of the intelligence of a crocodile and a dog.

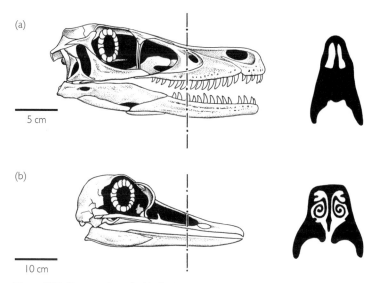

Figure 15.7. Cross-sections (in black) through the nasal regions of (a) an extinct dinosaur (*Velociraptor*) and (b) a living bird (*Rhea*); skulls and positions of the cross-section shown to left. The nasal cavity of the bird shows convolute respiratory turbinates, while that of the dinosaur does not. (Redrawn from Dingus and Rowe, 1997.)

What about dinosaurs? Although a number appear to have had olfactory turbinates (indicative of a well-developed sense of smell), apparently none – as far as we know – had respiratory turbinates to allow it to recoup respired moisture. Hillenius and Ruben argued, therefore, that on this basis dinosaurs could not have been endothermic in the way that most mammals and birds are today.

Histology

One of the truly exciting and original lines of evidence for dinosaur endothermy comes from the science of bone histology, or the study of bone tissues. Fossil bone may preserve fine anatomical details that are visible under a microscope. With a rock-saw, one can slice across a slab of fossil dinosaur bone in much the same way as a butcher uses a band-saw to cut across a leg of beef. To see the full details of the bone, a thin slice can be mounted on a glass slide, and ground down to a thickness that allows light to be transmitted through it (about 30 micrometers (μm)). Then the fine-scale structure can be studied under a microscope (Figure 15.8).

The idea of bone histology indicating levels of metabolism and its development is largely the work of A. J. de Ricqlès of the University of Paris VII (France). Ricqlès' key insight was the observation of so-called Haversian bone in birds, mammals, and dinosaurs.

Haversian bone

Bones grow by remodeling, which involves the resorption (or dissolution) of bone first laid down – primary bone – and redeposition of a kind of bone called secondary bone. Secondary bone is laid down in the form of a series of vascular canals called Haversian canals, and

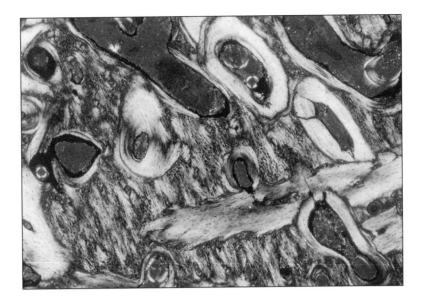

Figure 15.8. Ceratopsian bone under a microscope. Compare with Figure 15.9; note the presence of Haversian bone. Each cell is about 0.3 mm in diameter.

resorption and redeposition of secondary bone can occur repeatedly during remodeling. When this process repeats, a type of bone known as dense Haversian bone is formed. This bone has a distinctive look about it (Figures 15.8 and 15.9).

Among living endotherms, dense Haversian bone is found in many mammals and birds. Among extinct vertebrates, dense Haversian bone has been observed in dinosaurs, pterosaurs, and therapsids (including Mesozoic and Cenozoic mammals). It was not too great a leap of faith to propose that dinosaurs, too, must have been endotherms.

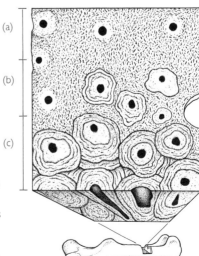

Figure 15.9. Primary bone in the process of being replaced by Haversian bone in the leg of a hadrosaur. Longitudinal canals (at the top of the figure) in primary lamellar bone (a) are resorbed (b) and then reconstituted as Haversian bone (c).

The significance of Haversian bone

It turns out that dense Haversian bone is due to a variety of factors, one of which may be endothermy. Haversian canals are known to be correlated with size and age, and are possibly correlated with the type of bone they replace, the amount of mechanical stress undergone by the bone in which they are found, and nutrient turnover (the interaction between soft tissue and developing bony tissue). Ricqlès wrote:

> In dinosaurs, as in living animals, Haversian bone is assumed to have had a complex physiological role … Indeed, it is likely that most of the processes which may contribute to Haversian replacement, including responses to individual age and growth, bio-mechanical demands, the functions of blood vessels, and the replacement of bone cells, would be accelerated by higher rates of metabolism. Just such an acceleration seems to be indicated by dinosaur bone structures, with their abundant Haversian replacement, which stands in marked contrast with the bone tissues of most other fossil and living reptiles … [A] high rate of metabolism cannot be inferred from the mere *occurrence* of remodeled [dense Haversian] bone, but is rather reflected by the rate and extent of bone replacement. For instance, a 50-year old turtle and a 5-year old dog might show the same amount of Haversian reconstruction, but the physiological significance of these tissues would be very different, particularly in regard to metabolic rate. (de Ricqlès, 1980, p. 125.)

And what does this mean for endothermic dinosaurs? It means that, assuming dense Haversian bone formed in dinosaurs at rates comparable to those in mammals, dinosaurs probably lived for lifespans approximating to those of Recent mammals, and dinosaurs could have had rates of bone growth similar to those found in mammals. If such rates really occurred, they would be in good agreement with conditions that might be expected to be found with an endothermic metabolism.

Bringing up baby

But what is really known about rates of dinosaur growth? Preliminary work was done by Jill Peterson, at that time an undergraduate student of J. R. Horner at Montana State University. Peterson correlated dense Haversian bone development with overall limb development. In doing so, she was able to establish developmental stages of growth in the hadrosaurid *Maiasaura*; she could take a cross-section of bone, and tell from the development of the Haversian canals which growth stage the animal had reached. Using rates of bone growth obtained from living organisms, Peterson and Horner estimated that *Maiasaura* grew an astounding 3 m/year. This work was continued by Horner, de Ricqlès, and K. Padian, who refined Peterson's original estimates by recognizing six different growth categories and assessing the stage of bone development in each stage. Again using modern rates of bone growth, they estimated that in the early years (nestling through juvenile stages) *Maiasaura* grew exceedingly fast, slowing markedly as adult body sizes were reached. This makes the growth pattern of *Maiasaura* distinctly different from that seen in living non-dinosaur reptiles, and much closer to that seen in modern birds.

Further work by Horner, de Ricqlès, and Padian reveals clear patterns in the development of living vertebrates. Unique in both young birds and other dinosaur juveniles, the bone has a distinctly porous quality, and bone cell development differs significantly from that seen in juvenile lizards, turtles, crocodiles, and mammals. Horner and colleagues linked the porosity to vascularization (an extensive network of blood vessels carrying nutrients), itself linked to the rate of deposition of the bone. With regard to the vascularization they inferred from the porosity in dinosaur bone, they wrote:

> Because these dinosaurs represent embryonic [stages in their growth histories], it is unlikely that the high vascularization of their bone could be explained by exercise or other biomechanical stresses. To the contrary, it is most probably that the high vascularization of these young bones is correlated with a very rapid growth rate that was provided by high, sustained metabolic rates.   (Horner *et al.*, 2001, p. 50.)

The message is clearly one of growth rates closer to those seen in modern birds than those seen in lizards, snakes, turtles, and crocodiles.

Horner and his colleagues had another point to make as well: the paleontological evidence – dinosaur juveniles in nests with differing stages of development (see Chapter 9) – suggested to them that parental care was involved in raising a dinosaur. Such behavior would be unlike that seen in conventional snakes, turtles, crocodiles, and lizards, but similar to that seen in birds. Indeed, C. Farmer (University of California at Irvine) suggested that parental care might have been a force in developing an endothermic metabolism. She proposed that the incubation of eggs required an endothermic metabolism. All of this work suggests that there is surely a relationship between growth rates, parental care, and metabolism.

Zonal growth

In 1979, P. A. Johnston, a paleontologist now at the Royal Tyrrell Museum of Palaeontology in Alberta, Canada, observed concentric growth rings in the teeth of saurischian and ornithischian dinosaurs. Among modern tetrapods, such growth rings are typically found in ectotherms. The rings are believed to represent seasonal cycles. During times such as dry seasons in the tropics, or cold seasons in more temperate latitudes, the metabolisms of many ectothermic tetrapods are slowed, and consequently growth is stymied. During more humid seasons in the tropics and warmer seasons in more temperate climates, bone and tooth growth resumes. Thus recorded in the teeth is a pattern of ring-like deposits representing annual cycles of growth and stasis.

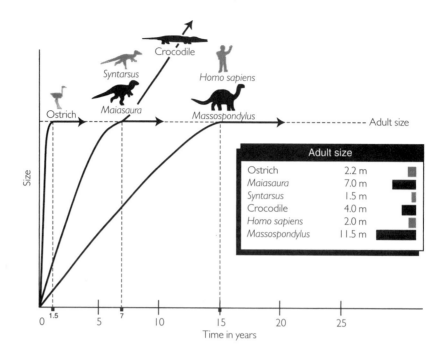

Figure 15.10. Estimated growth rates of some dinosaurs, *Alligator*, and a human. The graph is based upon guesses of how long it takes for the tetrapods to reach adult size. Note that because the sizes of these organisms vary extensively, the growth rates also vary. Unlike the other tetrapods presented, *Alligator* grows continuously throughout its life; hence, it has no fixed "adult size." Sexual maturity usually comes within six to eight years. (Estimates for *Syntarsus* and *Massospondylus* from the work of A. Chinsamy; estimates for *Maiasaura* from the work of J. Peterson and J. Horner.)

Among endotherms, on the other hand, such patterns are rare, because the relatively constant, elevated metabolic rates ensure growth at a constant rate. Johnston noted that the rings found in the dinosaur teeth were very much like those found in crocodile teeth from the same deposits. His conclusion, therefore, was that dinosaur teeth suggested an ectothermic metabolism for their owners.

Further work along these lines was undertaken by R. E. H. Reid and his student, A. Chinsamy, now at the South African Museum of Natural History in Capetown. These researchers recognized lines of arrested growth (LAGs), or bands in the bone, which they took to represent yearly increments. Using the sauropodomorph *Massospondylus* and an early theropod, the small, light-bodied theropod *Syntarsus*, both from South Africa, Chinsamy estimated rates of growth, and proposed that *Massospondylus* took 15 years to reach 250 kg (17 kg/year), while *Syntarsus* took 7 years to reach an estimated 20 kg (3 kg/year). These estimated rates that were somewhat slower than Peterson and Horner's estimate of *Maiasaura* growth rates (Figure 15.10). The appearance of secondary bone in the thigh bones of these organisms more closely approximated that of modern birds than that of a crocodile. Chinsamy concluded that the sauropodomorph and the theropod she studied showed an apparent mix of features characteristic of endothermic and ectothermic metabolisms: relatively slow rates of development that are characteristic of an ectothermic metabolism, but the dense Haversian bone that looked more like that of living endotherms (birds) than living ectotherms (crocodilians).

LAGs have come to play significant roles in the debate on dinosaur endothermy. LAGs occur when bone growth stops, usually the result of temperature fluctuations caused by seasonality. Living endotherms – birds and mammals – grow independently of temperature and seasonality, and thus do not possess well-developed LAGs. LAGs occur in many different kinds of dinosaur (notably among the best candidates for endothermy, *Coelophysis*, *Allosaurus*, and *Troodon*), suggesting to researchers that growth in dinosaurs was more susceptible to external climatic influences than had been predicted by the simple homeothermic endotherm view of dinosaur metabolism.

Chinsamy also studied LAGs in three enantiornithine birds (see Chapter 14), two small flyers and one large flightless bird (*Patagopteryx*). In all of these organisms, Chinsamy found LAGs, leading her to the conclusion that the early bird's metabolism was subject to seasonal growth, even though it clearly bore feathers. Her conclusion – contrasting with the view that feathers evolved to insulate a small, active endothermic dromaeosaur – was that a modern endothermic metabolism occurred after the development of feathers. Ultimately, she ascribed intermediate physiologies to these organisms; that is, she viewed them as endotherms that had not quite attained the level of endothermy seen in living birds (see below).

Chinsamy's observations in enantiornithines contrasted with her studies of the ornithurine birds (e.g., *Hesperornis* and *Ichthyornis*). In these, the bone tissue looks very similar to that in modern birds. So too is the bone tissue of *Confuciusornis*, a primitive Early Cretaceous bird from China.

LAGs leave researchers with a few important questions. For example, it is not necessarily evident that LAGs represent yearly increments. Indeed, one hadrosaurid fossil is reported to have different numbers of LAGs on its arms from those on its thighs! Moreover, the appearance (or not) of LAGs has never been tightly correlated with climate. Finally, the LAGs reported by Chinsamy in enantiornithine birds have been ascribed by other workers to "tide lines"; that is, lines that are associated with changes in mode of bone remodeling (rather representing cessation of bone growth).

Ecology

Of predators and their prey

Tantalizing evidence bearing upon dinosaur endothermy comes from paleoecology, the study of ancient interactions among organisms. The idea, from R. T. Bakker, was rooted in the fact that, as we have seen, endotherms require more energy than ectotherms. The total weights of the organisms involved – their biomass – was assumed by Bakker to directly reflect energy. Thus, if predators were endothermic, they should require more energy than if they were ectotherms, and this should be in turn reflected in the weight proportions in the community of predator and prey. Bakker wrote:

> [A] meter of heat production in extinct vertebrates is the predator–prey ratio: the relation of the "standing crop" of a predatory animal to that of its prey. The ratio is a constant that is a characteristic of the metabolism of the predator, regardless of the size of the animals of the predator–prey system. The reasoning is as follows: The energy budget of an endothermic population is an order of magnitude larger than that of an ectothermic population of the same size and adult weight, but … the yield of prey tissue available to predators … is about the same for both an endothermic and an ectothermic population … The maximum energy value of all the carcasses a steady-state population of lizards can provide is about the same as that provided by a prey population of birds or mammals of about the same numbers and adult body size. Therefore a given prey population, either ectotherms or endotherms, can support an order of magnitude greater biomass of ectothermic predators than of endothermic predators, because of the endotherms' higher energy needs. (Bakker, 1975, pp. 128–129.)

With that premise, Bakker obtained modern predator:prey biomass[3] ratios, with the intent to compare them with the ancient ones. The modern ones were relatively accessible from published accounts. It turned out that size played a role; a community of large herbivores might yield 25% of its total biomass as prey, while a community of small herbivores might produce 600% of its total biomass as prey.[4] How then to factor in the issue of size, which would obviously affect the final ratio? Bakker proposed that endpoints in the size spectrum of predators and prey effectively cancel each other out, and the result is that uniform predator:prey biomass ratios exist for endotherms, regardless of size. Bakker calculated that predator:prey biomass ratios for ectothermic organisms are around 40%, while predator:prey biomass ratios for endothermic organisms are 1–3%. Here, then, was an order of magnitude difference in the biomass ratios that ought to be recognizable in ancient populations.

Now he was armed with a tool from modern ecosystems that he believed could reveal the energetic requirements of ancient ecosystems. But how can one determine predator:prey biomass ratios in ecosystems that vanished from the face of the earth 65 (or more) Ma? By counting

---

3 He called them "predator:prey ratios," but in fact they were estimated predator *biomass*:estimated prey *biomass* ratios. Here for convenience, we'll refer to them as predator:prey biomass ratios.

4 How could a community produce six times as much biomass as there was community in the first place? The answer is time-averaging. Bakker (1975, p. 129) noted: "[A] herd of zebra yields from about a fourth to a third of its weight in prey carcasses a year, but a 'herd' of mice can produce up to six times its weight because of its rapid turnover, reflected in a short life span and high metabolism per unit weight."

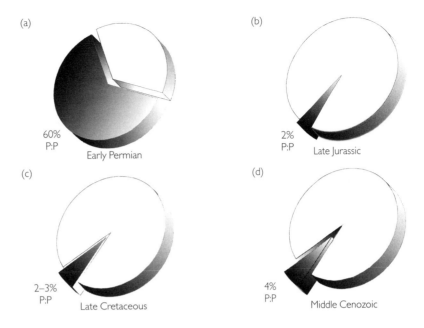

Figure 15.11. The proportions of predators to prey (P:P) in selected faunas in the history of life, as reconstructed by R.T Bakker. (a) Early Permian of New Mexico; (b) Late Jurassic of North America; (c) Late Cretaceous of North America; (d) Middle Cenozoic of North America. The Cenozoic fauna (d) is mammalian and obviously endothermic, providing clear guidelines on what predator:prey ratios are in an endothermic fauna. Predators are shaded; prey are white.

specimens of predators and presumed prey in major museums and by estimating the specimens' living weights, he attempted reconstructions of the biomasses of fossil assemblages. His results seemed unequivocal: among the dinosaurs, the predator:prey biomass ratios were very low, ranging from 2% to 4%. He interpreted this low number as indicative of endothermic dinosaurian predators (Figure 15.11).

Bakker's study, for all its creativity and originality, raised a howl of protest. J. O. Farlow, reviewing Bakker's conclusions, noted that even the initial premise – that there is an order of magnitude difference between ectothermic predator:prey biomass ratios and endothermic predator:prey biomass ratios – may not hold in all cases. Bakker's assumption that prey are approximately the same size as the predators is clearly not correct (consider a bear eating a salmon), and has drastic effects on the resultant ratio. Moreover, we really don't know who the prey of many carnivorous dinosaurs were. If we assume that such prey were about the same size as the carnivore, we can find suitable candidates. On the other hand, carnivorous dinosaurs may have consumed prey from a range of sizes.

The predator:prey calculation assumes that the number of predators is limited by the amount of food (in this case prey) available. However, a variety of spatial, social, behavioral, and health factors also constrain predator population size. More significantly, the predator:prey biomass calculation also assumes that all deaths are the result of

BOX 15.5

Weighing in

There are two commonly used ways to estimte the weight of a dinosaur. The first is based upon a relationship between limb cross-sectional area and weight (Figure B15.5.1). This relationship has some validity, because obviously, as a terrestrial beast becomes larger, the size (including cross-sectional area) of its limbs must increase. The question is "Does it increase in the same manner for all tetrapods?" If so, a single equation could apply to all. It is clear, however, that it cannot. As noted by J. O. Farlow, weight is dependent upon muscle mass and muscle mass is really a consequence of behavior. Therefore, weight estimates of dinosaurs are in part dependent upon presumed behavior. For example, reconstructing the weight of a bear would involves assumptions of muscle bulk and gut mass very different from those obtained in reconstructing the weight of an elk. Indeed, the cross-sectional area of their limb bones might be identical, but they might weigh very different amounts. Moreover, our knowledge of dinosaurian muscles and muscle mass is rudimentary. This method, although convenient and used by a number of workers (including R. T. Bakker), has the potential for serious misestimations of dinosaur weights.

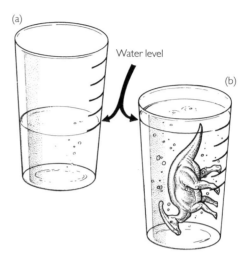

Figure B15.5.2. Estimating the weight of a dinosaur using displacement. For explanation of (a) and (b), see the text.

A second method, pioneered by American paleontologist E. H. Colbert in the early 1960s, involves the production of a scale model of the dinosaur, and then the calculation of its displacement in water (Figure B15.5.2). That displacement could then be (1) multiplied by the size of the scale model (if the model was 1/32 of the original, the weight of the displaced water would have to be multiplied by 32) and (2) further modified by some amount to a number corresponding to the specific gravity of tetrapods. But what is the specific gravity of a tetrapod? On the basis of studies with a baby crocodile, Colbert determined that baby crocodiles, at least, have a specific gravity of 0.89. Unfortunately, there is no uniform specific gravity shared by all tetrapods. Studies with a large lizard (*Heloderma*) showed that the specific gravity of that lizard was 0.81. Among mammals, it would not be surprising to find the specific gravity of a whale differing from that of a cheetah, which might in turn differ from that of a gazelle. In short, while the displacement method is perhaps a bit more accurate than limb cross-sectional calculations, it is still dependent upon inferred muscle mass (and thus behavior), and therefore somewhat problematical.

Figure B15.5.1. Estimating the weight of a dinosaur by comparing the cross-sectional areas of bones.

predation; that there can be no mortality due to other causes. This, of course, is obviously not the case. When the predator:prey biomass ratio was calculated by Farlow to incorporate deaths by predation only, the maximum ectothermic predator:prey biomass ratio was 20%. In reality, of course, because death comes to populations by means other than predation, this number could be smaller.

There are serious and legitimate problems with using fossils for such calculations. Most obvious are difficulties in estimating the true weight of dinosaurs. Several estimates are available (see Box 15.5), but, even discounting great differences in these values, small differences in weight estimates can substantially alter the predator:prey biomass ratios.

Beyond this, preservation of dinosaur faunas can be very misleading. In the field, the preservation of dinosaur material is subject to a variety of biases. How can one ever be sure that the proportions of the living community are represented? Because we can't, paleontologists commonly talk about fossil assemblages. The word assemblage means a collection, and the proportions of organisms in a particular assemblage may have nothing to do with the proportions of the same animals in the community in which those animals lived.

Finally, Bakker obtained his data by counting specimens in museum collections. But this is going out of the frying pan and into the fire: dinosaur fossils are commonly collected by museums to obtain rare or particularly well-preserved specimens, not to attempt to reconstruct ancient communities. Museum collections thus more likely represent an assemblage of well-preserved organisms with an increased percentage of rare organisms than was present in the original fauna. Ultimately, predator:prey biomass ratios was a brilliant, if flawed, idea. Much uncertainty is introduced through this method.

Zoogeography The global distribution of dinosaurs far exceeds that of modern ectothermic vertebrates. Correcting for continental movements, Cretaceous dinosaur-bearing deposits have been found close to 80° north and south of the equator. In the Northern Hemisphere, Late Cretaceous sites have been found in Alaska, the Yukon, and Spitsbergen. In the Southern Hemisphere, Early Cretaceous sites are known from the southern part of Victoria, Australia, and from Antartica.

What were these high-latitude localities like? In North America, evidence derived from theoretical considerations and fossil plants suggests that the region experienced a mean annual temperature of 8 °C, and a mean cold month temperature of 2–4 °C. By comparison, on the basis of stable isotopes, the Australian Early Cretaceous site had a estimated mean annual temperature of 5 °C, with temperatures descending to as low as −6 °C. Overall, the climate of the Australian locality is thought to have been humid, temperate, and strongly seasonal. Certainly sites at both the northern and southern latitudes experienced extended periods of darkness, and we can be reasonably sure that, at least occasionally, temperatures in winter went below freezing.

The assemblage from North America includes hadrosaurids, ceratopsids, tyrannosaurids, and troodontids. The dinosaur assemblage from

Australia includes a large theropod (possibly *Allosaurus*) and some basal euornithopods (including many juveniles). Along with these dinosaurs are fish, turtles, pterosaurs, plesiosaurs, birds (known solely from feathers), and, incredibly, a improbable late-surviving temnospondyl (an amphibian group that apparently went extinct in the Early Jurassic everywhere else in the world; see Figure 16.7).

This mix of animals is not particularly easy to interpret in terms of endothermy or ectothermy. T. H. Rich and P. Vickers-Rich from Monash University, Victoria, Australia, and their colleagues, studied the southern faunas, and remarked on the large brains and well-developed vision of the euornithopods, which they suggested were helpful during periods of extended darkness. Other members of the southern assemblage, however, were not so well equipped. The Riches and colleagues noted that some of the animals could potentially have survived winter by burrowing. The dinosaurs in the assemblage could have survived by migration; however, where they might have migrated to is an open question, and migration was certainly not an option for the turtles and labyrinthodonts, and possibly not for the plesiosaurs.

In the case of the North American faunas, only *Troodon* is of a size that could make burrowing feasible. Migration was potentially a solution to inclement winter weather, although the dinosaurs would

Figure 15.12. Gregory Paul's reconstruction of a tyrannosaurid drinking at a nearly frozen Arctic pond at night. (Drawing courtesy of G. S. Paul.)

have had to migrate for tremendous distances before temperatures warmed sufficiently.

In 1987, G. S. Paul published an evocatively illustrated argument for the idea that dinosaurs living in the high latitudes must have been endothermic (Figure 15.12). He noted that a large ectotherm, the Late Cretaceous crocodilian *Phobosuchus* (= *Deinosuchus*, a crocodilian comparable in size to the Alaskan and Canadian hadrosaurids and tyrannosaurids) is found in temperate latitudes and yet is absent in the northernmost deposits. He suggested too that the smaller, ectothermic members of the assemblage burrowed. On this basis of theoretical calculations of energy requirements, Paul rejected migration as a viable option for the dinosaurs in the northern assemblage. Finally, he suggested that the Arctic dinosaurs must have maintained some insulatory covering (such as fur or feathers) at least seasonally.

Biologist J. R. Spotila and colleagues reassessed the energy requirements for migration. Spotila, a 20 year veteran of studies of dinosaur endothermy and behavior, concluded on theoretical grounds that migration was a possibility and even a likelihood for medium-sized to large dinosaurs. Spotila's view was based on energy calculations from a large ectotherm, the leatherback turtle (*Dermochelys*).

Phylogeny

As we have seen, birds are dinosaurs (see Chapter 13) and modern birds are surely endothermic. Our question must be, therefore, at what point during theropod evolution did "avian" endothermy evolve? If we knew that it evolved very early in the evolution of dinosaurs, then it may be that the ancestor of dinosaurs (see Chapter 5) was endothermic and thus all dinosaurs would have inherited that metabolism. On the other hand, if it evolved with Avialae then perhaps birds and birds alone are the only endotherms in Dinosauria.

An important clue comes with insulation. All small- to medium-sized modern endotherms are insulated with fur or feathers.[5] Indeed, pterosaurs are suspected endotherms, in part because they are known to have been covered with a fur-like coat. There is a certain sense to this; if an ectotherm depends upon external sources for heat, why develop a layer of protection (insulation) from that external source? And can a small endotherm, constantly eating to maintain its metabolism, afford to lose heat? *Archaeopteryx*, with its plumage, is therefore usually considered to have been endothermic.[6] The discovery of non-avian, feathered theropods from China (see Chapter 13) reinforces the notion that the primary function of feathers was first for insulation and only secondarily did feathers evolve a flight function. These same theropods also give us a clue that endothermy occurred well within Deinonychosauria, and perhaps at an even more basal level within Theropoda.

Geochemistry

Remarkably, fossil vertebrates carry around their own paleo-thermometers. Recall from Chapter 3 that the proportion of the rare stable isotope of oxygen (^{18}O) varies as the temperature varies; therefore, if a substance contains oxygen, one can learn something about the temperature at which that substance formed by the ratio $^{18}O{:}^{16}O$. In the case of bone, the oxygen is contained in phosphate (PO_4) that forms part of the mineral matter in the bone. Thus, knowing the oxygen isotopic composition of the bone, one can learn something of the temperature at which the bone formed (Figure 15.13).

University of North Carolina geochemists W. Showers, R. E. Barrick and their co-workers analyzed samples from a variety of modern organisms (to get a sense of what variations are like in organisms of known metabolic strategies), and then from a Cretaceous lizard (presumably a known ectotherm) and some dinosaurs. Their idea was that if dinosaurs were ectothermic, there should be a large temperature differ-

5 While all mammals are fur-bearing, large mammals in tropical climates have very sparse coats that do not serve in an insulatory capacity. Here a combination between temperature and the size of the organism is assuring heat retention.

6 In 1992, J. A. Ruben suggested that *Archaeopteryx* could have been an ectotherm. His idea was based upon the amount of energy needed for flight, and upon the amount of energy available from an ectothermic metabolism. Since the bones of *Archaeopteryx* show limited adaptations for sustained, powered flight, Ruben argued that an ectothermic metabolism would have been more than sufficient for the kind of limited flight that apparently characterized *Archaeopteryx*. While powered flight may be possible in an ectothermic tetrapod, none (save perhaps *Archaeopteryx*) ever evolved it. Moreover, it seems to us that the presence of insulation (feathers) in *Archaeopteryx* is incompatible with an ectothermic metabolism.

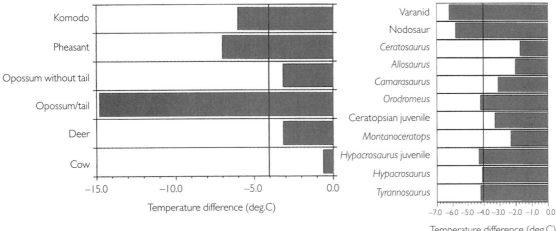

Figure 15.13. Estimated maximum temperature variations between bones located in the core of the body and those located at the extremities, in living and extinct vertebrates, reconstructed by use of oxygen isotopes. Living vertebrates are represented by the komodo dragon (an ectothermic lizard) and a selection of mammals (endotherms). The extinct vertebrates are all dinosaurs with the exception of the varanid (a type of lizard). Note that the greatest variation between core and extremeties occurs in the opossum (conventionally considered to be an endotherm), reinforcing the idea even endotherms can undergo significant fluctuations variations between core and extremities. The authors concluded that all dinosaurs tested, except the ankylosaur, matched the definition of homeotherms. (Data from Showers *et al.*, 1997.)

ence between parts of the skeleton located deep within the animal (i.e., ribs and trunk backbone) and those located toward the exterior the animal (i.e., limbs and tails). If, however, dinosaurs were endothermic, there should be little temperature difference between bones deep within the animal and those more external because the body would be maintaining its fluids at a constant temperature. The difference in temperatures – or lack thereof – should be reflected in the proportion of ^{18}O to ^{16}O.

Their studies revealed that bones of some of the dinosaurs tested – *Tyrannosaurus*, *Hypacrosaurus*, *Montanoceratops*, and a juvenile *Achelousaurus*) showed little temperature variation between core extremities, suggesting that they were formed under homeothermic conditions. The small euornithopod *Orodromeus* and a nodosaurid ankylosaur that they tested, on the other hand, had an isotopic variation (hence, a temperature variation) that pushed the limits of conventional homeothermy. The Jurassic dinosaurs *Ceratosaurus* and *Allosaurus* all showed an ectotherm-like variability in their core regions (the pelvis, in this case). The extremities of virtually all these dinosaurs – plus *Allosaurus* and *Ceratosaurus* – fell "near the range of homeothermy." In general, the extremites of these dinosaurs fell within 4 degrees Celsius (deg.C) of the cores. The authors concluded that virtually all of these dinosaurs were homeotherms that experienced some "regional heterothermy." In short, no clearly endothermic or ectothermic signal came from this type of work.

Opinions on dinosaur endothermy

Nobody has ever seen a living non-avian dinosaur. The consensus view held today is part of a continuum of important insights into the question, and a historical approach yields instructive insights into the originality and importance of current views.

1842

In 1842, Sir Richard Owen coined the term "Dinosauria," basing this group upon three very incompletely known dinosaurs (*Iguanodon*, an ornithopod; *Hylaeosaurus*, an ankylosaur; and *Megalosaurus*, a large theropod that remains, to this day, poorly known). Incredibly, his initial assessment of Dinosauria was that its members were close to achieving avian and mammalian levels of endothermy. Postulating a progressive thinning of the atmosphere through the Phanerozoic, he wrote:

> If we may presume to conjecture that atmospheric pressure has been diminished by a change in the composition … of the air, the beautiful adaptation of the structure of birds to a medium thus rendered lighter and more invigorating … must be appreciable by every physiologist. And it is not without interest to observe, that the period when such change would be thus indicated by the appearance of birds in Wealdon strata [based upon footprints, now thought to have been made by dinosaurs] is likewise characterized by the presence of those Dinosaurian reptiles which in structure most nearly approach Mammalia.
>
> The Dinosaurs, having the same thoracic structure as the Crocodiles, may be concluded to have possessed a four-chambered heart; and, from their superior adaptation to terrestrial life, to have enjoyed the function of such a highly-organized centre of circulation in a degree more nearly approaching that which now characterizes the warm-blooded Vertebrata [i.e., Mammalia and Aves].
>
> A too-cautious observer would, perhaps, have shrunk from such speculations, although legitimately suggesting themselves from necessary relations between the organs and media of respiration; but the sincere and ardent searcher after truth, in exploring the dark regions of the past, must feel himself bound to speak of whatever a ray from the intellectual torch may reach, even though the features of the object should be but dimly revealed.
>
> (Owen, 1842, p. 204.)

The first 60 years of the twentieth century

The first 60 years of the twentieth century brought about an expansion and consolidation of our basic understanding of dinosaurs and their diversity. Because dinosaurs were classified as "reptiles," it was assumed that they functioned like large crocodiles. Owen's original suggestion was forgotten, for there was certainly no evidence for a less dense atmosphere. In retrospect, it seems impossible that scientists could have ignored so pivotal a question for so long, but a look at publications through this period suggests that it was not even an issue.[7] A typical portrait of the "reptilian" qualities of the group is provided by R. C. Andrews, who, in 1953, wrote his immensely popular *All About Dinosaurs* for Landmark Books. His opening statement (ibid., p. 3) began, "Dinosaurs were reptiles, cold-blooded animals related to crocodiles, snakes, and lizards."

In 1946, paleontologist E. H. Colbert and his colleagues attempted to estimate the speed with which a large ectothermic dinosaur could regain heat once lost (e.g., perhaps during a cool, rainy night). The results didn't auger well for dinosaurs being physiologically like modern reptiles. Colbert and his colleagues determined that a rise of temperature of 1 deg.C in a 10 tonne dinosaur would take "more than 1 hour and something less than several days." They concluded:

> [T]emperature fluctuations in the larger dinosaurs must have been less abrupt than they are in most Recent reptiles. Therefore body temperatures in the large dinosaurs must have been relatively constant, with the result that these animals may have had some of the physiological advantages that characterize mammals and birds. Would this explain, to some degree, the long period of dominance enjoyed by the dinosaurs?
>
> (Colbert *et al.*, 1947, p. 142)

A few paleontologists, notably G. R. Wieland of Yale University, likewise shared a vision of some kind of dinosaur homeothermy. However, it was not until the 1960s that things began to change dramatically.

1960–1980

This 20 year period witnessed some of the most radical changes in ideas about dinosaur physiology. L. S. Russell, a paleontologist with the Royal Ontario Museum in Canada, noted that dinosaurs are in many respects intermediate in their anatomy between crocodiles and birds, and suggested that perhaps their metabolism was likewise intermediate. Indeed, he supposed that they might be more like birds than crocodiles, and stated that the concept of warm-blooded dinosaurs was "possible, even plausible."

However, it was J. H. Ostrom's 1969 description of *Deinonychus anthirropus* that truly changed everything. Here was a predatory dinosaur (a dromaeosaurid) obviously built for extremely high levels of activity. Its skeletal design simply made no sense considered otherwise. This view and that of Ostrom's then-student R. T. Bakker, that dinosaurs were fully endothermic, opened up the debate. Bakker's initial arguments were based primarily upon the skeletal morphology of dinosaurs; it was slightly later that he developed his ecological ideas.

---

7 The treatment of dinosaurs in the 1933 edition of A. S. Romer's *Vertebrate Paleontology*, the classic general treatment of the field, refers to dinosaurs as reptiles; thus physiological deviations from living "reptiles" are never mentioned. Interestingly, in the same edition, pterosaurs are considered as potential candidates for endothermy. In the 1966 edition of *Vertebrate Paleontology*, dinosaur physiology still does not merit a mention.

But nothing caught the attention of paleontologists quite like two papers published in 1974. The first, by Ostrom, was a brilliant, exacting study of the earliest-known bird, *Archaeopteryx*. Reviving the ideas of T. H. Huxley, a contemporary of Darwin, Ostrom concluded that the relationship between dinosaurs and birds was inescapable. "All available evidence," he wrote, "indicates unequivocally that *Archaeopteryx* evolved from a small coelurosaurian dinosaur and that modern birds are surviving dinosaur descendants." This suggested strongly that dinosaur physiology should be considered more along bird than crocodilian lines.

The second paper was by Bakker and P. M. Galton, a paleontologist from the University of Bridgeport, Connecticut. Bakker and Galton, basing their ideas on their work and Ostrom's, proposed to remove dinosaurs from "Reptilia" and establish them and birds as a new class of vertebrates, Dinosauria. The basis for this proposal were the "key advancements of endothermy and high exercise metabolism." The radical proposal fired imaginations and incensed detractors.

1980 and beyond: toward a modern consensus

The debate over dinosaur endothermy climaxed with the 1980 publication of an American Association for the Advancement of Science special volume that covered the 1978 proceedings of a symposium devoted solely to dinosaur endothermy. In general, the contributions in the book struck a compromise between Bakker's concept of dinosaurs having bird/mammalian levels of endothermy and the prevalent view of dinosaurs as oversized, uninspired crocodiles. Most authors acknowledged aspects of dinosaurian anatomy that appear to be closer to the crocodile model; nonetheless, all seemed to lean toward, at a minimum, some kind of homeothermy.

The most conservative viewpoint was that dinosaurs were mass homeotherms; that is, their very bulk all but precluded fluctuations in their core temperatures. This argument failed for small dinosaurs such as all hatchlings and juveniles, as well as small ornithopods and theropods (where surface area:volume ratios would have been relatively large and thus heat would have been efficiently dissipated), even if it makes sense for some of the larger dinosaurs. Would small dinosaurs have been something else?

Modern views on dinosaur metabolism

While the data do not convincingly demonstrate crocodile-style ectothermy or mammal-style endothermy in dinosaurs, it would be a mistake to infer that the data send no clear signal at all. On the contrary, what is to be learned from them is very clear: that dinosaurs were neither conventional mammalian or bird-like endotherms, nor "reptile"-style ectotherms. The trick is to determine exactly what they were.

One size fits all?

R. E. H. Reid envisioned all dinosaurs with a uniform physiology "intermediate" between complete endothermy and complete ectothermy. This viewpoint has historically challenged physiologists, because it is not clear exactly what an intermediate metabolism might have been like. Reid speculated:

> The simplest … reading of the evidence is that all dinosaurs shared one type of physiology, which allowed them to form bone continuously when or where growth was rapid, but also allowed them to to form "growth rings" when or where it was slower. This suggests, in turn, that while [dinosaurs were] more like birds and mammals physiologically than modern reptiles are, they were also more like modern reptiles than mammals and birds are … [T]he case for intermediate dinosaurs rests here. It cannot be proved conclusively, but it fits current evidence better than the view that they were endotherms … They could instead have been sub-endothermic "super-reptiles," or "supergiganto-therms," with a more advanced circulatory system than any modern reptile, and no true modern physiological counterparts.
>
> (Reid, 1997, p. 470–471.)

Different strokes for different folks?

Thinking along different lines, smaller dinosaurs may have been metabolically very different from larger dinosaurs. J. R. Spotila applied the concept of gigantothermy to large dinosaurs. Gigantothermy is a kind of modified mass homeothermy that mixes large size with relatively low metabolic rates, and control of blood circulation in peripheral tissues. Spotila and his colleagues noted that, among endotherms, size is inversely correlated with metabolic rate; that is, the larger the endotherm, the slower its relative metabolic rate. This makes a great deal of sense in the context of mass homeothermy: because large animals (with low surface area:volume ratios) retain heat, high levels of metabolic activity (and associated heat generation) would heat the animal beyond its ability to cool off. In other words, it would suffer the biological equivalent of a melt down.

To avert this thermal debacle, Spotila and colleagues argued, large dinosaurs might have resorted to gigantothermy. Because of its size, heat produced by the animal – even at low metabolic rates – can warm it quite efficiently, and control of circulation to the extremities would modify the rate of heat dissipation. Gigantothermy in combination with migration, they argued, would be as viable for these organisms as for a typical migrating large mammalian endotherm.

And what of the smaller dinosaurs? Spotila and his colleagues argued that if they maintained mammal-like metabolic rates, they would have had to have maintained an insulatory coating to prevent too rapid a loss of heat (like a modern bird or mammal). If they did not maintain a mammalian/avian level of endothermy, Spotila and colleagues suggested, perhaps small dinosaurs burrowed during cold months. Intermediate-sized dinosaurs perhaps enjoyed the best of two worlds: when very young, they maintained mammalian/avian levels of endothermy, but, as they got older and larger, they could have switched over to a more gigantotherm-type of metabolism.

Our view

Our best call is that dinosaurian physiology appears to have been a complex amalgam of various strategies, relating to size, behavior, and perhaps environment. That they weren't reptiles in the crocodile sense of physiology is evident, but 65 million years since the last one roamed earth has obscured our ability to completely resolve their metabolic strategies.

Still, it seems to us that the late 1990s discoveries of feathered dinosaurs from China (see Chapter 13) bear directly on this question. In these groups, at least, the question of "dinosaur" theromoregulation becomes effectively equivalent to the question of the origin of avian endothermy. And that, of course, depends upon what one chooses to call a "bird." As we noted in the section entitled "Phylogeny" above, living birds are clearly endotherms and have two key indicators of their metabolism: pneumatic foramina and feathers. More basal avialians (such as *Archaeopteryx*) have feathers but no pneumaticity, leading many researchers to posit an endothermy not quite as advanced as that seen in living birds. The fossil record does not allow us to distinguish between the presumed type of endothermy seen in Chinese deinonychosaurs – "dinosaurs" in the popular sense of the word – and that seen in *Archaeopteryx* (popularly considered a "bird"). But this leaves us with the conclusion that, in at least one group of dinosaurs, some kind of internal temperature regulation goes back at least as basally as Coelurosauria.

Here, we'll let Spotila and his colleagues have the last word, their viewpoint on dinosaur endothermy being much in accord with our own:

> The concept of dinosaurs as avian- or mammalian-style endotherms has intuitive appeal to the layperson, and is infinitely preferable to the image of dinosaurs as the cultural icons of stupidity, sluggishness, extinction, and above all, failure ... An eager public has been extremely receptive to popularizers willing to offer a readily comprehensible account of dinosaurs as highly active, racing, dancing endotherms ... However, the view of dinosaurs that the public has embraced so eagerly is partisan and flawed ... [This view] has derived from a philosophically ... suspect concept (that endothermy is superior to ectothermy ...), and from doctrinaire [views] that insist that all dinosaurs were endothermic that have failed to keep abreast with relevant developments in biophysics. (Spotila *et al.*, 1991, pp. 223–224.)

Important readings

Andrews, R. C. 1953. *All About Dinosaurs*. Random House, New York, 146pp.

Bakker, R. T. 1975. Dinosaur Renaissance. *Scientific American*, **232**, 58–78.

Bakker, R. T. 1986. *The Dinosaur Heresies*. William Morrow and Company, New York, 481pp.

Bakker, R. T. 1987. Return of the dancing dinosaur. In Czerkas, S. J. and Olsen, E. C. (eds.), *Dinosaurs Past and Present*, vol. I. Natural History Museum of Los Angeles County, Los Angeles, pp. 38–69.

Bakker, R. T. and Galton, P. M. 1974. Dinosaur monophyly and a new class of Vertebrates. *Nature*, **248**, 168–172.

Barreto, C., Albercht, R. M., Bjorling, D. E. and Horner, J. R. 1993. Evidence of the growth plate and the growth of long bones in juvenile dinosaurs. *Science*, **262**, 2020–2023.

Barrick, R. E., Stoskopf, M. K. and Showers, W. J. 1997. Oxygen isotopes in dinosaur bone. In Farlow, J. O. and Brett-Surman, M. K. (eds.), *The Complete Dinosaur*. Indiana University Press, Bloomington, IN, pp. 474–490.

Colbert, E. H., Cowles, R. B. and Bogert, C. M. 1947. Rates of temperature increase in the dinosaurs. *Copeia*, pp. 141–142.

de Ricqlès, A. 1980. Tissue structures of dinosaur bone: functional significance and possible relation to dinosaur physiology. In Thomas, R. D. K. and Olson, E. C. (eds.), 1980. *A Cold Look at the Warm-Blooded Dinosaurs*. AAAS Selected Symposium no. 28, 103–139.

de Ricqlès, A., Padian, K. and Horner, J. R. 2001. The bone histology of basal birds in phylogenetic and ontogenetic perspectives. In Gauthier, J. A., and Gall, L. F. (eds.), *New Perspectives on the Origin and Early Evolution of Birds*, Proceedings of the International Symposium in Honor of John H. Ostrom. Yale Peabody Museum Press, New Haven, CT, pp. 411–426.

Desmond, A. 1975. *The Hot-Blooded Dinosaurs*. The Dial Press, New York, 238pp.

Dingus, L. and Rowe, T. 1997. *The Mistaken Extinction*. W. H. Freeman and Company, New York, 332 pp.

Farlow, J. O. 1990 Dinosaur energetics and thermal biology. In Weishampel, D. B., Dodson, P. and Osmolska, H. (eds.), *The Dinosauria*, 1st edn. University of California Press, pp. 43–55.

Fisher, P. E., Russell, D. A., Stoskopf, M. K., Barrick, R. E., Hammer, M. and Kuzmitz, A. A., 2000. Cariovascular evidence for an intermediate or higher metabolic rate in an ornithischian dinosaur. *Science*, **288**, 503–505.

Feduccia, A. 1973. Dinosaurs as reptiles. *Evolution*, **27**, 166–169.

Gauthier, J. A. and Gall, L. F. (eds.), *New Perspectives on the Origin and Early Evolution of Birds*, Proceedings of the International Symposium in Honor of John H. Ostrom. Yale Peabody Museum Press, New Haven, CT, 613pp.

Hopson, J. A. 1977. Relative brains size and behavior in archosaurian reptiles. *Annual Review of Ecology and Systematics*, **8**, 429–448.

Hopson, J. A. 1980. Relative brain size in dinosaurs: implications for dinosaurian endothermy. In Thomas, R. D. K. and Olson, E. C. (eds.), *A Cold Look at Warm-Blooded Dinosaurs*. American Association for the Advancement of Science Symposium no. 27, pp. 287–310.

Horner, J. R., de Ricqlès, A. and Padian, K. 2000. Long bone histology of the hadrosaurid dinosaur *Maiasaura peeblesorum*: growth dynamics and physiology based upon an ontogenetic series of skeletal elements. *Journal of Vertebrate Paleontology*, **20**, 115–129.

Horner, J. R., Padian, K. and de Ricqlès, A. 2001. Comparative osteohistology of some embryonic and perinatal archosaurs: developmental and behavioral implications for dinosaurs. *Paleobiology*, **27**, 39–58.

Jerison, H. J. 1973. *Evolution of the Brain and Intelligence*. Academic Press, New York, 482pp.

Johnston, P. A. 1979. Growth rings in dinosaur teeth. *Nature*, **278**, 635–636.

Ostrom, J. H. 1969a. Terrestrial vertebrates as indicators of Mesozoic climates. In *Proceedings of the North American Paleontological Convention*, pp. 347–376.

Ostrom, J. H. 1969b. Osteology of *Deinonychus anthirropus*, an unusual theropod from the lower Cretaceous of Montana. *Bulletin of the Peabody Museum of Natural History, Yale*, **30**, 1–165.

Ostrom, J. H. 1974. Reply to "Dinosaurs as Reptiles." *Evolution*, **28**, 491–493.

Ostrom, J. H. 1976, *Archaeopteryx* and the origin of birds. *Biological Journal of the Linnean Society of London*, **8**, 91–182.

Paul, G. S. 1988. Physiological, migratorial, climatological, geophysical, survival, and evolutionary implications of Cretaceous polar dinosaurs. *Journal of Paleontology*, **62**, 640–652.

Paul, G. S. 1997. Dinosaur models: the good, the bad, and using them to estimate the mass of dinosaurs. In Wolberg, D. L., Stump, E. and Rosenberg, G. D. (eds.), *Dinofest International*. Academy of Natural Sciences, Philadelphia, pp. 129–154.

Reid, R. E. H. 1997. Dinosaurian physiology: the case for "intermediate" dinosaurs. In Farlow, J. O., and Brett-Surman, M. K., (eds.), *The Complete Dinosaur*, Indiana University Press, Bloomington, IN, pp. 449–473.

Rich, P. V., Rich, T. H., Wagstaff, B. E., McEwen Mason, J., Douthitt, C. B., Gregory, R. T. and Felton, E. A. 1988. Evidence for low temperatures and biologic diversity in Cretaceous high latitudes of Australia. *Science*, **242**, 1403–1406.

Romer, A. S. 1933. *Vertebrate Paleontology*. University of Chicago Press, Chicago, 491pp. (3rd edition, 1966, 468pp.)

Ruben, J. A. 1991. Reptilian physiology and the flight capacity of *Archaeopteryx*. *Evolution*, **45**, 1–17.

Ruben, J. A., Jones, T. D. and Gesit, N. R. 2003. Respiratory and reproductive paleophysiology of dinosaurs and early birds. *Physiological and Biochemical Zoology*, **76**, 141–164.

Russell, L. S. 1965. Body temperature of dinosaurs and its relationship to their extinction. *Journal of Paleontology*, **39**, 497–501.

Spotila, J. R., O'Connor, M. P., Dodson, P. and Paladino, F. V. 1991. Hot and cold running dinosaurs: body size, metabolism, and migration. *Modern Geology*, **16**, 203–227.

Thomas, R. D. K. and Olson, E. C. (eds.), 1980. *A Cold Look at the Warm-Blooded Dinosaurs*. AAAS Selected Symposium no. 28, 514pp.

Wieland, G. R. 1942. Too hot for the dinosaur! *Science*, **96**, 359.

CHAPTER 16

Patterns in dinosaur evolution

Throughout much of this book, we have considered dinosaurs as individual taxa; who they were, what they did, and how they did it. But of course dinosaurs were never isolated in nature. They were parts of complex ecosystems and owe their identities as much to their ecological context as to their phylogenetic relationships. The problem is that phylogenetic relationships are considerably easier to reconstruct than the vagaries of ecosystems now defunct over 65 Ma. Our initial goal here will be to take a first pass at understanding the great rhythms of dinosaur evolution, and then to integrate these into a large-scale ecological context, to obtain a feel for what dinosaurs as a group did over time, and to what kinds of global event they were responding.

Preservation

Table 16.1 shows the distribution of dinosaurs among the continents through time. The paucity of dinosaur remains from Australia and Antarctica, however, is surely more a question of preservation and inhospitable conditions today for finding and collecting fossils than one of where dinosaurs actually lived. Into this mix must be factored geological preservation; some time intervals simply contain more rocks than others. The most notorious example of this is the Middle Jurassic: clearly this was a time of great significance in the evolution of dinosaurs, but terrestrial rocks that could contain their remains

Table 16.1. Distribution of dinosaurs on continents during the Mesozoic Era. Filled areas indicate dinosaurs known

	Asia	Africa	South America	North America	Europe	Australia	Antarctic
Late Triassic	■	■	■	■	■	■	
Early Jurassic	■	■	■	■	■		■
Middle Jurassic	■	■	■	■	■	■	
Late Jurassic	■	■	■	■	■		
Early Cretaceous	■	■	■	■	■	■	■
Late Cretaceous	■	■	■	■	■	■	

serendipitously happen to be quite rare. The result is that the Middle Jurassic appears to be a time of very low tetrapod diversity (Box 16.1). Without the rocks to preserve the fossils, we have nothing to work with. The Late Cretaceous is rather the opposite; continental rock sequences of the Upper Cretaceous, thousands of meters thick, are preserved in both Asia and North America. The result is that we have a rich record of Late Cretaceous dinosaurs. Methods of estimating the completeness of fossil preservation, and thereby mitigating these problems, have been developed (Box 16.2), but these have not been applied to most dinosaur faunas.

Temporal distribution

The University of Pennsylvania's Peter Dodson has estimated that each dinosaur species existed between 1 and 2 million years. This means that every 1–2 million years, one species of a particular genus died out and another took its place. Occasionally a new genus and species took the place of an old genus and species. Thus, not only did new species appear as old ones disappeared, but new genera as well. One to two million years is a relatively high rate of taxonomic turnover, and it may have been accelerated by mass extinction events (see Chapter 18) that periodically left earth depleted and vacant for colonization by new dinosaurian faunas.[1]

Ultimately then, the manner in which these faunas are handled here condenses many millions of years and large geographical regions. It

---

1 The high rate of dinosaur species turnover has interesting consequences. Obviously, fossils are found within sedimentary rocks. Sedimentary rocks are commonly grouped into bodies called formations, which are simply rocks that can be mapped and which share similar features. Most terrestrial formations have taken longer than 1 million years to deposit, which means that a dinosaur species found in one formation will probably not occur in the one above or below it; too much time would have elapsed. Thus successive formations contain different species; there is virtually never an instance of a particular species found in one formation and then in the one above or below it. It is therefore very difficult to track the fates of individual species; terrestrial deposition being episodic as it is, species associated with one formation are long gone before the time of the next formation.

─ BOX 16.1 ──────────────────────────────

The shape of tetrapod diversity

Over the past 20 or so years, Bristol University's M. J. Benton has been compiling a comprehensive list of the fates of tetrapod families through time (Figure B16.1.1). We see several interesting features of the curve that results from this compilation. Note the drop in families during Middle Jurassic time. This, as we have seen, is an artifact; that is, a specious result. This particular one comes from the lack of localities more than from a true lack of families during the Middle Jurassic. Then, notice the huge rise in families during the Tertiary. Some of this may be real, and perhaps attributable in part to Tertiary birds and mammals (both of whom are tremendously diverse groups) but some of it might be another artifact, due to what is called the "pull of the Recent." The pull of the Recent is the inescapable fact that as we get closer and closer to the Recent, fossil biotas become better and

better known. This is because more sediments are preserved as we get closer and closer to the Recent, and a greater amount of sedimentary rocks preserved means more fossils. The big spike at the end of the Jurassic is the Morrison Formation, a unit that preserves an extraordinary wealth of fossils.

So a curve such as Benton's requires skill to understand and to factor out the artifacts. Nonetheless, we can see that, generally, dinosaurian diversity increased throughout their stay on earth and, as they progressed through the Cretaceous, dinosaurs continued to diversify. The increase in diversity shown in Benton's diagram may reflect the increasing global endemism of the terrestrial biota, itself driven by the increasing isolation of the continental plates.

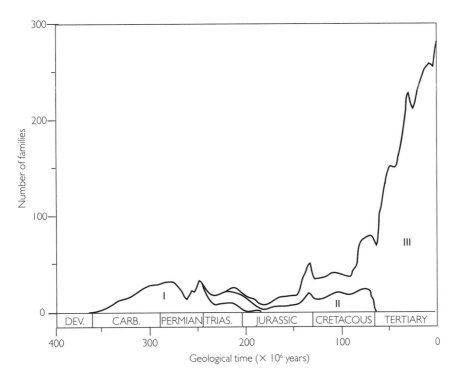

Figure B16.1.1. M. J. Benton's (1999) estimate of vertebrate diversity through time. On the x-axis is time; on the y-axis is diversity as measured in numbers of families. Benton recognized three successive vertebrate faunal assemblages (I, II, III).

BOX 16.2

Counting dinosaurs

There are several ways in which the true diversity of dinosaurs can be estimated. In this box we introduce one of them and exemplify it with Ceratopsia.

Although we have heretofore emphasized the use of cladograms for reconstructing evolutionary relationships, they can also be seen as indicating the relative sequence of evolutionary events. For animals with a fossil record – such as dinosaurs – this relative sequence can be compared to the sequence of appearance in the stratigraphic record. The phylogenetic sequence ought to mirror the stratigraphic sequence.

In addition, the combination of phylogeny and stratigraphy has a lot to say about the presence and meaning of gaps in the fossil record. Suppose Dinosaur X and Dinosaur Y were each other's closest relative; suppose also that X comes from rocks dated at 100 Ma and Y comes from rocks of 125 Ma. This means that the ancestor of X and Y has to be at least 125 million years (myr) old. And, if that is true, then there must be some not-yet-sampled history between this ancestor and Dinosaur X – to the tune of 25 myr. That 25 myr gap is called a "minimal divergence time" (MDT). Lineages which must have existed, but for which there is no known record, are called "ghost lineages" by paleontologists M. Norell and M. Novacek. MDTs are measures of the duration of ghost lineages (Box B16.2.1).

Ceratopsians counted

It has sometimes been claimed that ceratopsians have one of the best fossil records of all dinosaurs. How can we test this? We begin with the raw data (32 species; less than are known in theropods sauropodomorphs, and ornithopods, but more than in ankylosaurs, stegosaurs, and pachycephalosaurs), averaged over the duration of the higher taxon (Ceratopsia) and learn from this that ceratopsians produced new species at the rate of 1/1.9 myr (by comparison sauropodomorphs produced

them at 1/1.4 myr and stegosaurs produced them at 1/5.6 myr).

To estimate the overall diversity of a particular group, however, we must include ghost lineages. For ceratopsians, MDT values range from 0 to 30 myr, with an average of just over 5 myr. These are among the smallest MDTs for all Dinosauria. Moreover, actual ceratopsian species counts are nearly 70% of all ceratopsians after ghost lineages have been added to the diversity total, suggesting that, indeed, the fossil record of Ceratopsia lives up to its billing.

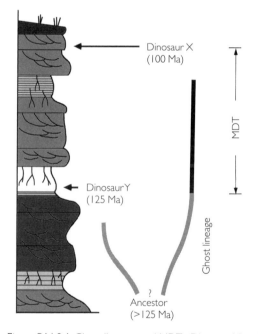

Figure B16.2.1. Ghost lineages and MDTs. Dinosaur X and Dinosaur Y are preserved 25 million years apart. If they are closely related, they both share a common ancestor that is at least as old as Dinosaur Y. Thus an estimate of the minimum divergence time (MDT) of the two lineages is as old as, or older than, Dinosaur Y (125 Ma). The record of that divergence is unpreserved and is therefore called a ghost lineage (curved gray lines on the figure).

is presented in this way to summarize the 160 or so million years of global dinosaur evolution. Considered on that scale, it is a reasonable approach. But recall that dinosaur species turn over at rates of 1–2 million years. If this is so, when we present, for example, the huge fauna from the last 30 million years of the Cretaceous, we are combining in time, and possibly in space, animals that certainly did not live together. So, considered on a global scale of 10s of millions of years, we can recognize large-scale evolutionary patterns from the lists in this chapter. But, considered on a smaller scale, the relationships of the individual faunas to each other may be illusory.

We can think of these geographical and temporal distributions as pages in a notebook, in which each succeeding page represents a new time interval, with new continental arrangements, and new and different dinosaurs populating the continents. Considered this way, the sequence of dinosaurs through time is like a grand pageant through earth history, in which each interval of time has a characteristic fauna that takes on a characteristic quality (Figure 16.1).

Finally, the fact that we are North American scientists in combination with the fact that European and North American fossil localities have been studied longer than those elsewhere and the fact that those studies are more accessible to us than those elsewhere all conspire to give this chapter a slightly Western slant. This is unfortunate. We have said this before and we will undoubtedly say this again: the answers to many of the questions posed in this book certainly reside in the localities and fossils of Asia, South America, Antarctica, Australia, and Africa. In many respects, these are the places that represent the future of our discipline.

A *caveat* is necessary: our summary cannot be fully comprehensive. Many places have produced a wealth of indeterminate material, as well as eggshells and ichnotaxa. We have not attempted to incorporate these finds into our discussion.

Dinosaurs through time

Figure 16.2 is a good place to begin: a compilation of dinosaur generic diversity over time. A graph such as this allows us to track large-scale, global fluxes in dinosaur populations through the approximately 160 million years that they were on earth.

Late Triassic (227–205 Ma)

Figure 16.2 shows that dinosaurs radiated quickly in the Late Triassic. We already know from Chapter 5, however, that this early step remains tantalizingly shrouded in the mists of antiquity: exactly how dinosaurs came to be the dominant terrestrial vertebrates in the Late Triassic is still unclear. We do not know whether dinosaurs moved quickly into ecospace abandoned by other vertebrates, as suggested by M. J. Benton, or whether dinosaurs, possessing presumably superior adaptations for terrestrial locomotion (and what else?), as suggested by A. J. Charig, outcompeted pre-existing forms and took charge, ecologically speaking (although we incline toward the former interpretation; see Chapter 5).

These potentially dramatic ecological steps are not so easily revisited, because the diversity of early dinosaurs is small, and the times

Figure 16.1. The great historical pageant of dinosaurs through time. The paleoenvironments of each time interval – and the dinosaurs that populated them – all have different qualities that characterized each successive ecosystem.

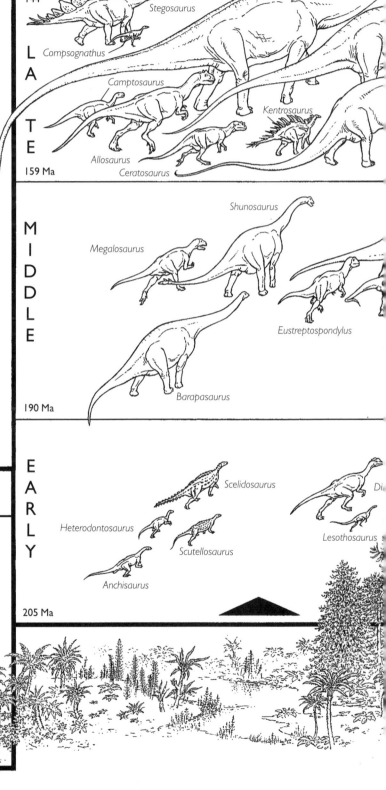

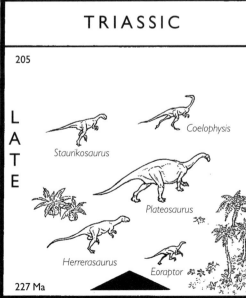

CRETACEOUS

LATE

65

Maiasaurae

Edmontonia

Oviraptor

Parasauralophus

Triceratops

T. rex

Euoplocephalus

Troodon

Gallimimus

Saltasaurus

Edmontosaurus

Protoceratops

Velociraptor

Centrosaurus

Pachycephalosaurus

Therizinosaurus

Alamosaurus

Carnotaurus

99 Ma

EARLY

Carchardodontosaurus

Psittacosaurus

Suchomimus

Deinonychus

Utahraptor

Ouranosaurus

Protarchaeopteryx

Polacanthus

Amargasaurus

Baryonyx

Sinosauropteryx

Wuerhosaurus

Iguanodon

Hypsilophodon

Caudipteryx

Brachiosaurus

144 Ma

Apatosaurus

Ornitholestes

plodocus

rchaeopteryx

Cetoisaurus

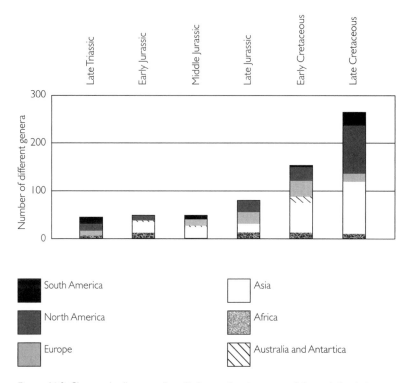

Figure 16.2. Changes in dinosaur diversity by continent measured through the Late Triassic–Late Cretaceous time interva. Each vertical bar shows the total number of different genera known from that particular time interval. Viewed from this perspective, dinosaurs appear to have steadily increased in diversity as the Mesozoic progressed. (Data from Fastovsky *et al.*, 2004.)

of their appearances are not particularly well known. The earliest dinosaurs known are *Herrerasaurus* and *Eoraptor* from the Ischigualasto Formation of Argentina, reliably dated at 228 Ma. Indeed, phylogenetic and biogeographical perspectives point to South America as the cradle of Dinosauria. Later Late Triassic dinosaurs include prosauropods (*Plateosaurus* and *Thecodontosaurus*), primitive ornithischians such as *Pisanosaurus*, and theropods such as *Liliensternus* and *Coelophysis*. The geographical distributions of dinosaurs rapidly enlarged: these dinosaurs are North America, South Africa, and Europe. Their appearances are too sparse, and their distribution is too wide for us to track the increase in diversity that must have occurred between the earliest appearance of dinosaurs and their clear ecological importance by Early Jurassic time. Even the dates – for those units for which we have them – are problematical. For example, the well-known, highly studied Late Triassic Chinle Formation – the source of *Coelophysis* – was dated in the early 1980s at 239 Ma, then in the early 1990s at 207 Ma, and most recently (2003) our best date (to date) is a reliable – finally (!) – 213 Ma.

Still, what we can be sure of is that Late Triassic terrestrial vertebrate faunas were not dinosaur dominated; rather they were an eclectic

Figure 16.3. Late Triassic therapsids and a very early mammal. (a) A large herbivore (the dicynodont *Kannemeyeria*); (b, c) two carnivorous cynodontians (*Cynognathus*); and (d) an early mammal (*Eozostrodon*).

mixture, including therapsids (advanced synapsids), archosaurs, primitive turtles, and some crocodile-like amphibians called "labyrinthodonts."

Members of the therapsid clade were of two major types: squat, beaked, tusked herbivorous therapsids called dicynodonts and a variety of carnivorous and herbivorous animals that must have looked and acted very much as mammals did. The earliest mammals, tiny, shrew-sized, omnivorous or insectivorous creatures, were also present. As it turned out, their appearance on earth was approximately coincident with – or even slightly preceded – that of dinosaurs. Examples of representatives of these clades are shown in Figure 16.3.

The earliest turtles also appeared during the Late Triassic, belying the idea that turtles are extremely primitive animals. From even the earliest and most primitive example (*Proganochelys* from the Late Triassic of Germany), characteristic turtle features are present (although the group did undergo significant evolutionary change early in their history), suggesting significant evolution prior to their first appearance (Figure 16.4).

And as we have seen from Chapter 5 (Figure 5.8) archosaurs played a major role in these early faunas as well. Among the most common members of the fauna were the crocodile-like phytosaurs (long-snouted, aquatic, fish-eaters). The early crocodyliform *Protosuchus* seems to have been more fully terrestrial than today's representatives of the group. Then there were a cloud of carnivorous primitive archosaurs, some large, quadrupedal, and heavily armored, and some small, lightly built, and

(a)

(b)

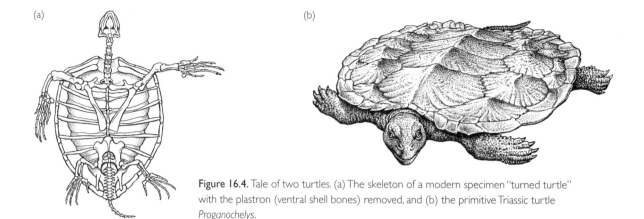

Figure 16.4. Tale of two turtles. (a) The skeleton of a modern specimen "turned turtle" with the plastron (ventral shell bones) removed, and (b) the primitive Triassic turtle *Proganochelys*.

not armored. Finally, there were aetosaurs, a group of large, armored, quadrupedal archosaurs that were vaguely crocodile-like but strictly herbivorous, with strange snubbed noses. Examples of these primitive archosaurs are shown in Figure 16.5.

Our reckoning of Late Triassic archosaurs would not be complete without some mention of pterosaurs (Figure 16.6). The very first pterosaurs appeared during the Late Triassic very shortly after the earliest (known) dinosaurs (see below). Already by this time they were highly specialized for flight, suggesting that, like turtles, significant evolution occurred before the Late Triassic. Be that as it may, no more primitive pterosaurs that indicate how they evolved their flight apparatus are known, and, once equipped with the morphology with which we find them, they were there to stay until the end of the Cretaceous.

(a)

(c)

(b)

(d)

Figure 16.5. Assorted primitive archosaurs. (a) A phytosaur (*Rutiodon*); (b) an aetosaur (*Staganolepis*); (c) a rauisuchian (*Postosuchus*); and (d) the primitive crocodilian (*Protosuchus*).

Figure 16.6. The primitive rhamphorhynchoid pterosaur *Dimorphodon.*

Other tetrapod clades at the time included a variety of amphibians, including the previously mentioned crocodile-like labyrinthodonts (Figure 16.7). These flattened aquatic creatures with upward directed eyes and immense mouths were clearly ambush carnivores, with a long and distinguished Paleozoic fossil record.

Contintental distributions and the Late Triassic fauna

What kinds of evolutionary force might have been driving the distinctive Late Triassic faunas? We have talked about competition and extinction

Figure 16.7. A "labyrinthodont" amphibian (metoposaur) grabbing a snack.

(see Chapter 5), but might other forces be involved? It is becoming clear that the very distributions of the continents themselves may play a role in the composition of global faunas.

Consider this example from the modern world: the large herbivore fauna of Africa is rather different from that of North America. And both differ from that of India. There are no physical connections among these continents that would allow the fauna of one to spread to the other. Therefore, each of these faunas – in fact, the ecosystems of which they are a part – has developed in relative isolation and is distinct with its own characteristics. This type of distinctness is called endemism. A fauna that is closely allied with a particular geographical region (large or small) is called indigenous or endemic. A region that is populated by distinct faunas unique to it is said to show high endemism. High endemism can be caused by evolution on widely separated continents, because there is no opportunity for faunal interchange.

Alternatively, if faunas appear very similar to each other, then it is likely that some land connection was present to allow the fauna of one continent to disperse to the other continent. Thus, we can imagine a region characterized by low endemism, because the continents are closely allied with each other, and there are extensive opportunities for faunal interchange.

It is now clear that, during the Triassic and Early Jurassic, global terrestrial vertebrate faunas were characterized by unusually low endemism. The supercontinent of Pangaea still existed during this time, and land connections existed among all continents (see Figure 2.5). The interesting mixtures of faunas outlined here look similar on a global scale during these time intervals. While Pangaea remained united, land connections existed and endemism was low. Here, then, is an excellent example of large-scale, abiotic events driving and modifying large-scale patterns of biotic evolution.

Although it was not so obvious in the Late Triassic, it is clear from our Recent vantage point that the Triassic fauna we have described consisted of "forward-looking" and "backward-looking" organisms. Turtles, pterosaurs, dinosaurs, and mammals represented the future of terrestrial vertebrate life during the Mesozoic. Labyrinthodont amphibians and therapsids, on the other hand, were representatives of the past, doomed to disappear into oblivion as the Jurassic got under way.

Early Jurassic (205–190 Ma) The Early Jurassic was the first time on earth when dinosaurs truly began to dominate terrestrial vertebrate faunas. Although they continued to share the limelight with some relict non-amniotes (see Chapter 4), a few of the very highly derived, mammal-like therapsids (as well as with the rather diminuitive contemporary mammals), turtles, and, of course, pterosaurs and the newly evolved crocodilians, many of the players in the terrestrial game were by now dinosaurs. Interestingly, the Early Jurassic faunas inherited some of the quality of low endemism that characterized the Late Triassic world. The unzipping of Pangaea was in its very earliest stages (see Figure 2.6), and it had not gone on so long that the fragmentation of the continents was yet reflected through increased global endemism. That was to await the Late Jurassic.

By the Early Jurassic, dinosaurs were no longer mainly theropods and prosauropods, although of course both these groups were well represented. Instead, ornithischians had by now made their appearance. These ornithischians were rather primitive, having just begun differentiation into one or two of the five major groups of ornithischians each of which forms a chapter in Part II of this book. Although prosauropods were still abundant during the Early Jurassic, true sauropods also made their appearance during this time interval.

Middle Jurassic (190–159 Ma)

The Middle Jurassic has historically been an enigmatic time in the history of terrestrial vertebrates. As noted in Chapter 2, Middle Jurassic terrestrial sediments are quite uncommon. When we look at the total diversity of tetrapods through time (see Box 16.1), the curve all but bottoms out during the Middle Jurassic. Did vertebrates undergo massive extinctions at the end of the Early Jurassic? Probably not. More likely, the curve is simply reflecting the absence of terrestrial Middle Jurassic sediments on earth. Without a good sedimentary record to preserve them, we can have little knowledge of the faunas that came and went during that time interval.

Regardless, the Middle Jurassic must have been an important time in the history of dinosaurs. We know that the dismemberment of Pangaea was well underway by this time. By the Late Jurassic, dinosaurs had diversified tremendously, and endemism was on the rise. Many of the diverse faunal elements that characterized earlier faunas – the advanced therapsids, for example, were largely out of the picture. The insignificant exception to this, of course, were the mammals, who were still around, but hanging on by the skin of their multi-cusped, tightly occluding teeth. Thus, by the Late Jurassic, dinosaurs had truly consolidated their hold on terrestrial ecosystems. The evolution that preceded this must have commenced sometime during the Middle Jurassic, which must therefore have been a kind of pivot point in the history of dinosaurs. It's a shame that we cannot know more of this crucial time.

Late Jurassic (159–144 Ma)

By the Late Jurassic, global climates had stabilized and were generally warmer and more equable (less seasonal) than they presently are (see Chapter 2). Polar ice, if present, was reduced. Sea levels were higher than today. The fact that earth's plates had by this time migrated to more disjunct positions (see Figure 2.8) than they had in the Late Triassic suggests that we should expect to find greater endemism in the terrestrial faunas of the Late Jurassic, as indeed we do.

For North Americans, and arguably for the world, the Late Jurassic has been called the Golden Age of Dinosaurs. Many of the dinosaurs that we know and love were Late Jurassic in age. Somehow that special Late Jurassic blend of equable climates, small brains, and massive sizes epitomizes dinosaur stereotypes and exerts a compelling fascination on humans.

In fact, the Late Jurassic produced a remarkable variety of forms. Many were large: gigantic sauropods (*Brachiosaurus*, *Diplodocus*, *Camarasaurus*, among others) as well as theropods that, if not *T. rex*-sized, still reached upwards of 16 m. Of course, not that all theropods were large; small, gracile ones abounded, and, as we have seen (see Chapter 12), it was during the Late Jurassic that the first unambiguous birds appeared. Moreover, this was the time of stegosaurs, ornithopods, and even a few ankylosaurs.

Who else was out there? Turtles had by this time explored both the terrestrial and marine realms. Pterosaurs had undergone much evolution and, although primitive forms with tails continued through the Late Jurassic, more derived tail-less forms had evolved. Crocodilians underwent a remarkable radiation; one flashy Late Jurassic model with fins was fully sea-going, among other more conventional kinds. Some of the early representatives of lizards appeared during this time. By the Late Jurassic, dinosaurs had consolidated their dominance of terrestrial vertebrate faunas.

Early Cretaceous (144–99 Ma)

The first part of the Cretaceous was a time of enhanced global tectonic activity. With this came increased continental separation (see Figure 2.9), as well as greater amounts of CO_2 in the atmosphere. This in turn probably produced warmer climates through the "greenhouse effect." Climates from the Early through mid-Cretaceous (to about 96 Ma) were therefore warmer and more equable than today.

Who enjoyed these balmy conditions? There were representatives of groups including all of our old friends from earlier times as well as of a number of new groups of dinosaurs. The Early Cretaceous marks the rise of the largest representatives of Ornithopoda. Ankylosaurs also became a significant presence among herbivores of the Early Cretaceous times, as did the earliest of ceratopsians.

Moreover, the balance of the faunas seems to have changed. During the Late Jurassic, sauropods and stegosaurs were the major large herbivores, with ornithopods being represented primarily by smaller members of the group. Now, in the Early Cretaceous (and in fact, throughout the Cretaceous), ornithopods first make their mark. Sauropods and stegosaurs were still present, but the significance of these groups seems to have been greatly reduced. Was the spectacular Cretaceous ascendency of Ornithopoda due to the feeding innovations developed by the group? Perhaps; however, as we have seen, such a question is not easily answered by the fossil record. The parallel success of Ceratopsia in Late Cretaceous time and the independent invention by that group of similar feeding innovations suggest that sophisticated oral food-processing probably didn't hurt. Then, too, the Early Cretaceous witnessed a revolution in carnivorous theropods, most notably, the deinonychosaurs of both North America and China (see Chapters 12 and 13).

Lastly, the Early Cretaceous was a time of increased endemism. The days of truly global faunas were coming to a close by the end of the Early Cretaceous, and instead distinctive regional faunas were beginning to appear, such as the psittacosaur fauna of southern Mongolia and northern China. This pattern was most marked by Late Cretaceous, but look for it in the Early Cretaceous.

Late Cretaceous (99–65 Ma)

The Late Cretaceous was a great time to be a dinosaur. Perhaps never before in their history were they so diverse, so numerous, and so incredible. The Late Cretaceous boasted the largest terrestrial carnivores in the history of the world (tyrannosaurids, carcharodontosaurids, and giganotosaurs), a host of sickle-clawed brainy (and brawny) killers worthy of any nightmare, a marvellous diversity of duck-billed and horned herbivores, armored dinosaurs, and dome-heads.

Climate seems not to have affected diversity. In fact, although diversity increased, Late Cretaceous climates deteriorated from former, equable mid-Cretaceous time onward. By this we mean that, generally, climates were less stable, and that seasonality increased. As we have seen (see Chapter 2), this occurred concomitantly with a marine regression, which undoubtedly played a role in the destabilization of climate. Conditions were not dreadful; they certainly were no worse than they are today. The deterioration was not a sudden drop near the end (or beginning) of the Late Cretaceous; it was a gentle increase in seasonality that marked the entire 30 or so million years of the time interval. At the very end of the Mesozoic, there is no evidence for a sudden drop in temperatures, or any significant modification of climate.

Endemism

Endemism among dinosaurs reached an all-time high during the Late Cretaceous as the continents had rarely been less connected in earth history (see Figure 2.10). Southern continents tended to maintain the veteran Old Guard: a large variety of sauropods, some ornithopods, and ankylosaurs, and in South America the unusual abelisaurid theropods. In northern continents, however, new faunas appeared that were very different. Among herbivores, sauropods were still present, although very rare. Stegosaurs took their final curtain call (stegosaurs were already extinct globally with the exception of one equivocal record in the Late Cretaceous of India). But in their place lots of new creatures roamed, including pachycephalosaurs, ceratopsids, and hadrosaurids. As we have seen, many among these new inhabitants of earth traveled in herds (or flocks), and large-scale migrations have been proposed for these gregarious animals. Likewise, a northern Late Cretaceous panorama would show herds (flocks) of ornithomimosaurs roaming, a sight not common in previous times.

Finally there is the magnificent diversity of Late Cretaceous theropods. Although carnivores of nearly equivalent size had existed previously, nothing shaped quite like tyrannosaurids had ever been seen, or has existed since. However compelling tyrannosaurids may be, however, the Late Cretaceous theropod story might be better told in the diversity of smaller forms: oviraptorosaurs, dromaeosaurids, troodontids, and therizinosaurs.

The Late Cretaceous record of dinosaurs is largely divided between North America and Asia. This is not to suggest that the other continents are not represented; as Table 16.1 indicates, they clearly are. It simply means that we have a much better understanding of the northern faunas than the southern ones, and that our story is likely to change as more and better dinosaur material from southern continents is unearthed. North America and Asia, however, share not only a rich record but an extremely strong similarity between their respective faunas. These two continents share hadrosaurids (also known from Europe and South America; although rare), pachycephalosaurs (earliest known from Europe), ceratopsians, tyrannosaurids, and ornithomimosaurs. This faunal similarity has been recognized for some years; however, in the late 1980s and early 1990s, P. Currie of Canada's Royal Tyrrell Museum along with the University of Chicago's P. C. Sereno undertook a detailed analysis of faunal distributions and determined that many of the North American faunas had their origins in Mongolia. They proposed that there may have been multiple migrations across a Bering land bridge (much as humans are thought to have migrated to North America from Asia some 64 million years later!). Once again, we have an example of geological forces (tectonics) modifying evolution.

Out with a whimper or a bang?

One of the great enigmas of dinosaur paleobiology is of course the nature of the end of the dinosaurs. And while this subject is treated in detail in Chapter 18, analyses such as that shown in Figure 16.2 bear significantly upon this subject. Clearly there are several apparent episodes in which dinosaur diversity was unusually high: the Late Jurassic and the latter part of the Late Cretaceous being the most spectacular. But as we have seen, some of the contours in this figure are simply issues of preservation: because there are few terrestrial Middle Jurassic sedimentary rocks exposed on earth, the diversity of Middle Jurassic dinosaurs was probably far greater than it appears to be. The latter part of the Late Cretaceous, however, seems to have been a time of unusually high diversity – unusually, we say, by comparison with times of more "average" diversity. Below we consider some explanations of why the diversity might have been so high. But here, we consider a somewhat different question: did the diversity of dinosaurs decrease just before their eventual extinction?

Considered over the large scale (see Figure 16.2), dinosaur diversity increases strikingly and steadily over the course of the Late Triassic–Late Cretaceous interval. At this scale dinosaurs as a group appear to be doing better and better (e.g., their diversity is increasing) until they go extinct.

We can (and should), however, look at these data a bit more closely. Considering diversity solely within the Late Cretaceous, however, the time interval running 83 to 71 Ma (called the Campanian Stage) appears to contain by far the greatest dinosaur diversity, and the time interval 71 to 65 Ma (called the Maastrichtian Stage) shows an apparent decrease in diversity (Figure 16.8). Thus it seems as if, at the very end of the Cretaceous, dinosaurs were beginning to dwindle in diversity. These and

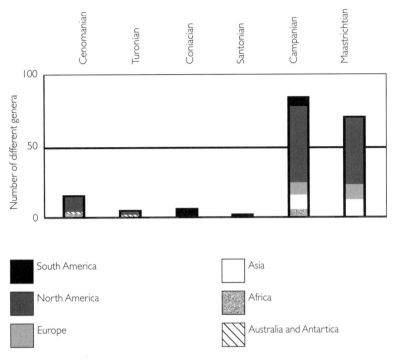

Figure 16.8. Changes in dinosaur diversity by continent measured through the Late Cretaceous. On the x-axis are names representing subdivisions of Late Cretaceous time. Each vertical bar shows the total number of different genera known from that particular subdivision. The paucity of samples from the Turonian, Coniacian, and Santonian are clearly artifacts of poor sampling in those time intervals. Viewed at this scale, diversity appears to decrease during the last stage of the Maastrichtian, suggesting to some paleontologists that dinosaurs began to decrease before the end of the Cretaceous. That decrease, however, must be considered in the context of the large increase in diversity during the penultimate stage (Campanian) of the Cretaceous (see the text). (Data from Fastovsky *et al.*, 2004.)

other data led University of California at Berkeley paleontologist W. A. Clemens and colleagues to conclude (with apologies to T. S. Eliot):

> This is the way Cretaceous life ended
> This is the way Cretaceous life ended
> This is the way Cretaceous life ended
> Not abruptly but extended.   (Clemens *et al.* 1981, pp. 297–298)

We have seen, however, that a literal reading of the fossil record may not always be the right one (remember the Middle Jurassic). Certainly the Campanian was a time of extremely high diversity, but this is in part because (1) the Campanian is a longer time interval than either its predecessors or its successor (the Maastrichtian), and (2) it contains some unusually fossiliferous deposits.

The Campanian spanned about 12 million years, double the 6 million year length of the Maastrichtian. Because the single number representing Campanian dinosaur diversity encompasses twice as much time as the duration of the Maastrichtian, it would be surprising if the

Campanian were not apparently more diverse than the Maastrichtian: more time, and thus more dinosaurs, are involved.

Moreover, preservational issues – analogous to those in the Middle Jurassic – may also be at work here. The Late Cretaceous of North America includes the greatest abundance and diversity of dinosaurs of all time. This is partly because Canada and the USA have been prospected by more or less well-funded field parties for about 150 years, but it is also simply the luck of the draw. Extensive Upper Cretaceous sediments are preserved on the eastern and southern coastal plain of the USA, as well as in the Western Interior of the continent, in a gigantic swath of sedimentary rocks running from the south (Mexico) all the way through much of the western part of the USA, to the central and even northern northern parts of the Canadian provinces of Alberta and Saskatchewan, to the northern slopes of Alaska. These rocks were deposited as a thick, east-facing wedge of sediment when the ancestral Rocky Mountains developed and were drained by east- and southeast-flowing river systems. This gigantic wedge of sediment produced by those streams includes some of the most famous Upper Cretaceous rocks of all time (Figure 16.9). In Canada, the Campanian is represented by the fluvial Dinosaur Canyon Formation and Milk River Formations, while the Maastrichtian is represented by the St Mary River Formation as well as the deltaic Horseshoe Canyon, Scollard, and Frenchman Formations. In the Western Interior of USA, the Campanian is represented by the fluvial Judith River and Two Medicine

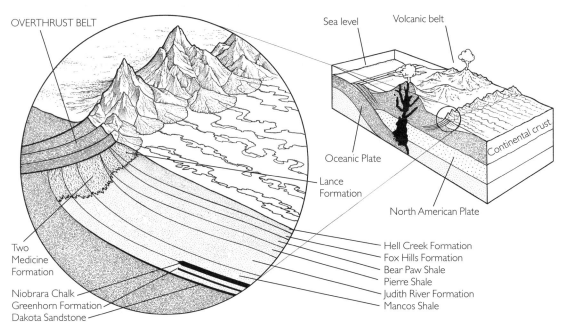

Figure 16.9. The thick wedge of sedimentation that was deposited in the western part of North America during the Late Cretaceous. As the ancestral Rocky Mountains were uplifted and then eroded by rivers, piles of sediment were deposited during floods into the geological basins adjacent to the mountains. In time this became the thick succession of sedimentary rocks that today preserve the famous dinosaur faunas from the upper Great Plains of the USA and Canada.

Formations; the Maastrichtian is represented by the St Mary River, North Horn, Laramie, Lance, and Hell Creek Formations. In the American southwest and southern Texas, the Kaiparowits, Fruitland, Kirtland, Aguja, and Javelina Formations have produced Campanian and Maastrichtian dinosaur faunas.

Who are these dinosaurs? A list would include tyrannosaurids (*Tyrannosaurus*, *Albertosaurus*, *Daspletosaurus*), ornithomimids (*Struthiomimus*, *Ornithomimus*), troodontids, dromaeosaurids (*Dromaeosaurus*, *Saurornitholestes* – not in North America!), oviraptorosaurs, a few small non-iguanodontian euornithopods (*Thescelosaurus*, *Parksosaurus*, *Orodromeus*), hadrosaurids (*Brachysaurolophus*, *Corythosaurus*, *Edmontosaurus*, *Gryposaurus*, *Lambeosaurus*, *Maiasaura*, *Prosaurolophus*, *Parasaurolophus*), pachycephalosaurs (*Stegoceras*, *Gravitholus*, *Pachycephalosaurus*, *Ornatotholus*, *Stygimoloch*), ceratopsians (*Achelousaurus*, *Anchiceratops*, *Arrhinoceratops*, *Avaceratops*, *Brachyceratops*, *Centrosaurus*, *Einiosaurus*, *Leptoceratops*, *Pachyrhinosaurus*, *Pentaceratops*, *Chasmosaurus*, *Styracosaurus*, *Montanoceratops*, *Torosaurus*, *Triceratops*), and ankylosaurs (*Ankylosaurus*, *Edmontonia*, *Panoplosaurus*, *Euoplocephalus*). In all of this, a few sauropods have been recovered, including the titanosaurid *Alamosaurus*.

A few finds have been made in the central and eastern parts of USA, including the theropod *Dryptosaurus*, an ornithomimid, the hadrosaurid *Lophorhothon* and *Hadrosaurus*. From Mexico comes the theropod *Labocania*, as well as a variety of other theropod, ankylosaur, hadrosaurid, and ceratopsian material referred to many of the genera described above.

On top of this wealth of preservation in North America, the Asian dinosaur faunas, particularly those from of the Djadochta, Barun Goyot, and Nemegt Formations of Mongolia, have also contributed to Campanian dinosaur diversity. Depending upon how one correlates these rocks with those in North America, dinosaurs such as *Protoceratops* and *Velociraptor* are surely Campanian; *Tarbosaurus* and *Saurolophus* are possibly Late Campanian or Early Maastrichtian. With ambiguous correlations as well as an unusually rich North American record, it would be easy to artifically inflate Campanian dinosaur diversity, to give an appearance that dinosaur diversity decreased from the Campanian to the Maastrichtian.

What does this all mean? Factoring out more rocks and longer, better-funded collecting, were dinosaurs in North America (and Asia) more diverse (and abundant) than in the rest of the world? Possibly, as indeed some authors have suggested that the Northern Hemisphere was simply where most of the progressive dinosaurs hung out. In terms of dwindling diversity at the end of the Cretaceous, however, it leaves an ambiguous picture with no clear story about final few million years of the dinosaurs on earth.

The Tertiary: after the ball is over

With the end of the Cretaceous, non-avian dinosaurs disappeared from earth forever. How suddenly or fast this occurred remains contentious (see Chapter 18), but the end of the Cretaceous literally was the end of an Era. Never again would the world see archosaur-dominated faunas;

Figure 16.10. Representative ferns, lycopods, and sphenopsids from the Mesozoic. Club moss: (1) *Pleuromeia* (Early Triassic). Ferns: (2) *Matonidium* (Jurassic–Cretaceous); (3) *Onychiopsis* (Jurassic–Early Cretaceous); (4) *Anomopteris* (Middle–Late Triassic); (5) Osmundaceae (Late Paleozoic–Recent). (6) Tree fern (Jurassic). Sphenopsids: (7) *Equisetum* (Late Paleozoic–Recent); (8) *Neocalamites* (Triassic–Lower Jurassic); (9) *Schizoneura* (Late Paleozoic–Jurassic).

mammals, once well entrenched as the dominant terrestrial vertebrates, would be no more likely to give up their place in Tertiary ecosystems to dinosaurs than dinosaurs were likely throughout the 160 million years of dinosaur incumbency to give up their place to mammals![2]

Plants and herbivores

As in most extant terrestrial mammalian communities, the majority of dinosaurs were herbivorous (a fact clearly appreciated in R. T. Bakker's assessment of dinosaur endothermy; see Chapter 15). It would not be surprising, therefore, to find some relationship between the

2 In 1987, J. K. Rigby, Jr, and colleagues of the University of Notre Dame, Indiana, reported a few relict specimens from eastern Montana that they claimed persisted past the end of the Cretaceous. While there is good reason to suspect that these authors are mistaken, the claim deserves no more than a footnote to the dinosaur extinction story, because as noted by P. Dodson, suppose it were true and a few stragglers staggered into the Tertiary. What difference would this make? "It's like driving along the freeway and seeing a Model T [Ford]," he said. "There may still be one or two relicts left driving, but that doesn't mean that anybody is still making the Model T!" As Model Ts are no longer parts of Second Millennial automotive ecosystems, so dinosaurs were not parts of early Tertiary ecosystems.

Figure 16.11. Representative cycads, gingkoes, gymnosperms, and angiosperms from the Mesozoic. (1) Cycadeoids (or bennettitaleans, as they are sometimes called; Triassic–Cretaceous); (2 and 5) *Williamsoniella* spp. (Triassic–Jurassic); (3) *Wielandia* (Jurassic); (4) *Williamsonia sewardiana* (Jurassic). Gingko: (6) *Gingkoites* (Triassic–Recent). Conifers: (7) Sequoia (mid Cretaceous–Recent); (8) *Araucaria* (Late Triassic–Recent); (9) *Pagiophyllum* (Triassic–Cretaceous). Angiosperms: (10) Magnoliaceae (magnolias; still small-flowered in the early days of their appearance on earth; Cretaceous?–Recent); (11) Nymphaeaceae (water lilies; Late Cretaceous–Recent). Inset: Seed (dicot) in cross-section. The cotyledons, shoot apex, and suspensor are all parts of the embryonic plant. The endosperm is a food source for the embryo as it develops, and the seed coat protects the embryo and its food source.

overall contours of herbivorous dinosaur evolution and their ultimate source of food: plant life. Several authors – notably B. H. Tiffney, D. B. Weishampel, D. B. Norman, S. Wing, P. Crane, and S. Lidgard – have struggled with this subject, and have obtained some interesting results.

Plants   Most paleobotanists – people who study extinct plants – recognize two major groupings of Mesozoic plants. The first is a non-monophyletic cluster of plants including ferns, lycopods, and sphenopsids (Figure 16.10). All of these plants tend to be low-growing and primitive, but like most land plants, they are vascular; that is, they possess specialized tissues that conduct water and nutrients throughout the plant.

The second major grouping of plants consists of the non-monophyletic "gymnosperms" and the angiosperms (a natural group better known as flowering plants). Together these two groups are united by the diagnostic character of possessing a seed (Figure 16.11). Seeds take varying configurations, but ultimately are nutrient-bearing pods apparently developed for the dissemination of gametes. "Gymnosperms" are best known as pines and cypress, and a lesser-known but Mesozoic-ly important group known as cycadophytes: plants with a pineapple-like stem and bunches of leaves springing out of their tops. Angiosperms today consist of magnolias, maples, grasses, roses, and orchids, among many other groups (Figure 16.11).

Several qualities distinguish these plant groups. In general, Mesozoic "gymnosperms" tended to be of two types: the more familiar conifers – epitomized, for example, by pines – were very tall and woody plants. These would have had relatively little nutritive value kilogram for kilogram, possessing coarse thick bark and cellulose-rich leaves. The modern representatives of these plants tend to secrete a variety of ill-tasting or poisonous compounds as a strategy to discourage their consumption; there is no reason to suppose that their Mesozoic counterparts were any different. The cycadophytes, on the other hand, tended to be fleshier and softer, with perhaps more nutritive value. Gingkoes would also have been plants available for dinosaur consumption (Figure 16.11).

Flowering plants appear to have evolved an entirely different approach to life. Far from discouraging herbivores from consuming them, they evolved a variety of strategies to actively court consumption by herbivores: bright tasty flowers, fruits, and tough seeds that can survive a visit to a digestive tract. Consumption by herbivores in the case of angiosperms appears to be a strategy for seed dispersal, not the destruction of the plant.

Dinosaur and plant co-evolution

Our analysis is built around Figure 16.12, which compares the record of Late Triassic through Late Cretaceous plant diversity with that of dinosaurian hervbivores. The lower part of the figure gives approximations of the global composition of plants through the time of the dinosaurs. The upper part of the figure is divided into various groups of herbivorous dinosaurs.

In terms of plants, Figure 16.12 shows some key patterns. Lycopods, seed ferns, sphenopsids, and ferns decrease in global abundance during the Late Triassic interval. From then until the end of the Mesozoic, they constitute a roughly constant proportion of the world's floras. Not so with the gymnosperms, which dramatically increase their proportion of the total global flora during the Late Triassic. And, it is clear that much of that increase is taken up by conifers, which constitute around 50% of the world's total floras through the rest of the Mesozoic.

Our best guess is that angiosperms were first invented in the very early part of the Cretaceous; however, it was during mid-Cretaceous times that they underwent a tremendous evolutionary burst. The uniquely efficient angiosperm seed dispersal mechanisms afforded by flowers were (and are) unparalleled in the botanical world, and consequently flowering plants have blossomed as no other group of plants has.

What is the relationship between dinosaurian herbivores and plants? It cannot be purely by chance that the rise of tall coniferous

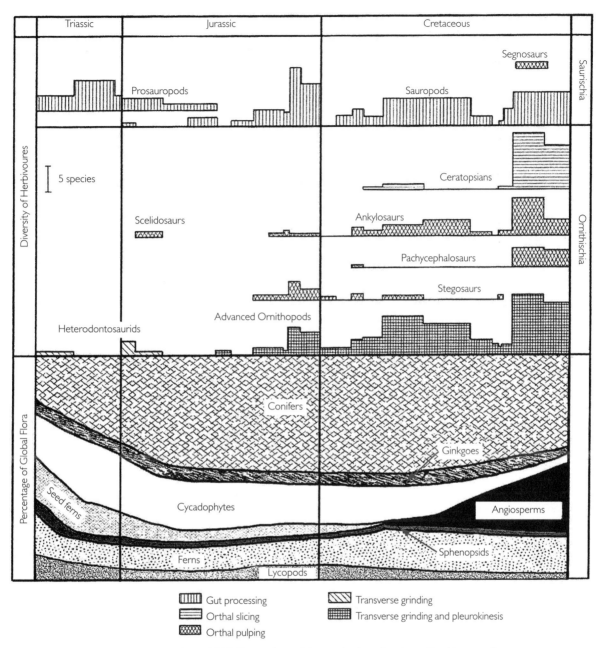

Figure 16.12. Comparison of changes in plant diversity and herbivorous dinosaur diversity during the Late Triassic through Late Cretaceous time interval. Upper part of the diagram shows the diversity of major groups of dinosaurs though time. Dinosaurs are divided into Ornithischia and Saurischia, and coded for food-processing method. Comparison between this diagram and that shown in Figure 16.2 suggests that one of the most important things driving dinosaur evolution and diversity was the development of new (or improved) ways to exploit the environments in which they lived (see the text). (Lower part of diagram redrawn from Tiffney (1997); upper part of diagram redrawn from Weishampel and Norman (1989). Entire figure redrawn from Fastovsky and Smith (2004).

BOX 16.3

Dinosaurs invent flowering plants?

In his popular book *The Dinosaur Heresies*, R. T. Bakker proposed that dinosaurs "invented" flowering plants. The germ behind Bakker's hypothesis is that Late Jurassic herbivores, epitomized by sauropods, were essentially high browsers, while Cretaceous herbivores, epitomized by ornithopods, ankylosaurs, and ceratopsians were largely low browsers. Bakker argued that Cretaceous low browsers put tremendous selective pressures on existing plants, so that survival could occur only in those plants that could disseminate quickly, grow quickly, and reproduce quickly. Angiosperms, he argued, are uniquely equipped with those capabilities. In his scenario, Bakker has Cretaceous low-browsing dinosaurs eating virtually all the low shrubbery, and plants responding by developing a means by which animals simply couldn't keep up with the growth, reproduction, and dissemination of the plants.

How likely is this? We are not sure. Troubling, of course, is the strongly North American and Asian cast of this hypothesis; southern latitude faunas seem to have had just the faunal compositions that Bakker claims would not have put intense selective pressure on contemporary low-growth floras. Yet, angiosperms were radiating worldwide by Late Cretaceous time. Still, what is significant about Bakker's hypothesis is that in it, he, as well as others such as S. L. Wing and B. H. Tiffney, are clearly recognizing and attempting to define the co-evolution between dinosaurs and plants.

forests is coincident with the appearance on earth of the world's first tall herbivores: prosauropods (and later sauropods). Within early populations, were those tall prosauropods that could take advantage of comparatively succulent leaves in the tree tops of conifers favored by natural selection? Alternatively, were conifers that were particularly tall favored by natural selection in response to the increasing height of prosauropods? Which is cause and which is effect is something we'll probably never know.

The figure also reveals another compelling relationship. The rise of the angiosperms occurs at approximately the same time as a major radiation in dinosaur groups. Were these groups – ceratopsians, pachycephalosaurs, hadrosaurids, and late-evolved ankylosaurs – groups that somehow took advantage of angiosperms as a food source and diversified? Is this a clue to what these dinosaurs were eating? Bakker had a different take: he inferred that, quite to the contrary, dinosaurs invented angiosperms (see Box 16.3). A more measured study on the evolutionary interactions between Mesozoic and Tertiary plants and herbivores was conducted by S. Wing of the Smithsonian Institution and B. Tiffney at the University of California at Santa Barbara. They described the probable diets of dinosaurian and other herbivores and the ways in which contemporary plants reacted to this kind of "predation." They found that, during the middle of the Mesozoic, few vertebrates fed selectively upon the relatively slow-growing conifers, cycads, and ginkgoes that formed the majority of terrestrial floras.

Even with the origin and spread of angiosperms during the early part of the Cretaceous, feeding consisted of low browsing and was rather generalized (much like a lawn-mower "grazes" over whatever is in its path, some have said). Because so many of these Mesozoic herbivores were also very large and may have lived in large herds, they probably also cleared expansive areas, trampling, mangling, uprooting, and otherwise disturbing areas that otherwise might be colonized by plants.

Such low-level, generalized feeding and disturbances of habitats tended to emphasize fast growth in plants, but discouraged any sort of forging of seed-dispersal relationships between plants and animals. Thus the picture of Mesozoic plant–herbivore interactions appears to be one in which (1) plants produced vast quantities of offspring to ensure the survival of the family line into the next generation and (2) herbivores took advantage of the rapidly and abundantly reproducing resource base to maintain their large populations of large individuals. Plant–herbivore co-evolution during the Mesozoic appears to have been based on habitat disturbance, generalized feeding, and rapid growth and turnover among plants.

Interestingly, with the extinction of the large dinosaurian herbivores at the end of the Cretaceous and the collapse of the plant–herbivore interactions that resulted, another suite of interactions developed. These emphasized more selective feeding by the much smaller mammalian herbivores, the development of seed dispersal complexes and evolution of fleshy fruits and seeds.

While the large-scale fossil record suggests this type of relationship, smaller-scale analysis leaves a more ambiguous picture. The "mummified" remains of hadrosaurids (*Edmontosaurus* and *Corythosaurus*) do not show the remains of angiosperms in the digestive tract, but rather the remains of coniferous plants. Likewise, K. Chin, who has studied the fecal remains (coprolites) of dinosaurs, reports that Late Cretaceous coprolites reliably attributed on the basis of size to dinosaurs (either ceratopsids or hadrosaurids) contained conifer fragments as well. If angiosperms were fueling this dinosaur radiation, where are the angiosperm pieces that we might hope to find?

The issue is further muddled by a closer look at the herbivorous adaptations of dinosaur groups that appeared in the great Late Cretaceous radiation. D. B. Weishampel and D. B. Norman, whose 1989 study is the source for the data in the upper part of Figure 16.12, classified groups of herbivorous dinosaurs on the basis of their chewing apparatus. They recognized gut processing (in which chewing is minimized; see Chapter 11), orthal pulping (in which chewing is rudimentary; see Chapters 6–8), orthal slicing, and transverse grinding with pleurokinesis (in which chewing is highly developed, and in some cases as efficient as anything invented by mammals; see Chapter 10). The overall pattern is clearly one in which increased chewing efficiency becomes more predominant in later dinosaurs. This is not to say that non-chewing dinosaurs were in a state of decline; as Figure 16.12 shows, sauropods – for whom chewing was a minimalist artform – were successful throughout the Cretaceous. Moreover, animals who indulged in rudimentary chewing – such as ankylosaurs and pachycephalosaurs – underwent strong evolutionary bursts during the latter part of the

Cretaceous. Still, ceratopsids and hadrosaurids – groups that elevated the art of chewing to new heights – are characteristic of the Late Cretaceous radiation. Hadrosaurids, of course, came from a long line of ornithopod chewers; they only refined the game a bit with pleurokinesis. Did advanced chewing mechanisms allow hadrosaurids and ceratopsids to take advantage of food resources not heretofore available to other dinosaurs?

We would be remiss if we did not mention therizinosaurs – the strange Late Cretaceous theropod foray into herbivory. By the chewing standards of their ornithopod brethren, these animals were mighty primitive. While speculation has analogized their behavior with mammals as diverse as sloths and the enigmatic, extinct, long-armed and heavy-clawed clicotheres, these dinosaurs remain fundamentally mysterious. Was there something about the rise of angiosperms that fueled this unusual radiation?

Carnivorous dinosaurs

And what about carnivorous dinosaurs (theropods)? Did the large-scale evolution of these mimic that of herbivorous dinosaurs? The short answer is a qualified "No." As we have seen, carnivorous dinosaurs are much rarer in life than their herbivorous cohort/counterparts (by some estimates, two orders of magnitude more rare). Moreover, the hollow bones of theropods don't preserve particularly well; transport (e.g., in a river) tends to shatter their thin walls. So any preservational problems are magnified in the case of theropods because of their rarity.

The basic theropod shape was set early on and changed little (see Chapter 12). Moreover, diverse lineages of theropods responded in similar ways to increasing body size: as they attained large sizes, they all tended to increase the size of the head, decrease the size of the arms, and lose fingers (particularly digits V and IV, and sometimes even III). Still, theropods – like their herbivorous counterparts – had a Cretaceous radiation. This is when we pick up forms as diverse as ornithomimosaurs, a high diversity of dromaeosaurs, and enigmatic light-bodied, sometimes-toothless theropods such as *Oviraptor*, *Ingenia*, and *Conchoraptor*. Since it is not clear what these animals did for a living, it is hard to attribute their evolution to a particular cause.

Theropod teeth appear from the outset as laterally compressed, serrated, and recurved. They have long been interpreted – especially via studies by J. O. Farlow – as multifunctional killing, slicing, and prey-restraint instruments. Because theropod behavior required all of these functions, there is no obvious trend in tooth evolution through the Mesozoic. There are distinctive differences in theropod teeth – compare, for example, the teeth of *Troodon* with those of the equivalently sized *Deinonychus*, the teeth of *Allosaurus* with those of *Carnotaurus*; the teeth of *Tyrannosaurus* with just about anything else (!) – but no particular trend in theropod tooth development, and hence no trend in theropod behavior, can be easily inferred throughout the Mesozoic. Even the pack behavior postulated for the Early Cretaceous *Deinonychus* cannot be ruled out for the Late Triassic *Ceolophysis*. Once a theropod, always a theropod.

Important readings

Bakker, R. T. 1986. *The Dinosaur Heresies*. William Morrow & Co., New York, 481pp.

Bakker, R. T. 1997. Raptor family values: allosaur parents brought giant carcasses into their lair to feed their young. In Wolberg, D. L., Stump, E. and Rosenberg, G. D. (eds.), *Dinofest International*, Proceedings of a Symposium Sponsored by Arizona State University. The Academy of Natural Sciences, Philadelphia, pp. 51–63.

Béland, P. and Russell, D. A. 1978. Paleoecology of Dinosaur Provincial Park (Cretaceous), Alberta, interpreted from the distribution of articulated remains. *Canadian Journal of Earth Sciences*, **15**, 1012–1024.

Benton, M. J. 1999. The history of life: large databases in palaeontoloy. In Harper, D. A. T. (ed.), *Numerical Palaeobiology: Computer-Based Modelling and analysis of Fossils and the Distributions*. John Wiley & Sons, Chichester and New York, pp. 249–283.

Brinkman, D. B. 1990. Paleoecology of the Judith River Formation (Campanian) of Dinosaur Provincial Park, Alberta, Canada: evidence from vertebrate microfossil localities. *Palaeogeography, Palaeoclimatology, Palaeoecology*, **78**, 37–54.

Chin, K. 1997. What did dinosaurs eat? Coprolites and other direct evidence of dinosaur diets. In Farlow, J. and Brett-Surman, M. K. (eds.), *The Complete Dinosaur*, Indiana University Press, Bloomington. pp. 371–382.

Chin, K. and Gill, B. D. 1996. Dinosaurs, dung beetles, and conifers: participants in a Cretaceous food web. *PALAIOS*, **11**, 280–285.

Chin, K., Torkaryk, T., Erickson, G. M. and Calk, L. C. 1998. A king-sized coprolite. *Nature*, **393**, 680–682.

Clemens, W. A., Jr, Archibald, J. D. and Hickey, L. J. 1981. Out with whimper not a bang. *Paleobiology*, **7**, 297–298.

Coe, M. J., Dilcher, D. L., Farlow, J. O., Jarzen, D. M. and Russell, D. A. 1987. Dinosaurs and land plants. In Friis, E. M., Chaloner, W. G. and Crane, P. R. (eds.), *The Origin of Angiosperms and their Biological Consequences*. Cambridge University Press, Cambridge, pp. 225–257.

Currie, P. J., Koppelhaus, E. B., and Muhammed, A. F. 1995. "Stomach" contents of a hadrosaur from the Dinosaur Park Formation (Campanian, Upper Cretaceous) of Alberta, Canada. In Sun, A. and Wang, Y. (eds.), *Short Papers of the Sixth Symposium on Mesozoic Terrestrial Ecosystems and Biota*, pp. 111–114.

Dodson, P. 1983. A faunal review of the Judith River (Oldman) Formation, Dinosaur Provincial Park, Alberta. *Mosasaur*, **1**, 89–118.

Dodson, P. 1987. Microfaunal studies of dinosaur paleoecology, Judith River Formation of Alberta. In Currie, P. J. and Koster, E. H. (eds.), *4th Symposium on Mesozoic Terrestrial. Ecosystems Short Papers of the Tyrell Museum of Palaeontology*, Drumheller, Alberta. pp. 70–75.

Dodson, P. 1990. Counting dinosaurs: how many kinds were there? *Proceedings of the National Academy of Sciences, USA*, **87**, 7608–7612.

Dodson, P., Behrensmeyer, A. K., Bakker, R. T. and McIntosh, J. S. 1980. Taphonomy and paleoecology of the Upper Jurassic Morrison Formation. *Paleobiology*, **6**, 208–232.

Fastovsky, D. E. 1989. Dinosaurs in space and time: the geological setting. In Culver S. J. (ed.). *The Age of Dinosaurs*. University of Tennessee, Knoxville, TN, pp. 22–33.

Fastovsky, D. E. 1995. Tetrapod recoveries from the Triassic–Jurassic and Cretaceous–Tertiary extinctions: comparing the incomparable. *Geological Society of America Abstracts with Programs*, **27**, A-164.

Fastovsky, D. E. 2000. Dinosaur architectural adaptations for a gymnosperm-dominated world. In Gastaldo, R. A. and W. A. DiMichele (eds.), *Phanerozoic Terrestrial Ecosystems*. The Paleontological Society Papers, vol. 6, pp. 183–207.

Fastovsky, D. E. and Smith, J. B., 2004, Dinosaur Paleoecology. In Weishampel, D. B., Dodson, P. and Osmólska, H. (eds.), *The Dinosauria*, 2nd edn. University of California Press, Berkeley, pp. 614–626.

Fastovsky, D. E., Huang, Y., Hsu, J., Martin-McNaughton, J., Sheehan, P. M. and Weishampel, D. B. 2004. The shape of dinosaur richness. *Geology*, **32**, 877–880.

Galton, P. M. 1985. Diet of prosauropod dinosaurs from the Late Triassic and Early Jurassic. *Lethaia*, **18**, 105–123.

Galton, P. M. 1986. Herbivorous adaptations of Late Triassic and Early Jurassic dinosaurs. In Padian, K. (ed.), *The Beginning of the Age of Dinosaurs*. Cambridge University Press, New York, pp. 203–222.

Hill, C. R. 1976. Coprolites of *Ptiliophyllum* cuticles from the Middle Jurassic of North Yorkshire. *Bulletin of the British Museum of Natural History*, **27** 289–294.

Johnson, K. R. 1997. Hell Creek Flora. In Currie, P. J. and Padian, K. (eds.), *Encyclopedia of Dinosaurs*. Academic Press, San Diego, pp. 300–302.

Johnson, K. R. 1999. The reconstruction of ancient landscapes: an example from the Late Cretaceous Hell Creek Formation of North Dakota. *Proceedings of the National Academy of Sciences, USA*, **53**, 134–140.

Lehman, T. M. 1987. Late Maastrichtian paleoenvironments and dinosaur biogeography in the western interior of North America. *Palaeogeography, Palaeoclimatology, Palaeoecology*, **60**, 189–217.

Lehman, T. M. 1997. Late Campanian biogeography in the western interior of North America. In Wolberg, D. L., Stump, E. and Rosenberg, G. D. (eds.), *Dinofest International*. Proceedings of a Symposium Sponsored by Arizona State University. The Academy of Natural Sciences, Philadelphia, pp. 223–239.

Lehman, T. M. 2001. Late Cretaceous dinosaur provinciality. In Tanke, D. H. and Carpenter, K. (eds.), *Mesozoic Vertebrate Life*. Indiana University Press, Bloomington, IN, pp. 311–328.

Lidgard, S. and Crane, P. R. 1988. Quantitative analyses of the early angiosperm radiation. *Nature*, **331**, 344–346.

Long, R. A. and Houk, R. 1988. *Dawn of the Dinosaurs*. Petrified Forest Museum Association, Petrified Forest National Park, 96pp.

Lull, R. S. 1953. The Triassic life of the Connecticut Valley. *Bulletin of the Connecticut Geology and Natural History Survey*, **81**, 1–336.

Norman, D. B. and Weishampel, D. B. 1985. Ornithopod feeding mechanisms: their bearing on the evolution of herbivory. *American Naturalist*, **126**, 151–164.

Norman, D. B. and Weishampel, D. B. 1987. Vegetarian dinosaurs chew it differently. *New Scientist*, **114**, 42–45.

Olson, E. C. 1971. *Vertebrate Paleozoology*. Wiley Interscience, New York, 839pp.

Parrish, J. M. 1989. Vertebrate paleoecology of the Chinle Formation (Late Triassic) of the southwestern United States. *Palaeogeography, Palaeoclimatology, Palaeoecology*, **72**, 227–247

Stokes, W. L. 1987. Dinosaur gastroliths revisited. *Journal of Paleontology*, **61**, 1242–1244.

Sun G., Dilcher, D. L., Zheng S. and Zhou, Z. 1998. In search of the first flower: a Jurassic angiosperm, *Archaefructus*, from northeast China. *Science*, **282**, 1692–1695.

Taggart, R. E. and Cross, A. T. 1997. The relationship between land plant diversity and productivity and patterns of dinosaur herbivory. In Wolberg, D. L., Stump, E. and Rosenberg, G. D. (eds.), *Dinofest International*. Proceedings of a Symposium Sponsored by Arizona State University. The Academy of Natural Sciences, Philadelphia. pp. 403–416.

Tiffney, B. H. 1989. Plant life in the Age of Dinosaurs. *Short Courses in Paleontology* no. 2, pp. 34–47.

Tiffney, B. H. 1992. The role of vertebrate herbivory in the evolution of land plants. *Palaeobotanist*, **41**, 87–97.

Tiffney, B. H. 1997. Land plants as food and habitat in the Age of Dinosaurs. In Farlow, J. and Brett-Surman, M. K. (eds.), *The Complete Dinosaur*. Indiana University Press, Bloomington, IN, pp. 352–370.

Weishampel, D. B., Barrett, P. M., Coria, R. A., Le Loeuff, J., Xu, X., Zhao, X., Sahni, A., Gomani, E. M. P. and Noto, C. R. 2004. Dinosaur distribution. In Weishampel, D. B., Dodson, P. and Osmolska, H. (eds.), *The Dinosauria*, 2nd edn. University of California Press, Berkeley, pp. 517–606.

Weishampel, D. B., and Norman, D. B. 1989. Vertebrate herbivory in the Mesozoic; jaws, plants, and evolutionary metrics. *Geological Society of America Special Paper no. 238*, 87–100.

Wing, S. L. 2000. Evolution and expansion of flowering plants. In Gastaldo, R. A. and DiMichele, W. A. (eds.), *Phanerozoic Terrestrial Ecosystems*. The Paleontological Society Papers, vol. 6, pp. 209–231.

Wing, S. L. and Sues, H.-D. 1992. Mesozoic and early Cenozoic terrestrial ecosystems. In Behrensmeyer, A. K., Damuth, J., DiMichele, W. A., Potts, R., Sues, H.-D. and Wing, S. L. (eds.), *Terrestrial Ecosystems through Time: Evolutionary Paleoecology of Terrestrial Plants and Animals*. University of Chicago Press, Chicago, pp. 326–416.

Wing, S. L. and Tiffney, B. H. 1987. The reciprocal interaction of angiosperm evolution and tetrapod herbivory. *Review of Palaeobotany and Palynology*, **50**, 179–210.

CHAPTER 17

Reconstructing extinctions: the art of science

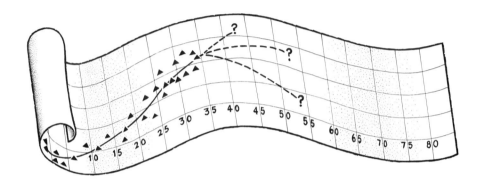

Sooner or later, everybody wants to know what happened to the dinosaurs. It is a question for which paleontology ought to be able to provide a satisfactory answer. But, as is so often the case, the question is deceptively simple.

Consider some of the problems with finding a "satisfactory answer." Suppose you think, as people have suggested, that dinosaurs died because the mammals ate their eggs. How do you test this hypothesis? Even if you were fortunate enough to find a fossil of a little mammal, snarling jaws wrapped around an innocent dinosaur embryo, would it prove that dinosaurs died because the mammals ate their eggs? It would be just another incredible, unique, fossil find, but certainly not adequate to infer the cause of dinosaur extinction. Likewise, the *au courant* idea that an asteroid killed the dinosaurs might be right, but you surely won't find even a common Late Cretaceous dinosaur like *Triceratops* flattened beneath a chunk of asteroid. At best, the evidence will be circumstantial.

It is not enough simply to find potential causes for the extinction. That is like looking at all potential murderers in the world to solve a particular murder case. Some relationship between the hypothetical cause and the extinction has to be established. For example, it is now pretty clear (as we shall see) that a large asteroid struck earth 65 Ma. So be it, but our question must be "What – if anything – does the asteroid have to do with this extinction?" Before we can blame the extinction on an asteroid

– or anything else – we need to find some kind of connection between the presumed culprit and the extinction event itself. In paleontology, as in a court of law, we affirm innocence until guilt is proven (or at least until the hypothesis is falsified).

Demonstrating what killed off the dinosaurs turns out to be non-trivial. It is easy to be seduced by the latest flashy theory, but instead one must dispassionately consider the data in a way that makes scientific sense. To do this, we need to develop a general understanding of mass extinction phenomena in general, and the dinosaur extinction in particular. Accordingly, this chapter and the next will forge an understanding of the kinds of inference that must be employed in order to reconstruct a mass extinction. As will become clear, the reconstruction of mass extinctions can involve as much art as science.

Extinctions

We all know that organisms have gone extinct. Certainly the present-day biota is not the same as one that existed 100 Ka (thousands of years ago), which in turn is not the same as one that existed 100 Ma, and so on. The biota has changed over time, and this is what produces the grand pagent of life described in Chapter 1. Extinction itself is straight-forwardly defined: when birth rates fails to keep up with death rates, we get extinction. But that definition begs the question. It says nothing about the nature or causes of the extinction, questions which turn out to be far more challenging and interesting.

Paleontologists generally divide extinctions into two categories. The first are the so-called background extinctions, isolated extinctions of species that occur in an on-going fashion. The second type are called mass extinctions. The latter certainly have caught the press and public's attention, but are they truly something qualitatively as well as quantitatively different from background extinctions?

Background extinctions

Although background extinctions are less glamorous than mass extinctions, they are essential to biotic turnover: University of Tennessee paleobiologist M. L. McKinney has estimated that as much as 95% of all extinctions can be accounted for by background extinctions. Isolated species disappear for a variety of causes, including out-competition, depletion of resources in a habitat, changes in climate, the growth or weathering of a mountain range, river channel migration, the eruption of a volcano, the drying of a lake, the spraying of a pesticide, or the destruction of a forest, grassland, or wetland habitat.

Dinosaur populations had a species turnover rate of around 2 million years per species. This means that each species lasted about 2 million years, before a new one appeared and the old one disappeared.[1] Although some dinosaur extinctions coincided with earlier mass extinction events

---

1 This is a simple statistical average for all of Dinosauria. It was not necessary for an older species to disappear before its descendant appeared. In other words, we are not positing a simple one-for-one replacement of species.

(such as those at the Triassic–Jurassic and Cretaceous–Tertiary boundaries), most dinosaurs fell prey to background extinctions. By far the majority of favorite and famous dinosaurs – *Maiasaura, Dilophosaurus, Protoceratops, Deinocheirus, Styracosaurus, Cetiosaurus, Iguanodon, Ouranosaurus, Allosaurus* – to name but a tiny fraction – were the victims of background extinctions. Indeed, the ultimate dinosaur extinction didn't wipe out the total species accumulated over 160 million years, it killed only the latest-evolved representatives of the group.

Mass extinctions

Mass extinctions are compelling events to try to understand, and the intellectual rewards in understanding one are great indeed. Aside from providing insights into some of the great mysteries of antiquity (such as the disappearance of the dinosaurs), understanding how past ecosystems have rebounded gives us some inkling as to how earth will recover from current human-induced ecological crises. For these reasons, many geologists and paleontologists have devoted their careers to reconstructing mass extinctions. Indeed, as we write these words, the study of mass extinctions is one of the most provocative and stimulating topics in the entire field of the geosciences.

Mass extinctions are somewhat poorly defined. They are said to involve *large numbers* of species and *many types* of species undergoing *global*

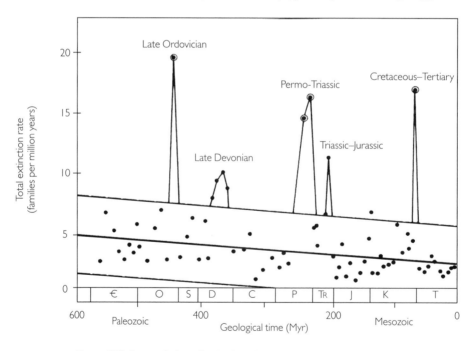

Figure 17.1. A compilation of extinctions through time, taken from the work of D. M. Raup and J. J. Sepkoski (1984). The five most significant were the Late Ordovician (438 Ma), the Late Devonian (380 Ma), the Permo-Triassic (245 Ma), the Triassic–Jurassic (205 Ma), and the Cretaceous–Tertiary (65 Ma). Є, Cambrian; O, Ordovician; S, Silurian; D, Devonian; C, Carboniferous; P, Permian; Tr, Triassic; J, Jurassic; K, Cretaceous; T, Tertiary.

extinction in a *geologically short* period of time. None of these has a truly precise definition, because there are no fixed rules for mass extinctions.

Indeed, how do we know that there even were mass extinction "events" and how can we recognize them? In 1983, University of Chicago invertebrate paleobiologists D. M. Raup and J. J. Sepkoski published a compilation of invertebrate extinctions through time (Figure 17.1). The compilation shows that, although extinctions characterize all periods (it is these that are termed background extinctions), there are intervals of time in which extinction levels are significantly elevated above background levels. Such intervals are said to contain the mass extinctions. Raup and Sepkoski recognized 15 intervals of mass extinction, of which five clearly towered above the others. These five are an early Paleozoic mass extinction (Late Ordovician; 438 Ma), a Middle Paleozoic mass extinction (Late Devonian; 380 Ma), a mass extinction that occurred right at the Paleozoic–Mesozoic boundary (Permian–Triassic boundary; 245 Ma), a Late Triassic mass extinction (205 Ma), and the K/T (Cretaceous–Tertiary) extinction (65 Ma).

Raup and Sepkoski divided all 15 extinctions into "minor," "intermediate," and "major" mass extinctions, based upon the amount of extinction that took place above background. In the entire history of life, only one extinction qualifies as "major": that is the Permian–Triassic (commonly called "Permo-Triassic") extinction. The remaining four of the Big Five – including dinosaur extinction – were all classified by Raup and Sepkoski as "intermediate." The rest of the 15 extinctions were considered "minor," although undoubtedly not to the organisms that succumbed during them.

Raup and Sepkoski's approach is worthy of note. They looked at the *pattern* of extinctions, and did not concern themselves with cause. The idea was to first establish the pattern, and, once this was understood, then to see whether the pattern of extinctions is in agreement with one or more of its potential causes. This will be a blueprint for our approach to the dinosaur extinction. Even establishing the pattern of the extinction, however, can be quite difficult. The Permo-Triassic extinction exemplifies this nicely.

The Permo-Triassic extinction is justifiably notorious. In it, as many as 96% of all species died, and as many as 52% of all taxonomic families disappeared from earth. Clearly, the magnitude of this extinction was extraordinary, measured either by counting species' disappearances or the disappearances of families. But 52% and 96% are very different numbers. Which most accurately characterizes what actually occurred 245 Ma? Was the world 50% empty, 90% empty, or neither? And which is relevant in terms of cause? Moreover, estimates of the duration of the Permo-Triassic extinction vary from essentially overnight to about 2 million years. Looking from a vantage point of 245 million years after the event, both are effectively geologically instantaneous, but the two are rather different in terms of probable causes.

Despite the mystery surrounding mass extinctions, paleobiologists' best guesses are that mass extinctions – although larger than background extinctions – may in most cases be caused by the same kinds of processes that cause smaller extinctions. Raup demonstrated that extinctions have a kind of periodicity, much like storms. Just as it is known that storms of a particular intensity occur at statistically regular intervals,[2] so it is that increasingly large mass extinctions occur increasingly infrequently. Raup calculated, for example, that the dinosaur extinction is a 100 million-year event, meaning that, on average, an extinction of that magnitude occurs once every 100 million years. In constructing the relationship extinctions and frequency, there is no particular jump to the category of extinction commonly considered a "mass extinction"; hence, Raup's conclusion was that mass extinctions are just more – much more – of the same.

Resolving the past

Fundamental to all problems of understanding the pattern of mass extinctions is the incompleteness of the fossil record, which means that, essentially, we must estimate the blanks and fill them in. These kinds of issue can lead reasonable, intelligent scientists of goodwill to intractably and vehemently defend mutually contradictory conclusions.

Resolution: the "R word"

The incompleteness of the fossil record is of two types: rocks do not record all of the time that has elapsed, and fossils are not preserved of every organism that has ever lived. In both cases, the incompleteness ultimately boils down to a question of resolution. In the study of optics, resolution is the degree to which one can distinguish the detail of things; the more one can distinguish detail, the better the resolution is said to be.[3] So it is in geology and paleontology. The better we can distinguish all of the components of an event that took place in the past and is preserved in the rock record, the better our resolution is said to be. Obviously, the business of understanding what took place in a mass extinction becomes in part the business of maximizing our resolution of the event.

Time resolution

We saw in Chapter 2 that rocks preserve the only tangible record of the time that has elapsed on earth. And we observed that the record of time is rather incomplete (see Figure 2.1). Estimates of the percentage of total earth time preserved in the rock record vary between 50% and 90%. This means that, globally, as much as 9 out of every 10 hours there are simply no rocks present recording the time that elapsed.

---

2 Storms are commonly divided into 5 year, 10 year, 50 year, and 100 year storms. The time designation is the statistical interval at which the storms reoccur; the intensity of the storms increases with their rarity. Sebastian Junger's book, *The Perfect Storm*, described a 100 year storm.

3 Televisions, computer monitors, comic books, and magazine pictures all rely upon the fact that human visual resolution is limited. If you had extremely well-developed eyesight, you would *resolve* a collection of dots instead of a picture. This was the underlying principle of the French impressionist Georges Seurat's *pointillisme* technique. Likewise, a magazine picture studied under a magnifying glass reveals the dots of color that our eye blends into an image. Recognizing art in the dots themselves, twentieth century American artist Roy A. Lichtenstein painted on a scale in which the dots and the images both were the visual objective.

Regardless of whether the rocks recording the time were eroded or were never deposited in the first place, the resolution is poor.

With resolution so poor, where we are able to study a particular event, even if it was originally global, becomes very restricted. We need to go wherever the time interval is present that preserves the event. In the case of the dinosaur extinction, many people – even professionals – are very surprised to discover that there are only a few places, most of which are concentrated in the upper Great Plains of North America, that preserve the last days of the dinosaurs. With the amount of rock that contains the last dinosaurs so locally restricted, care must be exercised in reaching global conclusions.

Paleontological resolution

Simply because something once lived does not guarantee that it will be preserved. Indeed, most fossil localities yield but a small fraction of the biota that once lived in the environment now represented by the locality. In part we know this because of accidents of preservation: extremely rare localities where tremendous numbers and many types of organism are preserved. By comparison with biotic diversity and abundance in modern environments, these special localities are probably preserving close to all of what lived in that place at that time. However, such marvels of preservation are very uncommon. By and large, localities preserve a very modest proportion of the original diversity. How modest? M. L. McKinney makes the following suggestion:

> There are an estimated 5 to 50 million species of life on earth today. Yet, since life first originated over 3.5 *billion* years ago, an estimated 1 to 3 billion species have come and gone. This means that over 99% of all species that ever existed are extinct. They will be found in the fossil record, if they are to be found at all. The question is, how complete is this record? ... Of the 1 to 3 billion species that have existed, only one to a few million, or less than 1% have probably been fossilized. Of those fossilized, only about 10% have so far been discovered and described ...
> (McKinney, 1993, pp. 123–124.)

Resolution and the reconstruction of extinction events

Many authors have likened the fossil record to a series of snapshots of the history of life. Another analogy might be a series of "stills," taken at irregular intervals, from a motion picture film. Consider a well-known extinction in terms of stills. Imagine a picture taken in eastern Montana, along Rosebud Creek, June 24, 1876. You'd see Lt. Col. G. A. Custer leading 250 well-equipped bluecoats of the Seventh Cavalry, flags flying and drums beating. A still photo taken as soon as the next day, would look utterly different; distributed across a west-facing slope just east of the Little Bighorn River would be the bodies of Custer and his annihilated troops. This landmark event occurred only 130 years ago, and yet – despite relentless investigations that began literally the day afterwards – we don't even know whether the Seventh Cavalry was destroyed in 15 minutes or 5 hours. Many aspects of the battle – including something as basic as its length – remain poorly understood; as with any extinction, the lack of resolution impedes understanding the events that took place.

Put most simply: *the poorer the resolution, the more abrupt any extinction event looks*. In one fossil locality (in our cinematic analogy, the last pre-extinction "still"), all the known members of the fauna are present; in the next younger fossil locality (the first post-extinction "still"), many of them are gone. Did they die out suddenly or did they die out over a long period of time? Considered uncritically, evidence from the localities shows an apparently dramatic extinction. But if we had more resolution and could see steps representing the time between the localities, we could better assess exactly how abrupt or gradual the extinction was.

Signor–Lipps effect

The difficulties in accurately characterizing the pattern of an extinction are increased by sampling problems, themselves related to the distribution of fossils in the rocks. The idea, first articulated in 1982 by paleobiologists P. W. Signor and J. H. Lipps (at that time both at the University of California at Davis) has profound implications in the reconstruction of mass extinctions. Suppose that, on average, fossils are distributed at some regular interval through the rocks[4] (whose thickness, you recall, represents some amount of time). Suppose, too, that we specify a moment in time after which there will be no fossils: we will call that an extinction boundary. As one approaches that boundary – that is, as one begins in older sediments and works upward (therefore through younger and younger sediments) to the boundary, the chances of finding a particular fossil become less and less in the remaining interval. This is because the area in which a fossil could potentially be found is decreasing. Indeed, chances are that there will be no fossils at exactly the moment of extinction, and that the "last" one (also the highest one) will occur somewhat before (or below) the boundary (Figure 17.2). Luis Alvarez, a Nobel prize-winning physicist, likened it to attempting to estimate the USA–Canada border by the home of the northernmost member of Congress. Because these homes are randomly distributed with respect to the border, the border must be estimated to be somewhat north of the northernmost member's house.

Here is another way to consider this problem. Think of a handful of different coins, which are tossed on the floor of a room toward a wall. As you get closer and closer to the wall (and the amount of area being

---

4 In fact terrestrial fossils such as dinosaurs are almost never distributed at regular intervals throughout sediments. Rather, the environment represented by the kind of rock that entombs them controls their distribution. Here, however, we assume a regular distribution to more easily explain the Signor–Lipps effect.

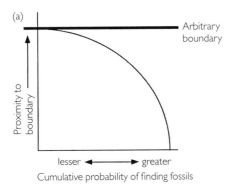

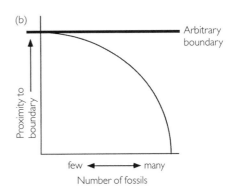

Figure 17.2. The Signor–Lipps effect. The fossil are randomly distributed throughout the rock sequence; however, as any arbitrary boundary is approached, the probability of maintaining (a) the cumulative fossil diversity of the rock sequence becomes smaller and smaller as the search area decreases, with the result that the extinction appears gradual even if (b) the distribution of fossils is in fact constant throughout the rock sequence.

searched becomes smaller and smaller), the probability of finding a particular type of coin – say, a nickel – becomes less and less in the remaining area. Indeed, as the area being searched becomes smaller and smaller (as one approaches the wall), the probability of finding any coin becomes less and less. So it is with fossils. As a boundary is approached, the cumulative probability of finding fossils is diminished because the area being searched becomes smaller.

Suppose you are truly omniscient, and you know that an extinction occurred suddenly, at a particular point in time. What would be the pattern of fossil distribution in the rocks? Because of the Signor–Lipps effect, as the place in the rock where the moment of extinction occurred (the boundary) is approached, the total variety of fossils found would gradually drop off. This is because the search area below the boundary is becoming smaller and smaller (as we get closer and closer to the zone of the extinction). The last occurrences of all fossils, therefore, would be some distance (and thus some amount of time) before the extinction boundary, because the rock is not simply a solid mass of fossils. With all such last occurrences at various distances from the boundary, *the extinction itself would appear to have occurred gradually when in fact it was instantaneous*. A real example of this is shown in Figure 18.12.

Higher taxa How we measure the magnitude of an extinction can have everything to do with how we perceive it. We saw that the Permo-Triassic extinction involved as much as a 96% extinction at the species level, but "only" a 52% extinction at the family level (52% still seems rather extreme to us). Both of these metrics are correct; the implication is that much of the species-level diversity is bound up within a relatively few families. But, as we asked before, what does this mean in terms of the magnitude of the extinction? The choice of taxonomic level clearly deserves careful consideration.

One might at first think that the most refined (lowest) taxonomic level would provide the most accurate reflection of an extinction. Indeed, it tells you in a detailed way who went extinct and who did not, but even this can lead to problems. Many fossils cannot be identified down to the species level. We have a choice, therefore, of either ignoring a potential piece of data (the fossil), or using it at a higher taxonomic level.

For example, suppose you are interested in proportions of fossils in an assemblage (e.g., as Bakker was when he formulated the predator:prey ratios; see Chapter 15). Unfortunately, preservation being what it is, suppose that you found 25 theropod fossils, of which you could only identify two below the level of Theropoda. Would it make sense to ignore the information about assemblage composition implied by 23 specimens, and consider only two specimens? Clearly not; you could include *all* the specimens in your database by simply designating them as "theropods." On the other hand, the significance of your data will be at least slightly blurred by the use of higher taxonomic categories.

The use of higher taxonomic categories (those above species), however, leads to a much more complex problem. Since such higher categories – that is, categories above the species level – have no precise definition (see Chapter 3), it is almost certain that in using higher categories for an analysis, one might be comparing incongruent taxa – even though they are both designated as "families" (or "genera" or "orders" or "classes" or whatever) – they may not truly be evolutionarily equivalents. But what is an "evolutionary equivalent?" There is no exact definition for what is meant by the the word "family" (or any other higher taxonomic category; see Box 17.1); who is to say that the groups that were used in calculating the 52% "familial" extinction at the Permo-Triassic boundary were really comparable (so that the extinction really was 52% at the *familial* level)? Maybe some of those taxa you called families are better grouped as subfamilies or superfamilies. Since there is no exact definition of "family," how would you ever know?

We can analogize the problem this way: suppose we are city planners and a developer wants to change local zoning from residential to commercial. We would like to know how dire the effects will be. One way to look at the problem is that 23 people will be made homeless. But if those 23 people really consist of four five-person families, each with three children, and three grandparents in separate apartments, it could be said that seven independent households will be made homeless. Suppose, however, that the fathers in two of the families are brothers, and the mothers in the other two families are sisters. Slick, the developer, tells the town council, "Look, only two families will be affected by this zoning change." Slick is deliberately referring to extended families. Here, and in biology, because the term "family" (or any other taxonomic designation above the species) is not precisely defined, it takes on a variety of meanings which change its significance, depending upon who the speaker is and what he hopes to convey.

The argument has been made that as long as the categories are monophyletic, they can still be useful (because they represent groups of

BOX 17.1

Full disclosure: the ugly truth about higher taxa

The issue under consideration is, yet again, exactly what is a family – or any taxon above the species level? Is it simply equivalent branch points in the phylogenetic history of the organisms (Figure B17.1.1)? This definition proves not to be terribly useful: because evolution consists (as we have seen) of repeated branchings of lineages, how could we ever be sure that the groups that we are comparing really represent exactly equivalent branch points? We obviously cannot.

Are higher taxa designated by some specified amount of diversity contained? That is, could a family be defined as containing, for example, not fewer than three but not more than six genera? Obviously not;

this depends upon the family – and it depends upon the choice of genera (themselves higher taxonomic categories). Indeed, some families are monospecific (they contain but one genus that has one species).

The possibility exists that higher taxa could be defined by some measured amount of difference (or similarity) among genomes. This, of course, would not work for fossil organisms, and nobody has yet suggested that it could or should be applied in the taxonomy of Recent organisms.

Could a higher taxon be defined as some amount of disparity – that is, its members have a given quantity of morphological distance between each other? Modern computerized methods have enabled

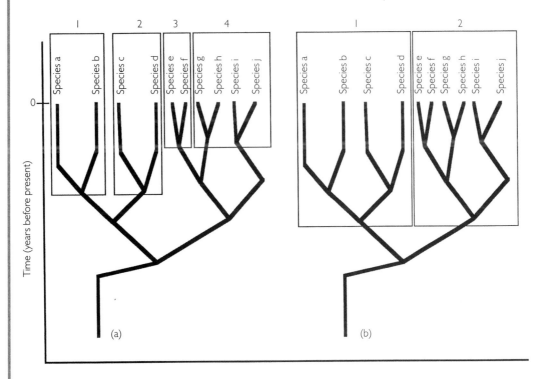

Figure B17.1.1. Two hypothetical trees showing all the speciation events leading to 10 species (of course, we can never really know all the speciation events that have occurred through time). In interpretation (a) the taxonomist has recognized four families. In interpetation (b), two families have been recognized. Because there is no absolute definition of "family," neither interpretation is wrong (or right); however, the taxonomic interpretation has made hypothesis (a) appear more diverse than hypothesis (b) even though both are interpretations of the same tree. If these families went extinct, should we say that two or four families went extinct? (continued on p. 422)

BOX 17.1 *(cont.)*

anatomists to digitally map morphology and is so doing quantify the differences between one shape and another. At this point, however, an organism's complete shape (measured in something called "morphospace") has yet to be defined (let alone measured), and the application of such quantities to taxonomy has thus never been attempted.

The problem ultimately lies with the higher taxonomic categories themselves. Linnaeus developed them in the middle 1700s, in an intellectual climate in which evolution (let alone evolution by natural selection) had barely been considered, let alone accepted. Because species were considered by Linnaeus to be essentially static, a static, hierarchical classification system made some sense. With the recognition of evolution as the

ultimate source of biotic diversity, the Linnaean system imperfectly fit what was and is now understood about biological diversity. In short, it remains convenient, but, as a tool for learning something about the history of life, it is fraught with significant problems. Because higher taxonomic categories are so poorly defined, many evolutionary biologists do not believe that taxonomic categories above the species level are useful in evolutionary reconstructions. Evolutionary biologist J. H. Woodger mordantly observed,

The doctrine of evolution is not something that can be grafted, so to speak, on to the Linnaean system of classification.
(Woodger, J. H., 1952, *British Journal of Philosophical Science.*)

evolutionary significance). This may be true, but it does not negate the problems raised above. It suggests that great care must be exercised in the way in which diversity in a mass extinction is measured, and that the assumptions that are embodied in the use of higher taxa be articulated and well understood.

Important readings

Donovan, S. K. (ed.) 1989. *Mass Extinctions: Processes and Evidence*. Columbia University Press, New York, 266pp.

McKinney, M. L. 1993. *Evolution of Life*. Prentice-Hall, Englewood Cliffs, NJ, 415pp.

Norell, M. A. and Novacek, M. J. 1992, The fossil record and evolution: comparing cladistic and paleontologic evidence for vertebrate history. *Science*, **255**, pp. 1690–1693.

Raup, D. M. 1991. *Extinction: Bad Genes or Bad Luck?* W. W. Norton and Company, New York, 210pp.

Raup, D. M. and J. J. Sepkoski, 1984. Periodicity of extinctions in the geological past. *Proceeding≥≥s of the National Academy of Sciences, USA*, **81**, 801–805.

Signor, P. W. and Lipps, J. H. 1982. Sampling bias, gradual extinction patterns and catastrophes in the fossil record. In L. T. Silver and Schultz, P. H. (eds.), *Geological Implications of Impacts of Large Asteroids and Comets on the Earth*. Geological Society of America Special Paper no. 190, pp. 291–296.

CHAPTER 18

The Cretaceous–Tertiary extinction: the frill is gone

The extinction to which the dinosaurs (non-avian dinosaurs, of course) finally succumbed after 160 million years on earth is called the "Cretaceous–Tertiary" extinction, commonly abbreviated K/T.[1] One of the most significant aspects of the K/T extinction – and one that commonly surprises people when they hear of it for the first time – is that it involved many events whose magnitude and significance far transcended the dinosaurs' extinction. Indeed, earth was redistributing its continents into a form very much as we find them today and, as has become clear in the past 25 years, a large extraterrestrial body – an asteroid – collided with earth. Moreover, the evidence is very strong that, for a period of time, the world's oceans were virtually "dead": that the great cycles of nutrients that form the complex food webs in the oceans temporarily shut down. By comparison with that, how important were the deaths of a few dinosaurs?

---

1 The extinction is thus said to have occurred at the Cretaceous–Tertiary (or K/T) boundary. The "T" in K/T obviously stands for the Tertiary Period. The "K" stands for the Latin word *creta*, which means chalk, which in German is the word *Kreide* – hence "K." The latest Cretaceous was first recognized at the well-known white chalk cliffs of Dover (England).

Geological record of the latest Cretaceous

Earth gets a makeover

Mountains and volcanoes

The Late Cretaceous was a time of active plate movement, mountain-building, and volcanism. The Rocky Mountains, Andes, and Alps were all entering important growth periods, fueled by extensive sea-floor spreading in the Pacific. With spreading rates significantly increased over previous times, the edges of the Pacific Basin became zones of subducted oceanic crust, lined by explosive volcanoes.

A unique episode of volcanism occurred between 65 and 60 Ma (from the very end of the Cretaceous into the Early Tertiary) in western and central India, consisting of a series of lava flows, called the Deccan Traps, which spewed molten rock over an area of 500,000 km^2. These flows were not the headline-grabbing Vesuvius/Krakatoa/Mt St Helens explosive volcanism that blasts clouds of gas and glass high into the atmosphere, making eerie red sunsets that are visible around the globe. Rather, they were pulses, in which immense volumes of basaltic lava sluggishly flowed from deep fissures in the earth's crust, cooling to form a broad plateau. The pulses occurred over several million years; sediments interlayered between the volcanic episodes attest to quiescent times of animal and plant habitation between flows. Volatile gasses – carbon dioxide, sulfur oxides, and possibly nitric oxides among the most prevalent – were emitted into the atmosphere, possibly affecting global temperatures and damaging the ozone layer.

Sea level

The Late Cretaceous was marked by a significant lowering of global sea level (called a regression), from high water stands enjoyed during mid-Cretaceous time (approximately 100 Ma). Evidence suggests that the regression maximum actually occurred slightly before the K/T boundary, and that global sea levels began to rise as 65 Ma came and went. It is clear that by the end of the Cretaceous, more continental land mass was exposed than had been in the previous 60 or so million years.

Seasons

The latter half of the Cretaceous seems to have been a time of gentle cooling from the highs reached in the mid-Cretaceous. While temperatures remained warm well into the Cenozoic, it is thought that by the Late Cretaceous, global mean temperatures had descended about 5 deg.C from mid-Cretaceous maxima. J. A. Wolfe, a paleobotanist with the U.S. Geological Survey, has argued that, in North America at least, climates through the Late Cretaceous were relatively equable, on the basis of plant fossils. Indeed, Wolfe suggested that the mean annual temperature range was just 8 deg.C at middle paleolatitudes (51–56° N). This is by contrast with modern conditions in approximately the same region, in which the mean annual temperature is about 15 °C. A latest Cretaceous dinosaur fauna north of the Arctic Circle is thought to have lived in a setting where the mean annual temperature was between 2 and 8 °C.

Asteroid impact   In the late 1970s, Walter Alvarez, a geologist at the University of California at Berkeley, was studying K/T marine outcrops now exposed on dry land near a town called Gubbio, in Italy. He was interested in learning how long it takes to deposit certain kinds of rock. He knew that cosmic dust (particulate matter from outer space) rains slowly and steadily down on earth. If he knew the rate at which the stuff falls, he could determine – by how much dust is present – how much time had elapsed during the deposition of a particular body of rock. So Alvarez ended up studying the amount of cosmic dust present in the rock.

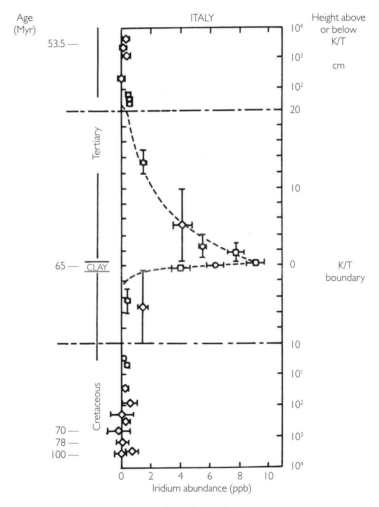

Figure 18.1. The iridium (Ir) anomaly at Gubbio, Italy. The amount of Ir increases dramatically at the clay layer to 9 ppb, and then decreases gradually above it, returning to a background count of about 1 ppb. On the right are numbers representing the thickness of the rock outcrop; on the left, the time intervals and rock types are identified. Note that the vertical scale is linear close to the K/T boundary but logarithmic away from the boundary, to show results well above and well below the boundary. (Redrawn from Alvarez et al., 1980.)

Gubbio is interesting. The lower half of the exposure is composed of a rock made up entirely of thin beds of the shells of Cretaceous micro-organsims (primarily microscopic creatures called dinoflagellates, coccolithophorids, and foraminifera). Abruptly, these beds are followed by a thin (2–3 cm) layer of clay, after which there begins (again, abruptly) the upper half of the exposure, made up almost exclusively of thin beds of the shells of Tertiary microorganisms. Obviously here was a K/T boundary, and right at the thin clay layer.

When Alvarez began his analysis for cosmic dust, he discovered unusually high concentrations of iridium (a rare, platinum-group metal) in the clay layer.[2] Iridium is normally found at the earth's surface in very low concentrations; it is found in higher concentrations in the core of earth and from extraterrestrial sources; that is, from outer space. Instead of the expected amount at earth's surface, about 0.3 ppb (parts per billion), the iridium content was an anomalous 10 ppb at Gubbio. So the iridium anomaly, as it came to be called, contained about 30 times as much iridium as Alvarez expected to find (Figure 18.1).

Was the source of the anomaly extraterrestrial or was it from the core of the earth? Discussion with Luis Alvarez, Walter's father and the Nobel Prize-winning physicist we met in Chapter 17, convinced Walter that the iridium had an extraterrestrial source. The case was sealed when the Alvarezes and two geochemist colleagues (Figure 18.2) at the Lawrence Berkeley Laboratories, Frank Asaro and Helen Michel, found iridium anomalies at two other K/T sites, one in Denmark and the other in New Zealand. The scientists concluded that at 65 Ma, an asteroid smacked into earth causing the K/T mass extinction. Luis Alvarez described the relationship between an asteroid impact and the iridium layer in this way:

When the asteroid hit, it threw up a great cloud of dust that quickly encircled the globe. It is now seen worldwide, typically as a clay layer a few centimeters thick in which we see a relatively high concentration of the element iridium – this element is very abundant in meteorites, and presumably in asteroids, but is very rare on earth. The evidence that we have is largely from chemical analyses of the material in this clay layer. In fact, meteoric iridium content is more than that of crustal material by nearly a factor of 10^4. So, if something does hit the earth from the outside, you can detect it because of this great enhancement. Iridium is depleted in the earth's crust relative to normal solar system material because when the earth heated up [during its formation] and the molten iron sank to form the core, it "scrubbed out" [i.e., removed] the platinum group elements in an alloying process and took them "downstairs" [to the core].

(Alvarez, 1983, p. 627.)

2 It is a common misconception that iridium is a deadly, toxic metal. In fact, like its chemical relatives gold and platinum, it is quite unreactive. For those with significant disposable incomes, costly fountain pens and watches made with iridium are available.

Figure 18.2. The team of University of California (Berkeley) scientists who first successfully proposed the theory of an asteroid impact at the K/T boundary. Left to right, geochemists Helen V. Michel and Frank Asaro, geologist Walter Alvarez, and physicist Luis Alvarez. (Photograph courtesy of W. Alvarez.)

On the basis of what was inferred to be a worldwide distribution (from the three sites sampled), the asteroid was estimated to be about 10 km in diameter.

Alvarez and Alvarez, Asaro, and Michel published their conclusions in the scholarly journal *Science* on June 6, 1980. Rarely has a scientific article provoked more controversy. Resistance to the idea came in two forms: some people challenged the conclusion that an asteroid hit at all, and others questioned what the work of the Alvarezes and their colleagues had to do with extinctions in general and dinosaurs in particular.

In the years since 1980, a tremendous amount of work has been done to explore the possibility of an asteroid impact 65 Ma. Most importantly, through the 1980s, the number of K/T sites with anomalous concentrations of iridium at the boundary has continued to climb, until today anomalous concentrations of iridium have been found in well over 100 K/T boundary sites around the world (Figure 18.3). In fact, as the Alvarezes continued to update a well-known diagram showing iridium-rich K/T localities on a world map, William A. Clemens, a University of California paleontologist, likened the diagram to the ubiquitous signs outside McDonald's restaurants: he wryly observed that the Alvarez diagram showed "over 100 iridium anomalies served!"

With the increase in anomalies came an increase in the understanding of background levels of iridium throughout the world. A particularly important demonstration of the ubiquity of the iridium anomaly came from U.S. Geological Survey paleobotanist Robert Tschudy and Los Alamos National Laboratories geochemist Carl Orth and their colleagues, who in 1983 discovered that the iridium anomaly occurred on

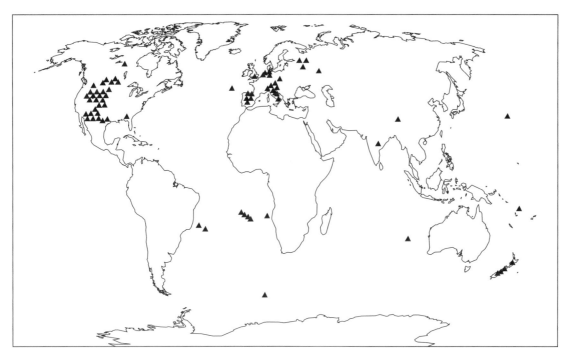

Figure 18.3. One hundred and three known iridium anomalies around the world.

land as well as in ocean sediments, in New Mexico and in eastern Montana (and later, in North Dakota, Wyoming, and southern Canada; Figure 18.4).

Figure 18.4. The iridium-bearing clay layer in Montana; one of the first localities on land where anomalous concentrations of iridium were discovered. (Courtesy of Milwaukee Public Museum.)

Figure 18.5. Shocked quartz. The etched angled lines across the face of a grain of quartz represent a failure of the crystal lattice along known crystallographic directions within the mineral. Grain is 70 μm (1μm = 10^{-6} m) across. (Photograph courtesy of B. F. Bohor, U.S. Geological Survey.)

Along with these discoveries, shocked quartz came to be recognized as part of the fingerprint left by the asteroid. Shocked quartz is the name given to quartz that has been placed under such pressure that the crystal lattice (a repeating sequence of atoms), which is normally organized in planes at 90° to each other, becomes compressed and deformed (Figure 18.5). In time, it generally came to be recognized that such lattice-deforming pressures could be generated at earth's surface only by impacts. The discovery of shocked quartz at the K/T boundary by U.S. Geological Survey mineralogist B. Bohor was therefore strong evidence that an extraterrestrial body was involved.

Another clue to the existence of an asteroid impact are micro-tektites, small, droplet-shaped blobs of silica-rich glass. They are believed to represent material from meteors that is thrown up into the atmosphere in molten state owing to the tremendous energy released when a meteor strikes earth. Quick cooling occurs while they're still airborne and then they plummet down on earth as a solid.

Volcanism

Were there any explanations possible for the coincidence of these apparently unusual geological features other than an asteroid impact? In 1983, volcanologist W. H. Zoller and colleagues observed that material being ejected from the vent of the Hawaiian volcano Kilauea was rich in iridium (which had evidently come from the core by way of the mantle). Moreover, Neville Carter of Texas A&M University and colleagues demonstrated that shock quartz metamorphism can occur through volcanic processes. Could the platinum-group metals, shocked quartz, and microtektites all have originated from more earthbound processes, such as the extensive volcanism that occurred at the Deccan Traps?

Finding the "smoking gun"

For many geoscientists, volcanism was not a terribly satisfactory explanation for the features observed at the K/T boundary. The Deccan Traps produced earthbound flood basalts, rather than the kind of explosive event necessary to blast impact materials into the stratosphere so that they could be distributed globally. Moreover, the timing was a bit late for the K/T boundary. To top it off, a variety of authorities, including U.S. Geological Survey planetary geologist Eugene Shoemaker and U.S. Geological Survey petrologist G. Izett, concluded that the shock metamorphic features produced by volcanoes are different from those produced by impacts.

The ultimate solution to the problem came out of the Caribbean. As early as 1981, geophysicist Glen Penfield had reported a bowl-shaped structure 180 km in diameter located in the Yucatán Penninsula of Mexico, in the region near the town of Chicxulub (Figure 18.6), that he

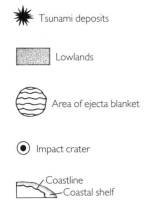

Tsunami deposits

Lowlands

Area of ejecta blanket

Impact crater

Coastline
Coastal shelf

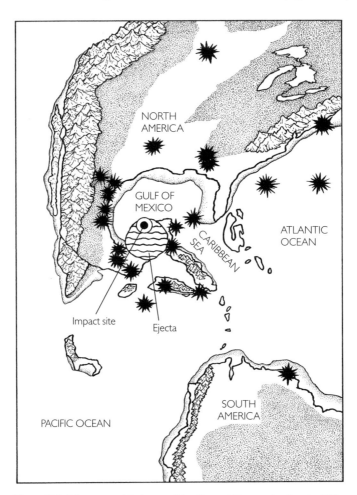

Figure 18.6. Paleogeographical map of the Ground Zero region for the K/T asteroid, Yucatán Penninsula, Mexico. The geography of the region as we know it today is superimposed over the geography of the region 65 Ma. (After Claeys et al., 1996.)

believed to be an impact crater. The Chicxulub structure (translated approximately as "devil's tail") is buried under many meters of more recent sediments. Ten years later, drill cores taken through the structure and reported in 1990 by geologist A. Hildebrand and colleagues, then at the University of Arizona, revealed shocked quartz. Moreover, the cores indicated that the structure was formed within a sequence of 1–3 km of evaporitic deposits, as well as some continental crust.

At about the same time, an approximately 1 m thick sequence of glass was discovered in Haiti, suggesting that the source of the glass had to be somewhere relatively nearby. Its chemical composition was shown by University of Rhode Island volcanologist Haraldur Sigurdsson and colleagues to come from a combination of molten continental crust and carbonate rock deposits. The glass that seemed to come from the carbonate deposits was rich in sulfur, such as might be expected if the carbonate originally formed as an evaporite; that is, a carbonate salt precipitated from water by evaporation.

The pieces of the puzzle were really starting to fall into place. The match between the chemical composition of the material in which the Chicxulub structure occurs and the composition of the Haitian glass was impressive. The continental crust and evaporitic deposits were geochemically exactly those expected from the thick glass deposits in Haiti. Moreover, the location was right: close enough to Haiti to produce the thick glass ejecta sequence, and close enough to Montana, North Dakota, Wyoming, and southern Canada to blast sand-sized material to the center of the North American continent. Indeed, several years earlier (1988) University of Washington sedimentologist Joanne Bourgeois and colleagues reported on evidence of a tsunami in K/T deposits in the Gulf Coast region of Texas. The Chicxulub site was well situated to produce the tsunami deposits Bourgeois recognized in the sedimentary record. Finally, convincing evidence for the relationship between the Chicxulub crater and global events (including the extinction) came when the impact that produced it was dated at 65 Ma. The smoking gun had been found.[3]

Further study of Chicxulub revealed much of interest. Study below the surface of the earth using sophisticated geophysical techniques showed a bullseye pattern with a circular peak and a large concentric ring around it, representing topography preserved in buried rocks below the surface (Figure 18.7). Interestingly enough, the northwest part of the outermost ring is broken through. Planetary geologist P. H. Schultz

---

3 The history of the discovery was not without irony. A. R. Hildebrand and colleague W. V. Boynton wrote, "In 1990, we, together with geophysicist Glen Penfield and other co-workers, identified a … candidate for the crater" (Hildebrand and Boynton, 1991). Penfield demurred, noting that 10 years earlier, he and colleague Antonio Camargo had prophetically written, "We would like to note the proximity of this feature in time to the hypothetical Cretaceous–Tertiary boundary event responsible for the emplacement of iridium-rich clays on a global scale and invite investigation of this feature in light of the meteorite impact–climatic alteration hypothesis for the late Cretaceous extinctions" (Penfield, G.T., 1991, Pre-Alvarez impact. *Natural History*, **6**, 4). The elusive crater had been found just after the initial Alvarez *et al.* (1980) publication, only nobody noticed it for almost 10 years!

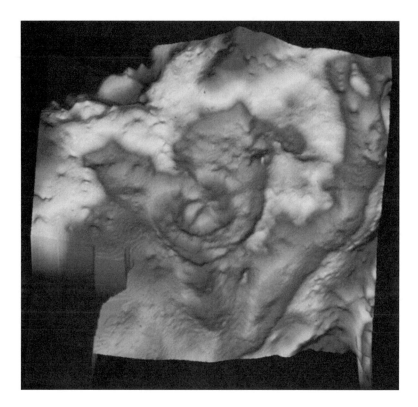

Figure 18.7. Three-dimensional geophysical reconstruction of the remnants of the Chicxulub crater. A gravimeter measures subsurface changes in gravitational attraction of rocks under the town of Chicxulub. These variations in gravitational attraction show a large-scale bullseye pattern of concentric rings, diagnostic of a meteor impact. North is toward top of page. (Image courtesy of M. Pilkington, Geological Survey of Canada, and A. Hildebrand, University of Calgary.)

of Brown University, who has spent a lifetime studying the patterns of impacts, used the distinctive pattern at Chicxulub to reconstruct the K/T asteroid event. He envisioned a large asteroid, 10–15 km in diameter (recall that previous estimates of size were based upon the global distribution of the ejecta; this estimate is based upon the morphology of the impact site) coming in from the southeast at a low angle of about 30°, which he believed is consistent with the morphology of the crater. The distribution of iridium, shocked quartz and micro-tektites across the Western Interior of North America (north and west of the crater) reinforce the idea of a low-angle, directional impact.

What did the asteroid do to earth when it struck? Numerous scenarios have been proposed, most of them inspired by post-nuclear apocalyptic visions.

- *Blockage of sunlight.* Based upon calculations of the size of the asteroid and the amount and size of material blasted into the atmosphere, it was initially hypothesized that sunlight would have been blocked for about three months. The blockage of sunlight, potentially causing a cessation of photosynthesis and a temperature decrease on earth, was termed an "impact winter." Estimates were calculated theoretically upon the amount of time that fine volcanic dust stayed in the air and the distance that such material can be dispersed based on modern explosive volcanic events such as the eruptions of Krakatoa, St Helens, and Pinatubo. Recently, however, K. Pope of Geo Eco Arc Research concluded that the amount of particulate matter kicked up from the impact would not have been sufficient either to stop photosynthesis or to cool the earth. He suggested that previous scenarios overestimated the amount of atmospheric dust by two to three orders of magnitude; he noted, however, that atmospheric aerosols and soot from wildfires (see below) might still have blocked sunlight as originally proposed.

- *Short-term global warming.* Tremendous amounts of energy in the form of heat must have been released immediately upon impact. Such heat release could have caused global temperatures to rise by as much as 30 deg.C for as long as 30 days. The initial global heat release at Ground Zero might have been 50 to 150 times as much as the energy of the sun as it normally strikes earth. One group of scientists likened this radiation of heat energy to an oven left on broil.

- *Global wildfires.* With so much instantaneous heat production, fires might have broken out spontaneously around the globe. Geochemist W. Wolbach of University of DePaul, and a variety of colleagues, described soot-rich horizons from five K/T sites in Europe and New Zealand. The amount of the element carbon (in the form of soot) was enriched between 100 and 10,000 times over background, a fact which they attributed to wildfires that overtook the globe as a result of the intensive heat release associated with the impact.

- *Acid rain.* In 1987, R. G. Prinn and B. Fegley of the Massachusetts Institute of Technology proposed that the initial "shock heating" of the atmosphere would produce nitric acid rain. The hypothesis was based on the fact that thermonuclear explosions, lightning, and modern, smaller meteors have all been shown to produce nitrous and nitric acids in the atmosphere. Sigurdsson and colleagues' work on the evaporite target rock of the impact, however, has suggested to some that sulfuric acid might be another candidate for the acid in the rain. Empirical evidence for significant acid rain, however, has never been discovered.

- *Long-term global cooling.* It is well known that dust from the comparatively miniscule eruption of Krakatoa caused a global decrease in temperature of about 1 deg.C. What would the asteroid have done? We have seen that not enough dust was ejected into the atmosphere to produce any kind of long-term global temperature changes. K. O. Pope noted that other aspects of the impact − perhaps soot or sulfate aerosols − might have caused global cooling as readily as dust. The heat transfer properties of water (see Chapter 2) would mitigate the extreme effects of the atmospheric cooling in the oceans, but far more severe effects could have been experienced on land.[4]

---

4 Cooling scenarios developed from this type of reasoning were instrumental in focusing thought on a problem that, unfortunately, is still timely: nuclear winter.

- *Long-term global warming.* Equally plausible is the possiblity of an enhanced "greenhouse effect," causing global warming. In this scenario,
the extermination of much of the world's biomass (see below) decreases
the amount of organic carbon taken up by the biota. This, in turn, would
produce more CO_2 in the atmosphere, which would absorb heat energy
radiated from earth, and *voilà!* – "greenhouse" conditions are induced.

The problem with all of these hypotheses of environmental modification is that the geological record doesn't really bear them out. For
example, Wolfe has estimated, on the basis of fossil plants, that North
American terrestrial deposits underwent an increase in the mean
annual temperature of about 10 deg.C in the first 500,000 to 1,000,000
years after the K/T boundary. However, the ocean record is more
equivocal. Some records, based upon stable isotopes, indicate a slight
warming after the boundary; others do not. The entire problem,
however, can be plausibly explained away by the claim that the
asteroid-induced climate change was simply too short to be clearly recognized in the fossil record.

Regardless, it is clear that, in a long-term sense, and by "long-term"
we mean on a million-year time scale, climates were little affected by the
asteroid impact. What we know is that climates in the latest Cretaceous
did not differ significantly from those in the early Tertiary.

Biological record of the latest Cretaceous

No amount of comets, volcanoes, asteroids, meteors, cooling, warming,
ice ages, or natural catastrophes can be used to explain any extinction
until we understand the anatomy of the extinction itself. The pattern
of extinction and survival becomes an important issue in understanding an extinction and determining potential causes.

The S-word resurfaces

It all boils down to stratigraphy. Suppose that the extinctions were
diachronous (*dia* – separate; *chronos* – time); that is, different groups
went extinct at different times. In this case, the K/T boundary would
occur in one place at one time and in another place at another time.
That would be unfortunate for attempting to reconstruct a pattern of
extinction, since it was the disappearance of the Cretaceous fossils and
the appearance of the Tertiary fossils (biostratigraphy) that are being
used to tell time.

What is needed here is a means of assessing time that is apart from
the fossils, against which the stratigraphic distributions of the fossils can
be compared. Most people immediately think of radiometric dates,
which allow one to tell time without using fossils (see Chapter 2). These,
when applicable, show that the K/T boundary is isochronous worldwide
to within about 500,000 years. This means that the record of extinctions
on land can be compared with that in the oceans, as long as it is understood that there is a half-million year margin for error. While this doesn't
sound particularly promising, there are other ways of linking the
oceanic and terrestrial records with greater precision. The problem is

that a lot can happen during 500,000 years that doesn't involve earth-shattering, heart-palpitating global catastrophes.

Is there a more refined way, *independent* of fossils, of identifying the K/T boundary globally? The answer is "Yes," and here is where the iridium anomaly assumes a scientific importance beyond its catastrophic allure. *If the iridium anomaly was caused by an asteroid impact of global proportions, then it represents a global isochron, and where it is preserved precisely and exactly identified that moment in time 65 Ma that we call the K/T boundary, independent of fossils.* If our ideas about its origin are correct (and the evidence strongly inclines us to believe that they are), then anywhere one finds the iridium layer he or she is looking at the same approximately *three months*. Considered in this way, the iridium anomaly is the most significant stratigraphic marker in all earth history.

Oceans
Continental seas and shelves

The marine regression of the last 2–3 million years of the Cretaceous exposed large expanses of the continents that previously had been innundated by ocean waters. This means that continental sea deposits can tell us very little about the last 2–3 million years of the Cretaceous. Because many groups of organisms lived and died in shallow continental seas, we lack data for many groups.

How well or badly bony fish fared remains largely conjectural, although it is apparent that, as a group, they did not suffer the kind of wholesale decimation seen in other groups. Not so in the case of plesiosaurs, the long-necked, Loch Ness-type, fish-eating diapsids of the Jurassic and Cretaceous (Figure 18.8c), which may or may not have been a group waning in abundance and diversity by the end of the Cretaceous. What we do know is that there are no credible accounts of post-K/T plesiosaurs. The whale- and dolphin-like marine diapsids called ichthyosaurs (Figure 18.8a) are thought to have disappeared well before the K/T boundary. The Late Cretaceous was a time of the appearance and radiation of an important group of marine-adapted lizards called mosasaurs (Figure18.8b). What is known of these suggests that, as a group, mosasaurs thrived until the very end of the Cretaceous. Because the best Late Cretaceous record of all these groups is found in continental sea deposits, however, exactly how these vertebrates fared during the regression remains unknown.

Among fossil invertebrates, perhaps the most famous group are the ammonites (Figure 18.8d). University of Washington paleontologist P. D. Ward has clearly demonstrated that ammonites lived right up to the K/T boundary, before finally going extinct.

Another important group of invertebrates are the bivalves. Obviously these did not all die out; there are many familiar bivalves today, including mussels, pectens (best known on the Shell Oil logo), and oysters (gullible morsels in Lewis Carroll's *The Walrus and the Carpenter*). During the Cretaceous, bivalves occupied the same important position in the ecosystem that they occupy today, although there was a somewhat different cast of characters. Careful study by D. M. Raup and D. Jablonski

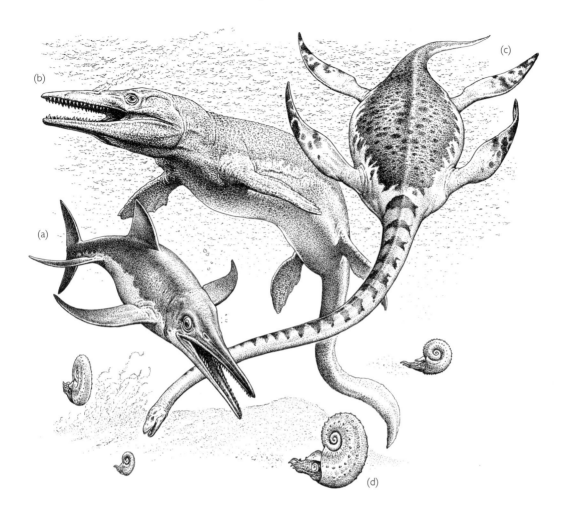

Figure 18.8. Some of the better-known inhabitants of Cretaceous seas. Vertebrates are: (a) ichthyosaur (*Platypterygius*); (b) mosasaur (*Tylosaurus*), and (c) plesiosaur (*Elasmosaurus*). The shelled, tentacled invertebrates toward the bottom of the drawing are cephalopod mollusks called ammonites (d).

has shown that, with one exception (which went extinct much earlier), 63% of all bivalves went extinct sometime within the last 10 million years of the Cretaceous. The record is unfortunately not more precise than this, but it does show that the extinction took place without regard for latitude: bivalves in temperate regions were just as likely to go extinct as those in the tropics.

Marine microorganisms

Because of their ubiquity, the record of foraminifera that are either planktonic (living in the water column) or benthic (living within sediments) has dominated discussions of K/T boundary events (we first met these in Chapter 2; see Figure 2.11). Micropaleontologists studying foraminifera, especially J. Smit of the Netherlands' Geologisch Instituut and S. L. D'Hondt of the University of Rhode Island argued persuasively, since as early as the late 1970s (and in many studies thereafter), that the planktonic foraminifera extinction was abrupt, with only a few species crossing the boundary into the Paleocene. G. Keller, of Princeton University, and N. MacLeod, of the Natural History Museum (London), disagree. Rather than an abrupt extinction, they see a smear, in which some forms go extinct before the boundary, some at the boundary, and some after the boundary. Their work has been strongly criticized by Smit, D'Hondt, and many other colleagues,[5] for not taking into account preservational biases, taxonomic inconsistencies, and stratigraphic problems. Florida State University's J. J. Pospichal, working with an entirely different group of microorganisms (calcareous nannofossils), observed an abrupt extinction. It is safe to say that most paleontologists working with marine microorganisms are inclined toward a catastrophic view of the extinction.

Some of the most exciting results came from a series of studies done by paleoceanographers J. C. Zachos, M. A. Arthur, and W. E. Dean. Zachos and his colleagues studied stable isotopes of foraminifera across the K/T boundary, with an eye toward characterizing primary productivity (the amount of organic matter synthesized by organisms from inorganic materials and sunlight) across the boundary. Obviously, primary productivity is the base of both the oceanic and terrestrial food chains, and with oceans covering 75% of earth's surface (or even more during the many high sea levels experienced during earth history), it would not be an exaggeration to state that earth's ecosystems are largely dependent upon oceanic primary productivity. Today, oceanic productivity can be measured and characterized by the $^{13}C:^{12}C$ ratio between surface and deep waters. Such a ratio can be obtained in ancient sediments by comparing the isotopic signatures of planktonic foraminifera with those of benthic foraminifera. Zachos and his colleagues did this using foraminifera from ocean bottom sediments taken from the North Pacific. At the K/T boundary they observed a "rapid and complete breakdown" in primary productivity, to less than 10% of what it had been. For the succeeding 1.5 million years, primary production remained at levels well below those preceding the original drop. They called this apparently moribund ocean a "Strangelove Ocean" after Dr Strangelove, a brittle, grotesque character (from the eponymous film), clearly unconcerned about a scorched, post-nuclear world.[6] D'Hondt and others subsequently interpreted the Strangelove Ocean to reflect a breakdown in nutrient cycling in the world's oceans.

Terrestrial record Virtually all of what we know of the K/T boundary on land also comes from Western Interior of North America (Figure 18.9). There, a well-

5 Of these, The Smithsonian Institution's B. Huber, University of North Carolina paleontologist T. Bralower, and Rutgers University's R. K. Olsson are among the most vocal.
6 The name was actually coined by oceanographer K. Hsü, who in 1982 (years before the work of Zachos and colleagues) speculated that the oceans after an asteroid would be a bit like the earth after Dr Strangelove.

Figure 18.9. The K/T boundary in eastern Montana. The boundary is midway up the butte, at the dark band. The gray rocks below are Cretaceous sediments; above, the lighter colored rocks are Tertiary sediments.

studied, complete section (see Chapter 16) has provided the best insights available into the dynamics of the extinction. In the following section, we outline some of the key indicators of the extinction, particularly as they are known from North America.

Plants

The plant record in the Western Interior has two major components, a palynoflora (spores and pollen) and a megaflora (the visible remains of plants, especially leaves; Figure 18.10). After a decade of intensive scrutiny, both records agree nicely with each other, and both records indicate that a major extinction occurred geologically instantaneously at the K/T boundary.

The pollen record, as studied by D. J. Nichols, F. Fleming, C. Hotton, and R. J. Tschudy of the U.S. Geological Survey, and A. Sweet of the Canadian Geological Survey, shows a distinctive pattern that is coincident with the iridium anomaly in Montana, North Dakota, and Wyoming. Pollen that is typical of early Paleocene time does not immediately follow the extinction of the Cretaceous pollen. Instead, as first documented by Tschudy, there is a high concentration of fern spores just after the iridium anomaly, suggesting that immediately after the extinction of the Cretaceous plants (represented by the pollen), there was a "bloom" of fern growth, interpreted by Tschudy as a pioneer community growing on a devastated post-impact landscape. In time, the fern flora gave way to a more diverse flora of flowering plants characteristic of the early Paleocene.

The situation reported by Sweet and his colleagues in southern and central Canada is somewhat different. These researchers have been able

Figure 18.10. Plant fossils. (a) Late Cretaceous leaf. The leaf is an as-yet unnamed angiosperm species that became extinct at the K/T boundary. The specimen is from Mud Buttes, Bowman County, North Dakota. (Photograph courtesy of K. R. Johnson, Denver Museum of Natural History.) (b) Pollen grains belonging to the genera *Proteacidites* (1) and *Aquilapollenites* (2), both important genera in measuring the moment of the terrestrial K/T extinction. *Proteacidites* is about 30 μm across; *Aquilapollenites* is about 50 μm. (Scanning electron micrographs courtesy of D. J. Nichols, U.S. Geological Survey.)

to track a series of five major changes in pollen composition across the boundary at a variety of latitudes. They correlated these changes to latitudinally related climate changes, as well as to the extinctions. Sweet and his colleagues concluded that no single extinction occurs at the K/T boundary, but rather that a series of extinction events occurred before, during, and after the K/T boundary.

Outside North America, an interesting but very preliminary pollen and spore record is slowly being built from a southern high-latitude flora from New Zealand. There, V. Vajda of Lund University, Sweden, and her colleagues from New Zealand documented an abrupt pollen and spore extinction, as well as the fern spike. Their conclusion, given the location of their study in comparison with the North American sites, was that the terrestrial K/T boundary was characterized by "global deforestation."

The megafloral record has been definitively documented by K. R. Johnson and L. J. Hickey, at that time both at Yale University. Johnson and Hickey tracked floras through the last 3 million years of the Cretaceous in North Dakota, South Dakota, Montana, Wyoming, and Colorado. Basing their study upon 25,000 plant specimens, they found that, while some environmental changes caused extinctions earlier than the K/T boundary, a major extinction took place precisely at the boundary. This, of course, is exactly correlated with the pollen extinction and iridium anomaly. Indeed, they observed an extinction of 79% of the dicot flowering plant taxa, suggesting that (as suspected) Tschudy's fern "bloom" may have been a response to the absence of flowering plants that would normally have occupied the ecosystem.

Vertebrates

Paleontologists J. D. Archibald (San Diego State University) and L. J. Bryant (U.S. Bureau of Land Management) made a thorough study of 150,000 vertebrate specimens (now housed in the University of California Museum of Paleontology) from eastern Montana asking the simple question "Of the vertebrates living in the region up to 3 million years before the K/T boundary, who survived the boundary and who did not?"

While their initial study claimed no recognizable pattern of survivorship, subsequent review of the data by P. M. Sheehan (Milwaukee Public Museum) and D. E. Fastovsky (University of Rhode Island), as well as by Archibald, himself, revealed that indeed there is a striking survivorship pattern: those organisms that lived in aquatic environments (i.e., rivers and lakes) showed up to 90% survival, whereas those organisms living on land showed as little as 10% survivorship. Thus the extinction seems not to have drastically affected aquatic organisms such as fish, turtles, crocodiles, and amphibians, but apparently wreaked havoc among terrestrial organisms such as mammals and, of course, dinosaurs. Archibald also noted subsequently that several other survivorship patterns appear in the data: small vertebrates are favored over large vertebrates, ectotherms over endotherms, and non-amniotes over amniotes (Figure 18.11).

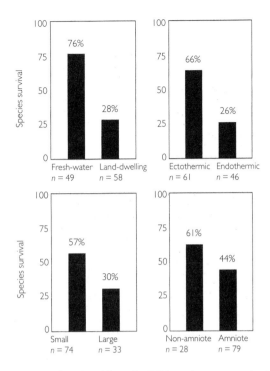

Figure 18.11. Patterns of survivorship at the K/T boundary as reconstructed by J. D. Archibald. Fastovsky and Sheehan reconstructed the aquatic and land-dwelling survivorship pattern as even more extreme, with land-dwelling organisms showing only a 12% survivorship, but aquatic organisms showing 90% survivorship. (From Archibald, 1996, Copyright Columbia University Press.)

Dinosaurs

Dinosaurs are difficult animals to study (see Box 18.1) and for many years, no scientific study of dinosaurs at the K/T boundary was ever carried out. Strangely (because no data were ever published to show this), it was (and still is, in some quarters) widely accepted that the last dinosaurs went extinct at the boundary, but that, as a group, they gradually began to die off about 10 million years before the boundary. Studies on patterns of dinosaur extinction were first performed only as recently as 1986 by University of Minnesota paleontologist R. E. Sloan and colleagues, who attempted to integrate data collected by them from a place called Bug Creek, Montana, with other North American data on dinosaur diversity. They concluded:

> Dinosaur extinction in Montana, Alberta, and Wyoming was a gradual process that began 7 million years before the end of the Cretaceous and accelerated rapidly in the final 0.3 million years of the Cretaceous, during the interval of apparent competition from rapidly evolving immigrant ungulates [hoofed mammals]. This interval involves rapid reduction in both diversity and population density of dinosaurs.   (Sloan et al., 1986, p. 629.)

BOX 18.1

Dinosaurs and mass extinctions

What are some of the problems with reconstructing changes in dinosaur populations over time? For one thing, dinosaurs are, by comparison with foraminifera for example, large beasts and, more importantly, not particularly common.[1] For this reason, the possibility of developing a statistically meaningful database is impractical, and rigorous studies of dinosaur populations are very hard to carry out. Just counting dinosaurs can be difficult. Mostly, one doesn't find complete specimens, and adjustments have to be made. For example, if you happen to find three vertebrae at a particular site, they might be from one, or two, or three individuals. The only way to be sure that they belong to a single individual is to find them articulated. Suppose they are not; then one must speak of a "minimum numbers of individuals" (often abbreviated to MNI), in which case the three vertebrae would be said to represent one individual: that would be the minimum number of individual dinosaurs that could have produced the three vertebrae. On the other hand, if one found two left femora, then the minimum number of individuals represented would be two.

It would be nice to use all the specimens that have been collected in the last 170 years of dinosaur studies in a study of changes in dinosaur diversity. Unfortunately, dinosaur specimens are commonly collected because they are either beautiful specimens or rare; hardly criteria for assuring that an accurate census of dinosaur populations has been performed. So any study that really is designed to get an accurate census of dinosaur abundance or diversity at the end of the Cretaceous must begin by counting specimens in the field, which is a labor-, time-, and cost-intensive proposition.

Then, of course, the taxonomic level at which to count dinosaurs can create problems. Recall from Chapter 17 the problems with higher taxa. Suppose that two specimens are found; one is clearly a hadrosaurid, and the other is an indeterminate ornithischian. The indeterminate ornithischian might be a hadrosaurid, in which case we should count two hadrosaurids. But then again it might not (because its identity is indeterminate), in which case calling it a hadrosaurid would give us more hadrosaurids in our survey than actually existed. On the other hand, calling both specimens "ornithischians" is quite correct, but not very informative, if we hope to track the survivorship patterns of *different types* of dinosaurs.

Finally, within the sediments themselves, problems of correlation exist. Suppose that we record the last (highest) dinosaur in the Jordan, Montana, area, and then record the last (highest) dinosaur in the Glendive, Montana, area, about 150 km away from Jordan. Can these two dinosaurs be said to have died at the same time? How could one possibly know? Suppose that in fact these dinosaurs died 200 years apart. An interval of 200 years, viewed from a vantage point of 65 million years is literally a snap of the fingers. Yet, 200 years is a long time when one is considering an instantaneous global catastrophe.

---

1 How rare are dinosaurs in this part of the world? Of course, we cannot know the density of dinosaurs within the rocks, but their surface density was calculated by sedimentologist P. White and colleagues, using the Sheehan *et al.* database. Fastovsky and Sheehan reported, "White and Fastovsky calculated that 0.000056 dinosaurs are preserved per m² of outcrop. Considered more realistically, in a statistical sense one must search a 5 m wide path of exposed rock that is 4 km long to find a single dinosaur fragment identifiable to family level (or lower)" (Fastovsky and Sheehan, 1997, p. 527.)

The work provoked a storm of interest and considerable criticism. It became clear that, statistically, the data did not support the conclusions. Moreover, the methods used for recognizing time in the Montana K/T rocks appeared to be largely conjectural. And as it turned out, the ancient sedimentary setting at Bug Creek was such that the locality bore an unwholesome jumble of Cretaceous and Tertiary fossils. With no stratigraphy, the sequence of extinctions could not be reliably reconstruced.

In 1986, paleobiologist P. M. Sheehan began asssembling a team of specialists to study the end-of-Cretaceous diversity of dinosaurs in eastern Montana and western North Dakota. The three-year study consisted of a quantitative census of dinosaur diversity during the last 1.6 million years of the Cretaceous. Sheehan and his colleagues were able to show that, within 150,000 years of the K/T boundary, the ecological diversity of the samples – that is, the proportion of the total dinosaur population taken by each of the eight families of dinosaurs preserved there – did not change. Change, if it occurred at all, had to have occurred sometime during the last 150,000 years of the Cretaceous. The study could not distinguish between whether the extinction took every day of that last 150,000 years of the Cretaceous, or whether it took only the last minute of that last 150,000 years.

The study by Sheehan and colleagues drew no less criticism than did that of Sloan and his colleagues. An oft-repeated criticism was that the

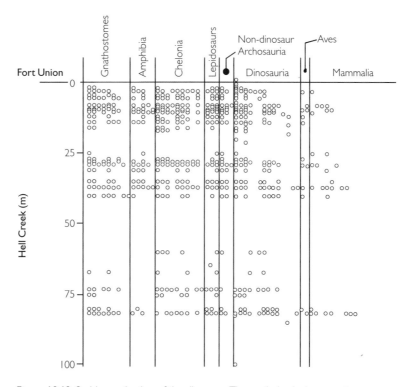

Figure 18.12. Sudden extinction of the dinosaurs. The vertical axis shows meters through the Hell Creek Formation, the uppermost unit in the Western Interior of the USA. "0" is the K/T boundary. The horizontal axis shows various vertebrate groups (including dinosaurs) that are found within the Hell Creek layers. Virtually all vertebrate groups are present throughout the thickness of the Hell Creek layers; there is no gradual decrease in the groups as the boundary is approached. The data indicate that the extinction of the dinosaurs and other vertebrates 65 Ma was geologically instantaneous. (Redrawn after Pearson et al., 2002.)

data were too sparse for any claim about the extinction – either catastrophic or gradual. What was needed in this acrimonious climate was
another approach to the problem, developed from a field-based study,
whose methods were exclusively designed to answer the key question
about the speed with which dinosaur populations went extinct.

Into that breach stepped an unlikely champion, North Dakota
amateur paleontologist D. A. Pearson of the Pioneer Trails Regional
Museum and a carefully assembled team of his professional colleagues.
Pearson and his associates collected vertebrate fossils from the eastern
Montana and western North Dakota for 10 years, painstakingly develop-
ing an immense database. What they found was clear evidence that, in
North America at least, dinosaur (and other vertebrate) populations
remained unchanging in either diversity or abundance right up to the
K/T boundary. Figure 18.12 reproduces their results.

In summary, the very limited data from the Western Interior of the
USA strongly indicate an abrupt end for the non-avian dinosaurs. Only
time and much further study will enable us to integrate dinosaur-
bearing localities from around the world into what is already known of
North America.

Extinction hypotheses

Adhering to a scientific standard

Much – indeed, most – of what has been proposed to explain
the extinction of dinosaurs does not even possess the basic prerequi-
sites for a viable, scientific theory. These minimal criteria are as
follows:

1   *The hypothesis must be testable.* As we have seen (Chapter 3), for a hypothesis
    to be considered in a scientific context, it must be testable, that is, it
    must have predictable, observable consequences. Without testability,
    there is no way to falsify a hypothesis, and in the absence of falsifiabili-
    ty, we are then considering belief systems rather than scientific
    hypotheses. If an event occurred and left no traces that could be
    observed (by whatever means available), science is simply not an appro-
    priate means by which to investigate the event.

2   *The hypothesis must explain all the events in question.* This criterion is rooted
    in the principle of parsimony (again, see Chapter 3). We would like to
    explain an event or series of events. If each step of the event (or events)
    requires an additional *ad hoc* explanation, our hypothesis loses strength.
    It is strongest when the most parsimonious explanation is used: that
    explanation which explains the most observations. For this reason, if we
    can explain all that we observe at the K/T boundary with a single hypoth-
    esis, we have produced the most parsimonious hypothesis and it has a
    good chance of being correct.

Extinction hypotheses

In Table 18.1 we present about 80 years of serious, published proposals
designed to explain the extinction of the dinosaurs (although see Box
18.2). The majority were published within the last 30 years. This
exhaustive list is abstracted from that published in 1990 by English
paleontologist M. J. Benton. Consider each; one does do not need to be
a professional paleontologist to reject most of them.

Table 18.1. Proposed causes for the extinction of the dinosaurs (after Benton, 1990)

I. Proposed biotic causes

A. Medical problems

(a) Slipped disks in the vertebral column causing dinosaur debilitation

(b) Hormone problems

    (1) Overactive pituitary glands leading to bizarre and non-adaptive growths

    (2) Hormonal problems leading to eggshells that were too thin, causing them to collapse in on themselves in a gooey mess

(c) Decrease in sexual activity

(d) Blindness owing to cataracts

(e) A variety of diseases, including arthritis, infections, and bone fractures

(f) Epidemics leaving no trace but wholesale destruction

(g) Parasites leaving no trace but wholesale destruction

(h) Change in ratio of DNA to cell nucleus causing scrambled genetics

(i) General stupidity

B. Racial senescence – this is the idea, no longer given much credence, that entire lineages grow old and become "senile," much as individuals do. Thus, in this way of thinking, late-appearing species would not be as robust and viable as species that appeared during the early and middle stages of a lineage. The idea behind this was that the dinosaurs as a lineage simply got old and the last-living members of the group were not competitive for this reason

C. Biotic interactions

(a) Competition with other animals, especially mammals, which may have out-competed dinosaurs for niches, or perhaps ate their eggs

(b) Overpredation by carnosaurs (who presumably ate themselves out of existence)

(c) Floral changes

    (1) Loss of marsh vegetation (presumably the single important source of food)

    (2) Increase in forestation (leading to loss of dinosaur habitats)

    (3) General decrease in the availability of plants for food with subsequent dinosaur starvation

    (4) The evolution in plants of substances poisonous to dinosaurs

    (5) The loss from plants of minerals essential to dinosaur growth

II. Proposed physical causes

A. Atmospheric causes

(a) Climate became too hot so they fried

(b) Climate became too cold so they froze

(c) Climate became too wet so they got waterlogged

(d) Climate became too dry so they desiccated

(e) Excessive amounts of oxygen in the atmosphere caused:

    (1) Changes in atmospheric pressure and/or atmospheric composition that proved fatal; or

    (2) global wildfires that burned up the dinosaurs

(f) Low levels of CO_2 removed the "breathing stimulus" of endothermic dinosaurs

(g) High levels of CO_2 asphyxiated dinosaur embryos

(h) Volcanic emissions (dust, CO_2, rare earth elements) poisoned dinosaurs one way or another

B. Oceanic and orographic causes

(a) Marine regression produced loss of habitats

(b) Swamp and lake habitats were drained

(c) Stagnant oceans produced untenable conditions on land

(d) Spillover into the world's oceans of Arctic waters that had formerly been restricted to polar regions, and subsequent climatic cooling

(e) The opening of Antarctica and South America, causing cool waters to enter the world's oceans from the south, modifying world climates

(f) Reduced topographic relief and loss of habitats

C. Other

(a) Fluctuations in gravitational constants leading to indeterminate ills for the dinosaurs

(b) Shift in earth's rotational poles leading to indeterminate ills for the dinosaurs

(c) Extraction of the moon from the Pacific Basin perturbing dinosaur life as it had been known for 1 40 million years (!)

(d) Poisoning by uranium from earth's soils

D. Extraterrestrial causes

(a) Increasing entropy leading to loss of large life forms

(b) Sunspots modifying climates in some destructive way

(c) Cosmic radiation and high levels of ultraviolet radiation causing mutations

(d) Destruction of the ozone layer, causing (c)

(e) Ionizing radiation as in (c)

(f) Electromagnetic radiation and cosmic rays from the explosion of a supernova

(g) Interstellar dust cloud

(h) Oscillations about the galactic plane leading to indeterminate ills for the dinosaurs

(i) Impact of an asteroid (for mechanisms, see the text)

BOX 18.2

The real reason the dinosaurs became extinct

Not every published hypothesis has been serious. In 1964, for example, E. Baldwin suggested that the dinosaurs died of constipation. His reasoning went as follows: toward the end of the Cretaceous, there was a restriction in the distribution of certain plants containing natural laxative oils necessary for dinosaur regularity. As the plants became geographically restricted, those unfortunate dinosaurs living in places where the necessary plants no longer existed acquired stopped plumbing and died hard deaths. The same year, humorist W. Cuppy noted that "the Age of Reptiles ended because it had gone on long enough and it was all a mistake in the first place," a view with which the goat in *Jurassic Park* I would probably have agreed.

The November, 1981 issue of the *National Lampoon* offered its explanation, entitled "Sin in the Sediment." The Christian right was the target:

> It's pretty obvious if you just examine the remains of the dinosaurs … Dig down into older sediments and you'll see that the dinosaurs were pretty well off until the end of the Mesozoic. They were decent, moral creatures, just going about their daily business. But look at the end of the Mesozoic and you begin to see evidence of stunning moral decline. Bones of wives and children all alone, with the philandering husband's bones nowhere in sight. Heaps of fossilized, unhatched, aborted dinosaur eggs. Males and females of different species living together in unnatural defiance of biblical law. Researchers have even excavated entire orgies – hundreds of animals with their bones intertwined in lewd positions. Immorality was rampant.

In 1983, the sedimentary geologist R. H. Dott Jr published a short note in which he vented his frustrations with the pollen season, suggesting that it was pollen in the atmosphere that killed the dinosaurs. He called his contribution "Itching Eyes and Dinosaur Demise."

The issues raised by the *National Lampoon* were compelling enough to again be raised in 1988 by the *Journal of Irreproducible Results*. There, in an offering called "Antileduvian Buggery: the Role of Deviant Mating Behavior in the Extinction of Mesozoic Fauna," L. J. Blincoe informed us:

> A thorough but cursory review of fossil specimens … has revealed a unique fossil found in the Cretaceous "beds" of Mongolia in 1971. The fossil featured two different species of dinosaur, one a saurischian carnivore (*Velociraptor*), the other a [sic] ornithischian herbivore (*Protoceratops*), in close association at the moment of their deaths. Prejudiced by their preconceived notions of dinosaur behavior, paleontologists have almost unanimously interpreted this find as evidence of a life and death struggle [see Figure 12.13] … However, an alternative theory has now been developed which not only explains this unusual fossil, but also answers the riddle of the dinosaurs disappearance. Quite simply, when their lives were ended by sudden catastrophe, these two creatures were locked together...in a passionate embrace. They were, in fact, prehistoric lovers.

> The implications of this startling interpretation are clear: dinosaurs engaged in trans-species sexual activity. In doing so they wasted their procreative energy on evolutionarily pointless copulation that resulted in either no offspring or, perhaps on rare occasions, in bizarre, sterile mutations (the fossil record is replete with candidates for this later category).
>
> (Blincoe, 1988, p. 24.)

For the ultimate causes of the extinction, however, we think O'Donnell's perspective published in the *New Yorker* says it all (see Figure B18.2.1).

BOX 18.2 (cont.)

EXTINCTION of the DINOSAURS FULLY EXPLAINED

Figure B18.2.1. One cartoonist's take on the causes of the extinction of the dinosaurs. (Drawing by O'Donnell ©1992 *The New Yorker Magazine*, Inc.)

Many of the ideas – parasites, blindness, epidemics – fail for leaving no visible trace: they are not within the realm of science to address. Beyond this, however, virtually all of these hypotheses do not explain all, or even most, of what is known of K/T events. Most (such as mammals causing the extinction by eating dinosaur eggs) are special pleading for dinosaurs, as if dinosaurs alone were the key components in the K/T boundary extinctions. But as we have seen, in the context of earth's ecosystems, dinosaurs were really only a small part of what took place. Any theory that purports to explain K/T events in a meaningful way must also explain these other events. With that in mind, the hypothesis that an asteroid impact caused the events at the K/T boundary becomes a much more interesting and plausible hypothesis.

Does the idea that an asteroid impact caused the K/T extinctions have predictable consequences?
Clearly, it does. First, if the asteroid produced global consequences, evidence for it should be visible globally. After 25 years of research, the

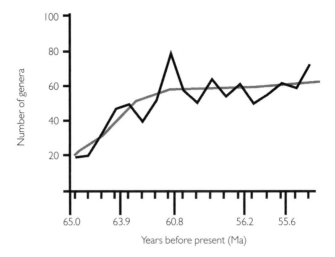

Figure 8.13. Radiation of mammals after the K/T boundary. The vertical axis shows species of mammals; the horizontal axis shows time. The black line is the exact counts of genera at any particular time; the gray line is the inferred, general shape of mammalian diversity. The interpretation (gray line) shows a rapid increase in the number of genera of mammals during the first 3 or so million years of the Tertiary, followed by a leveling off of diversity. (Redrawn after Maas, M. C. and Krause, D. W. 1994. Mammalian turnover and community structure in the Paleocene of North America. *Historical Biology*, **8**, 91–128; dates interpolated from Woodburne, M. O. and Swisher, C. C. 1995. Land mammal, high-resolution geochronology, intercontinental overland dispersals, sea level, climate, and vicariance. In Berggren, W. A., Kent, D. V., Aubry, M.-P. and Hardenbol, J. (eds), *Geochronology, Time Scales, and Global Correlation*. SEPM Special Publication no. 54, pp. 335–364.)

evidence for global influence of the K/T boundary asteroid impact is overwhelming (see Figure 18.3). That being the case, what kind of predictable consequences are there in terms of the extinction?

In the case of the bivalves and plants, the fact that the extinctions took place regardless of latitude is strong evidence that those extinctions were due to a global effect, which was apparently unrelated to climate. Had climate been involved as a causal agent, one might expect to see latitudinal changes in the patterns of extinction, but, as we have seen, such is not the case.

Other evidence comes from the rate at which the extinction took place. If the asteroid really caused the extinction, the event should have been what W. A. Clemens called a "short, sharp shock."[7] There should be no evidence of biotic abundance and diversity dwindling in the years preceding the boundary. Rather, the fossil record should indicate that it was "business as usual" for most lineages right up until the moment of the impact, at which point the extinctions should be global and isochronous. Therefore, patterns of *gradual* extinction should falsify the asteroid impact as a causal agent; patterns of *abrupt* or *catastrophic* extinction should corroborate the hypothesis.

---

7 Quoting Pooh-Bah in Gilbert and Sullivan's *Mikado*.

Evidence is mounting that the extinction was abrupt. This pattern is best seen in ammonites, in foraminifera, in plants (both in pollen and in the megaflora), and in dinosaurs. The bivalve data, however, are not sufficiently refined at present to distinguish between a gradual and an abrupt extinction. In the case of the mammals, a number of workers have published accounts of "stepwise" extinctions for the group, although all would agree that mammals – particularly marsupials – underwent significant extinctions at the K/T boundary.

Recovery

Catastrophic events tend to leave a distinctive mark: organisms that first colonize deserted ecospace tend to speciate rapidly, to be rather small, and to adopt generalist lifestyles (rather than developing a highly specialized behavior such as exclusively carnivorous or herbivorous behaviors). Such faunas – which tend to be uniformly small-sized, speciate rapidly, and pursue generalist life strategies – are termed "disaster faunas" and have analogues within vertebrate, invertebrate, and plant communities.

The mammals that evolved throughout the recovery at the K/T boundary were quite remarkable in that they had neither the size range nor the specializations of later mammalian faunas. As reconstructed by M. Maas and D. Krause (both then at the State University of New York (SUNY) at Stony Brook), and later by J. Alroy (University of California, Santa Barbara), the earliest Tertiary faunas were extremely small and generalist. They speciated rapidly, only leveling off after about 5 million years (Figure 18.13). Theirs is the pattern of a fauna that came through a catastrophic event and inherited deserted ecospace.

Clemens suggested that the mammalian radiation that characterized the Tertiary was actually well underway during the latest Cretaceous. Be that as it may, the resultant patterns of diversification of mammals – including small body size, generalist habit, and high rates of speciation – as well as the monumental revolution in the terrestrial vertebrate ecosystem (going from dinosaur-dominated to mammal-dominated) bear the imprint of the terrestrial ecosystem having undergone an ecological disaster.

Toward a consensus

Does the asteroid impact hypothesis explain all the data?
Initially, researchers struggling to evaluate the asteroid impact hypothesis as a causal agent for the K/T extinctions had some trouble explaining the selectivity of the extinctions. We have seen from Archibald's work that aquatic organisms were heavily favored over terrestrial organisms, smaller organisms were favored over larger organisms, non-amniotes were favored over amniotes, and ectotherms were favored over endotherms (see Figure 18.11).

Recently, a pattern has begun to emerge that seems to correlate extinction selectivity with the asteroid as a causal agent in the extinctions. The original idea was articulated for the invertebrate extinctions by Sheehan, who later extended the idea to the vertebrate record. Sheehan and a colleague, T. A. Hansen, noted that those marine creatures that suffered the most extinctions were those that depended directly upon primary productivity for their food source. Such creatures included not only the planktonic foraminifera and other planktonic marine microorganisms, but also ammonites, other cephalopods, and a variety of mollusks. On the other hand, Sheehan and Hansen noted that organisms that not only depended on primary productivity but could also survive on detritus – that is, the scavenged remains of other organisms – fared consistently better. In the marine rocks that formed the basis of their report, Sheehan and Hansen observed that detritus feeders were apparently unaffected by the extinction, and that, in some cases, even experienced an evolutionary diversification across the K/T boundary.

This idea received a real boost from the work of Zachos and his colleagues, who only a year after Sheehan and Hansen's study demonstrated that there was a dramatic drop in global primary productivity in the world's oceans at the K/T boundary. Now, the question was, could the primary productivity/detritus-feeding extinction selectivity be traced in the terrestrial realm as well?

Apparently so. Sheehan and Fastovsky noted that the strong selectivity between land-dwelling and aquatic vertebrate survival (see Figure 18.11) is very much paralleled by their feeding strategies: those aquatic vertebrates that lived in river environments and in floodplain lakes and wetlands tend to utilize detritus as a major source of nutrients, while land-dwelling veterbrates are far more dependent upon primary productivity. This is because river and lake systems can serve as a repository for detrital material; organisms living in such environments who can utilize this resource are therefore "buffered" or protected against short-term drops in primary productivity. Sheehan and Fastovsky's conclusion was reinforced by the work of D. A. Pearson and colleagues, who reaffirmed that those animals who survived across the boundary were primarily aquatic detritus-feeders.

So what happened at the K/T boundary? Scientists who view the asteroid impact as the cause of K/T events usually envision some kind of dramatic and short-term disturbance to the ecosystem. Such a disturbance could be, as we have seen, a dust cloud blocking sunlight for a few months, global wildfires, acid rain, or some combination of ills. It may never be possible to know exactly which factor(s) did which deed(s), and trying to reconstruct it in so exact a manner may be stretching the resolution of the fossil record well past its capability. Whatever the disturbance to the ecosystem, it seems to have had, at the very least, a deadly effect on global primary productivity, which in turn seems to have decimated organisms solely dependent upon upon primary productivity. In this view, extraterrestrial processes become an integral part of the very earthly process of organic evolution.[8] In the case of the mammals, it seems clear that the absence of dinosaurs after 160 million years of terrestrial importance was the event that allowed the mammals to evolve and occupy the place in the global ecosystem that they presently hold.

8 This subject is treated in detail in J. D. Powell's 1998 book, *Night Comes to the Cretaceous.*

The loyal opposition

It cannot be said that the idea of the asteroid as the cause of the extinctions is accepted in all quarters. Clemens and Archibald have both argued forcefully, on the basis of data from Cretaceous and early Tertiary mammals that, regardless of cause, the extinction and recovery of Mammalia during the K/T transition was stepwise (it occurred in small increments in which old species died out one by one, and new ones appeared to replace them). They concluded that the mammalian extinction and radiation began well in advance of the asteroid impact. The authors' point was that such a pattern is not in agreement with an abrupt extinction by an asteroid. More recently, Archibald tested the patterns of survivorship that he has reconstructed against the putative effects of an asteroid. His conclusion is that the match between the selectivity of the extinctions and the asteroid impact is imperfect at best.[9]

Instead, he suggested that the marine regression at the end of the Cretaceous was responsible for the demise of the dinosaurs, along with a series of causes, such as volcanism and even the asteroid. The regression, he has claimed, caused "habitat partitioning," by which he means that as the seas receded, the habitats in which the dinosaurs had been living were divided, and thus rendered ecologically inadequate for the creatures. Weakened by partitioned habitats, dinosaur populations were more susceptible to other environmental perturbations, such as the variety of climatic and environmental changes that might have been induced by volcanoes or an asteroid.

Our view

We are not persuaded by the ideas expressed by Archibald and his colleagues. First and foremost, there is no geological evidence for habitat partitioning in latest Cretaceous rocks in eastern Montana (where Archibald indicates that it occurred). Indeed, recent geological evidence suggests quite the opposite: that the boundary was actually marked by a transgression. Given that, the whole hypothesis of habitat partitioning loses its geological basis.

Moreover, the "fit" or lack thereof of the extinction patterns to the supposed effects of an asteroid impact is not of much concern. Modelers cannot agree on the exact effects of an asteroid impact, and so far no one has been able to predict how a modern (let alone a Cretaceous!) ecosystem would react to such an event. Our view is that – beyond the prediction that the extinction ought to be sudden – the fossil record may not allow us to construct mechanistic hypotheses detailing who went extinct due to which particular inferred killing mechanism.

The problem with a concatenation of causes (habitat partitioning, regression, an asteroid, volcanoes, and other environmental perburbations) to explain the extinction is that, in large part, it does not

---

9 His most complete exposition of this idea is in his 1996 book, *Dinosaur Extinction and the End of an Era.*

Figure 18.14. Reconstruction of an asteroid impact with earth. Planetary geologist P. H. Schultz and geobiologist S. L. D'Hondt suggest that the asteroid struck earth at an angle of about 30°, coming from the southeast.

constitute a unified hypothesis or suite of hypotheses that explain what is known of K/T boundary events. Even if supported by evidence, the habitat-partitioning-and-other-causes viewpoint explains only the vertebrate extinctions in North America – and it is our view that they fail to explain even those particularly well. There has not yet been formulated a theory alternative to the asteroid impact hypothesis that explains all – or even most – of the data as fully as the asteroid impact hypothesis. For this reason, the extinction of the dinosaurs and the many other organisms that went extinct at the K/T boundary remains best explained by causes directly or indirectly related to the impact of an asteroid with earth 65 Ma (Figure 18.14).

Important readings

Alvarez, L. W. 1983. Experimental evidence that an asteroid impact led to the extinction of many species 65 Myr ago. *Proceedings of the National Academy of Sciences, USA*, **80**, 627–642.

Alvarez, L. W., Alvarez, W., Asaro, F. and Michel, H. V. 1980. Extraterrestrial cause for the Cretaceous–Tertiary extinction. *Science*, **208**, 1095–1108.

Alvarez, W. 1997. *T. rex and the Crater of Doom*. Princeton University Press, Princetown, NJ, 185pp.

Archibald, J. D. 1984. Bug Creek Anthills (BCA), Montana: faunal evidence for Cretaceous age and non-catastrophic extinctions. *Geological Society of America Abstracts with Programs*, vol. 16, p. 432.

Archibald, J. D. 1996. *Dinosaur Extinction and the End of an Era*. Columbia University Press, New York, 237pp.

Archibald, J. D. and Bryant, L. J. 1990. Differential Cretaceous/Tertiary extinctions of nonmarine vertebrates; evidence from northeastern Montana. In Sharpton, V. L. and Ward, P. D. (eds.), *Global catastrophes in Earth History; An Interdisciplinary Conference on Impacts, Volcanism, and Mass Mortality*. Geological Society of America Special Paper no. 247, pp. 549–562.

Archibald, J. D. and Clemens, W. A., Jr. 1980. Evolution of terrestrial faunas during the Cretaceous–Tertiary transition. *Mémoires de la Société Geologique de la France*, 139, 67–74.

Clemens, W. A., Jr, Archibald, J. D. and Hickey, L. J. 1981. Out with a whimper not a bang. *Paleobiology*, **7**, 293–298.

Claeys, P., Kiessling, W. and Alvarez, W. 2002. Distribution of Chicxulub ejecta at the Cretaceous–Tertiary boundary. In Koeberl, C. and MacLeod, K. G. (eds.), *Catastrophic Events and Mass Extinctions: Impacts and Beyond*. Geological Society of America Special Paper no. 356, pp. 55–68.

D'Hondt, S., Donaghay, P., Zachos, J. C., Luttenberg, D. and Lindinger, M. 1998. Organic carbon fluxes and ecological recovery from the Cretaceous/Tertiary mass extinction. *Science*, **282**, 276–279.

Dodson, P. 1991. Maastrichtian dinosaurs. *Geological Society of America Abstracts with Programs*, **23**, no. 5, pp. A184–A185.

Fastovsky, D. E. and Sheehan, P. M. 1997. Demythicizing dinosaur extinctions at the Cretaceous–Tertiary boundary. In Wolberg, D. L., Stump, E. and Rosenburg, G. D. (eds.), *Dinofest International*. The Academy of Natural Sciences, Philadelphia, pp. 527–531.

Frankel, C. 1999. *The End of the Dinosaurs – Chicxulub Crater and Mass Extinctions*. Cambridge University Press, New York, 223pp.

Hartman, J., Johnson, K. R. and Nichols, D. J. (eds.) 2002. The Hell Creek Formation and the Cretaceous–Tertiary Boundary in the Northern Great Plains. *Geological Society of America Special Paper* no. 361, 520pp.

Hildebrand, A. R. and Boynton, W. V. 1991. Cretaceous Ground Zero. *Natural History*, **6**, 47–53.

Jablonski, D. 1986. Background and mass extinctions: the alternation of macroevolutionary regimes. *Science*, **231**, 129–133.

Jablonski, D. and Bottjer, D. J. 1991. Environmental patterns in the origins of higher taxa: the post-Paleozoic fossil record. *Science*, **252**, 1831–1833.

Johnson, K. R. and Hickey, L. J. 1990. Megafloral change across the Cretaceous/Tertiary boundary in the northern Great Plains and Rocky Mountains, USA. In Sharpton, V. L. and Ward, P. D. (eds.), *Global Catastrophes in Earth History; An Interdisciplinary Conference on Impacts, Volcanism, and Mass Mortality*. Geological Society of America Special Paper no. 247, pp. 433–444.

Koeberl, C. and MacLeod, K. G. (eds.). *Catastrophic Events and Mass Extinctions: Impacts and Beyond*. Geological Society of America Special Paper no. 356, 746pp.

Pearson, D. A., Schaefer, T., Johnson, K. R., Nichols, D. J. and Hunter, J. P. 2002. Vertebrate biostratigraphy of the Hell Creek Formation in southwestern North Dakota and northwestern South Dakota. In Hartman, J., Johnson, K. R. and Nichols, D. J. (eds.), *The Hell Creek Formation and the Cretaceous–Tertiary Boundary in the Northern Great Plains*. Geological Society of America Special Paper no. 361, pp. 145–167.

Pope, K. O. 2002. Impact dust not the cause of the Cretaceous–Tertiary mass extinction. *Geology*, **30**, 99–102.

Powell, J. L. 1998. *Night Comes to the Cretaceous*. W. H. Freeman and Company, New York, 250pp.

Russell, D. A. 1982. The mass extinctions of the late Mesozoic. *Scientific American*, **246**, 58–65.

Ryder, G., Fastovsky, D. E. and Gartner, S. (eds.). 1996. *The Cretaceous–Tertiary Event and Other Catastrophes in Earth History*. Geological Society of America Special Paper no. 307, 569pp.

Sharpton, V. L. and Ward, P. D. (eds.). *Global Catastrophes in Earth History; An Interdisciplinary Conference on Impacts, Volcanism, and Mass Mortality*. Geological Society of America Special Paper no. 247, 631pp.

Sheehan, P. M., Fastovsky, D. E., Hoffmann, R. G., Berghaus, C. B. and Gabriel, D. L. 1991a. Sudden extinction of the dinosaurs: Latest Cretaceous, upper Great Plains, U.S.A. *Science*, **254**, pp. 835–839.

Sheehan, P. M., Fastovsky, D. E., Hoffmann, R. G., Berghaus, C. B. and Gabriel, D.L. 1992. Reply. *Science*, **256**, pp. 160–161.

Sigurdsson, H., D'Hondt, S. L. and Carey, S. 1992. The impact of the Cretaceous/Tertiary bolide on evaporite terrain and generation of major sulfuric acid aerosol. *Earth and Planetary Science Letters*, **109**, 543–559.

Silver, L. T. and Schultz, P. H. (eds.). 1982. *Geological Implications of Large Asteroid and Comets on Earth*. Geological Society of America Special Paper no. 190, 528pp.

Sloan, R. E., Rigby, J. K., Van Valen, L. and Gabriel, D. L. 1986. Gradual dinosaur extinction and simultaneous ungulate radiation in the Hell Creek Formation. *Science*, **232**, 629–633.

Vajda, V., Raine, J. I. and Hollis, C. J. 2002. Indication of global deforestation at the Cretaceous–Tertiary boundary by New Zealand Fern spike. *Science*, **294**, 1700–1701.

Glossary

Absolute age. The age of a rock or fossil measured in years before present.

Acromion process. A broad and plate-like flange on the forward surface of the shoulder blade.

Actinopterygii. The clade of ray-finned fish.

Adenosine triphosphate. *See* ATP.

Advanced. In an evolutionary context, shared or derived (or specific), with reference to characters.

Aestivate. In zoology, to spend summers in a state of torpor.

Allometry. The condition in which, as the size of organisms changes, their proportions change as well. For example, if an ant were scaled up to the the size of a 747, its features – body, head, legs, etc. – would no longer have the same proportions relative to each other.

Altricial. Pertaining to organisms that are born relatively underdeveloped, requiring significant parental attention for survival.

Alveolus (pl. alveoli). A sac-like anatomical structure.

Amnion. A membrane in some vertebrate eggs that contributes to the retention of fluids within the egg.

Anaerobic. Without oxygen.

Analogue. In anatomy, structures that perform in a similar fashion but have evolved independently.

Analogous. Adjective form of analogue.

Anamniotic. An egg without an amnion.

Anapsida. The group that contains all amniotes with a completely covered skull roof.

Ancestral. In an evolutionary sense, relating to forebears.

Angiosperms. Flowering plants.

Ankylopollexia. A clade of ornithopods including *Camptosaurus, Iguanodon, Ouranosaurus,* and hadrosaurids.

Antagonistic muscle masses. Groups of muscles whose movements oppose each other; for example, muscles whose movements open a jaw are said to be antagonistic to muscles that close the jaw.

Anterior. Pertaining to the head-bearing end of an organism.

Antitrochanter. A downward-directed process on the upper edge of the ilium in hadrosaurids and ceratopsians.

Antorbital fenestra. An opening on the side of the skull, just ahead of the eye. This is a character that unites the clade Archosauria.

Arboreal. Pertaining to trees.

Archipelago. A group of islands.

Archosauria. A clade within Archosauromorpha. The living archosaurs include birds and crocodiles.

Archosauromorpha. The large clade of diapsids that includes the common ancestor of rhynchosaurs and archosaurs, and all its descendants.

Articulated. In paleontology, bones are said to be articulated when they are found positioned relative to each other as they were in life.

Artifact. In an experiment, an incorrect result caused by something in the nature of the data or the method of experimentation.

Ascending process of the astragalus. A wedge-shaped splint of bone on the astragalus that lies flat against the shin (between the tibia and fibula) and points upwards. Diagnostic character of Theropoda.

Assemblage. In paleontology, a group of organisms. The term is used to refer to a collection of fossils in which it is not clear how accurately the collection reflects the complete, ancient formerly living community.

Asteroid. A large extraterrestrial body.

Astragalus. Along with the calcaneum, one of two upper bones in the vertebrate ankle.

Atom. The smallest particle of any element that still retains the properties of that element.

Atomic number. The number of protons (which equals the number of electrons) in an element.

ATP (adenosine triphosphate). The compound that cells use to store energy. The reaction ATP→ADP (adenosine diphosphate) involves the breakage of a phosphate bond and the release of energy which is then converted to work by organisms.

Autotroph. An organism that uses energy from the sun as well as from inorganic nutrients to produce complex molecules for nutrition.

Background extinctions. Continually occurring, isolated extinctions of individual species. As distinct from mass extinctions.

Barb. Feather material radiating from the shaft of the feather.

Barbule. A small hook that links barbs together along the shaft of the feather.

Beak. Sheaths of keratinized material covering the ends of the jaws (synonym: rhamphotheca).

Benthic. With reference to the marine realm (oceans), living within sediments.

Bilateral symmetry. A kind of symmetry in which the right and left halves of the body are mirror images of each other.

Biogeographical. Pertaining to the distribution of organisms in space.

Biomass. The sum total of the weights of organisms in the assemblage or community being studied.

Biostratigraphy. The study of the relationships in time among groups of organisms.

Biota. The sum total of all organisms that have populated the earth.

Body fossil. The type of fossil in which a part of an organism becomes buried and fossilized – as opposed to trace fossil.

Bone. Calcified skeletal tissue.

Bone histology. The study of bone tissue.

Bonebeds. Relatively dense accumulation of bones of many individuals, generally composed of a very few kinds of organism.

Boss. A large mass or knob of bone, commonly used with reference to structures on the skull.

Brain. A centralized cluster of nerve cells.

Brain endocasts. Internal casts of braincases.

Braincase. Hollow bony box that houses the brain; located toward the upper, back part of the skull.

Cadence. In locomotion, the rate at which the feet hit the ground.

Caliche. Calcium carbonate nodules that form in soils.

Campanian. The penultimate Stage (subdivision) of the Cretaceous Period.

Caniniform. Canine-like; something that is caniniform has the shape of the elongate, pointed (approximately nine) teeth in dogs.

Carbohydrates. A family of 5- and 6-carbon organic molecules whose chemical bonds, when broken, release energy.

Carnian. The older of two stages (time subdivisions) of the Late Triassic.

Carnosaur. A large-bodied theropod, with a tendency toward small forelimbs and a large head.

Carpal. Wrist bone.

Carpometacarpus. Unique structure in all living, and in most ancient, birds, in which bones in the wrist and hand are fused.

Cast. Material filling up a mold.

Cellular respiration. The breakdown of carbohydrates through a regulated series of oxidizing reactions.

Cenozoic Era. That interval of time from 65 Ma to present.

Centrum. The spool-shaped, lower portion of a vertebra, upon which the spinal cord and neural arch rest.

Cerapoda. The ornithischian clade of Ceratopsia + Pachycephalosauria + Ornithopoda

Character. An isolated or abstracted feature or characteristic of an organism.

Choana (pl. choanae). In the skull, a passageway leading from the nasal cavity to the interior of the mouth.

Chondrichthyes. The gnathostome clade that includes sharks, skates, and rays.

Chronostratigraphy. The study of geological time.

Circumpolar currents. Cold water masses that circulate around the earth's poles.

Clade. Group of organisms in which all members are more closely related to each other than they are to anything else. All members of a clade share a most recent common ancestor that is itself the most basal member of that clade. Synonymous with "monophyletic group" and "natural group."

Cladistic analysis. Analysis of the ancestor–descendant (evolutionary) relationships among organisms using hierarchies of shared, derived characters.

Cladogram. A hierarchical, branching diagram that shows the distribution of shared, derived characters among selected organisms.

Clavicle. Collarbone.

Climate. The sum of all weather conditions. Usually one refers to particular climatic variables, such as precipitation or temperature.

Cnemial crest. A bony flange on the upper end of the front surface of the tibia.

Co-evolution. The idea that two organisms or groups of organisms may have evolved in response to one another.

Collect. To obtain fossils from the earth.

Competitive edge. Some aspect of an organism or group of organisms that enhances the ability of the possessor to compete successfully against those that do not.

Continental crust. Quartz-rich material of which continents are formed.

Continental effects. The effect on climate exerted by continental masses.

Convergent. In anatomy, pertaining to the independent invention (and thus, duplication) of a structure or feature in two lineages. The streamlined shape of whales, fish, and ichthyosaurs is a famous example of convergent evolution.

Coprolite. Fossilized feces.

Coracoid. The lower (and more central) of two elements of the shoulder girdle (the upper being the scapula).

Cosmic dust. Particulate matter from outer space.

Cranial. Referring to the skull (cranium).

Craton. A large body of continental crust that has been stable for millions of years.

Cretaceous–Tertiary boundary. That moment in time, 65 Ma, between the Cretaceous Period and the Tertiary Period. *See also* **K/T (boundary)**.

Crurotarsi. A clade of archosaurs including crocodilians and their close relatives.

Crust. A thin, chemically distinct rind on the earth's lithosphere.

Curate. To incorporate, preserve, and catalog specimens into museum collections.

Cursorial. Pertaining to running.

Cursorial locomotion. Running locomotion on land.

Cycadophytes. A bulbous, fleshy type of gymnosperm.

Deccan traps. Interbedded volcanic and sedimentary rocks in western and central India of Cretaceous–Tertiary age.

Deltoid crest. A large process at the head of the humerus.

Dense Haversian bone. A type of Haversian bone in which the canals and their rims are very closely packed.

Dental battery. A cluster of closely packed cheek teeth in the upper and lower jaws, whose shearing or grinding motion is used to masticate plant matter.

Deposition. Net addition of sediment to a land surface.

Derived. In an evolutionary context, pertaining to characters that uniquely apply to a particular group and are thus regarded as having been "invented" by that group during the course of its evolutionary history.

Detritus. Loose particulate rock, mineral, or organic matter; debris.

Diachronous. Occurring at different moments in time; not synchronous.

Diapsida. The large clade of amniotes that includes the common ancestor of lepidosauromorphs and archosaurs, and all its descendants.

Diastem(a). A gap.

Digitigrade. In anatomy, a position assumed by the foot when the animal is standing, in which the ball of the foot is held high off the ground and the weight rests on the ends of the toes.

Dinosauria. A clade of ornithodiran archosaurs.

Diphyletic. In evolution, having two separate origins.

Disarticulated. Dismembered.

Diversity. The variety of organisms; the number of kinds of organism.

Dominant. In an ecological sense, being the most abundant or having the greatest effect on a particular aspect of the ecosystem. A rather general term without well-constrained meaning.

Dorsal sacral shield. The upper portion of the sacrum, composed of the ilia, vertebrae, and sacral ribs.

Down. A bushy, fluffy, type of feather in which barbules and vanes are not well developed, used for insulation.

Dryomorpha. A clade of iguanodontians including *Dryosaurus*, *Camptosaurus*, *Iguanodon*, *Ouranosaurus*, and hadrosaurids.

Dynamic similarity. A conversion factor that "equalizes" the stride rates of vertebrates of different sizes and proportions, so that speed of locomotion can be calculated.

Ecological diversity. The proportion of an ecosystem that is occupied by a particular lifestyle, such as feeding type or mode of locomotion. For a simple example, one might study an ecosystem by dividing it into herbivores, carnivores, and omnivores.

Ectothermic. Regulating temperature (and thus, metabolic rate) using an external source of energy (heat). The opposite of endothermic.

Edentulous. Without teeth.

Elbow. The joint between the upper arm and the lower arm.

Electron. A negatively charged subatomic particle. Electrons reside in clouds around the nucleus of an atom.

Element. (1) In anatomy, discrete part of the skeleton; i.e., an individual bone. (2) In chemistry, a substance that cannot be broken down by chemical means into a simpler substance.

Encephalization. That condition in which an organism bears a head structure that is distinct from the rest of the body and that contains a brain.

Endemic. An organism or fauna is said to be endemic to a region when it is restricted to that region.

Endemism. The property of being endemic.

Endosymbionts. Organisms that live within another organism in a mutually beneficial relationship.

Endosymbiosis, theory of. The idea that eukaryotic cells evolved as a result of the ingestion, by prokaryotes, of other prokaryotes. The ingested prokaryotes eventually adopted specialized functions as organelles.

Endothermic. Regulating temperature (and thus, metabolic rate) using an internal source of energy. The opposite of ectothermic.

Epeiric sea. Relatively shallow (at most, a few hundred meters) marine water covering a craton (synonym: epicontinental sea).

Epioccipital. Bone ornamenting the rim of the frill in ceratopsians.

Epoch. Subdivisions of a period, several million years in duration.

Era. A very large block of geological time (hundreds of millions of years long), composed of periods.

Erect stance. In anatomy, the condition in which the legs lie parasagittal to (along side of) the body and do not extend laterally from it.

Erg. A large "sea" of sand, commonly associated with deserts.

Eukaryote. Complex cell that has a nucleus and a variety of internal chambers called organelles.

Eurypoda. The ornithischian clade Stegosauria + Ankylosauria.

Eustatic. Global.

Evaporite. Rock composed of minerals precipitated through desiccation.

Evolution. In biology, descent with modification.

External mandibular fenestra. An outward-facing opening toward the rear of the mandible (commonly between the dentary, surangular and angular bones) found in many archosaurs.

Extinction. When the birth rate fails to keep up with the death rate.

Extinction boundary. The moment in time when organisms or groups of organisms became extinct.

Extraterrestrial. From outer space.

Fauna. A group of animals presumed to live together within a region.

Femur. The upper bone in the hindlimb (thighbone).

Fibula. The smaller of the two lower leg bones in the hindlimb; the bone that lies alongside the shin bone (tibia).

Flight feather. Elongate feather with well-developed, asymmetrical vanes; usually associated with flight.

Flood basalt. Episodic lava flow from fissures in the earth's crust.

Flux. A measure of change; rate of discharge times volume.

Footprint. Trace fossil left by the feet of vertebrates.

Foramen magnum. The opening at the base of the braincase through which the spinal cord travels to connect to the brain.

Foraminifera (sing. foraminifer). Single-celled, shell-bearing protists that live in the oceans.

Fossil. Anything buried.

Fourth trochanter. A ridge of bone along the shaft of the femur (thighbone) for muscle attachment.

Fractionation. As discussed here, the separation of isotopes that occurs during physical or chemical processes.

Frill. In ceratopsians, a sheet of bone extending dorsally and rearward from the back of the skull, made up of the parietal and squamosal bones.

Furcula. Fused clavicles (collarbones); the "wishbone" in birds and certain non-avian theropods.

Gastralia. Belly ribs.

Gastrolith. Smoothly polished stone in the stomach, used for grinding plant matter.

Genasauria. The ornithischian clade of Thyreophora + Cerapoda.

General. In phylogenetic reconstruction, referring to a character that is non-diagnostic of a group; in this context, synonymous with primitive.

Geographical distribution. In biology, the spatial placement of organisms.

Ghost lineage. Lineage of organisms for which there is no physical record (but whose existence can be inferred).

Gigantothermy. Modified mass homeothermy, which mixes large size with low metabolic rates and control of circulation to peripheral tissues.

Gizzard. A muscular chamber just in front of the glandular part of the stomach.

Gnathostome. A vertebrate with a jaw (formal term: Gnathostomata).

Gondwana. A southern supercontinent composed of Australia, Africa, South America, and Antarctica.

Goyocephalia. A clade of highly derived pachycephalosaurs.

Gymnosperms. Paraphyletic group of seed-bearing, non-flowering plants, including pines and cypress.

Habitat partitioning. In biology, the division, by organisms, of available ecospace into non-overlapping domains.

Half-life. The amount of time that it takes for 50% of an unstable isotope to decay.

Hard part. In paleontology, all hard tissues, including bones, teeth, beaks, and claws. Hard parts tend to be preserved more readily than soft tissues.

Haversian canal. In bone histology, a canal composed of secondary bone.

Heterotroph. In biology, an organism that must ingest all nutrients necessary for survival.

Hierarchy. As applied here, the ordering of objects, organisms, and categories by rank. The military and the clergy are both excellent examples of hierarchies; in these, rank is a reflection of power and, one hopes, accomplishment. Another hierarchical system is money, which is ordered by value.

High pressure zone. In meteorology, a zone in the atmosphere where large, moist, and dense air masses accumulate.

Hip socket. A depressed area in the pelvis where the femur (thighbone) articulates.

Homeotherm. Organism whose core temperatures remain constant.

Homologue. A homologous feature.

Homologous. Two features are homologous when they can be traced back to a single structure in a common ancestor.

Humerus. The upper arm bone.

Hyposphene–hypantrum articulation. Extra articular surface on the neural arches connecting successive elements in the backbone.

Hypothesis of relationship. A hypothesis about how closely or distantly organisms are related.

Ichnofossil. Impression, burrow, track, or other modification of the substrate by organisms.

Ichnotaxa. Taxa established on the basis of trace fossils.

Ichthyosaurs. Dolphin-like marine reptiles of the Mesozoic.

Igneous. Rocks or minerals derived from molten material.

Ilium. The uppermost of three bones that make up the pelvis.

Impact ejecta. The material thrown up when an asteroid strikes the earth.

Incongruent. Not equivalent; conflicting.

Indigenous. Restricted to a particular geographical region.

Induration. The process by which rock is hardened.

In-place. Not reworked.

Interspecific. Among different species.

Intraspecific. Within the same species.

Iridium. A non-toxic, platinum-group metal, rare at the earth's surface.

Ischium. The most posterior of three bones that make up the pelvis.

Isochronous. Occurring at the same time (synonym: synchronous).

Isotopes. In chemistry, elements that have the same atomic number but different atomic masses.

Jacket. In paleontology, a rigid, protective covering placed around a fossil, so that it can be moved safely out of the field. Commonly made up of strips of burlap soaked in plaster.

Jugal. One of the bones in the cheek region of the skull.

K-strategy. The evolutionary strategy of having few offspring that are cared for by the parents. The symbol "K" stands for the carrying capacity of the environment.

K/T (boundary). Common abbreviation for that moment in time, 65 Ma, which marks the boundary between the Cretaceous and Tertiary periods.

Keel. A flange or sheet of bone, as in the keeled sternum of birds; named for its resemblance to the keel on a sailboat.

Knee. The joint between the upper hind leg (thigh) and lower leg (shin).

Lambeosaurines. The hollow-crested hadrosaurid dinosaurs.

LAG. *See* **Lines of arrested growth.**

Laurasia. A northern supercontinent made up of the Siberian craton and the Old Red Sandstone Continent (Europe and North America).

Lepidosauromorpha. One of the two major clades of diapsid reptiles; the other clade is Archosauromorpha.

Lesser trochanter. A crest-like ridge on the femur.

Lines of arrested growth (LAGs). Lines that are inferred to represent times of non-growth, visible in the cross-section of bones.

Lissamphibia. Modern clade of amphibians: frogs, salamanders, and caecilians.

Lithosphere. The rigid, outermost layer of the earth, 100 km thick.

Lithostratigraphy. The general study of all rock relationships.

Low pressure zones. Regions of greater vertical accumulation of dry air.

Lower temporal fenestra. The lower opening of the skull just behind the eye.

Lycopod. A primitive type of vascular plant.

Ma. Millions of years ago.

Maastrichtian. The last stage (time subdivision) of the Cretaceous.

Mandible. The lower jaw.

Marginocephalia. The clade of dinosaurs that includes the most recent common ancestor of pachycephalosaurs and ceratopsians and all of its descendants.

Mass extinctions. Global and geologically rapid extinctions of many kinds of, and large numbers of, species.

Mass homeotherm. An organism that has a relatively constant body temperature because of its large size.

Mass number. In chemistry, the total number of neutrons plus the total number of protons for a given element.

Mass spectrometer. An instrument able to separate and measure tiny amounts of isotopes.

Matrix. In paleontology, the rock that surrounds fossil bone.

Maxilla. The upper jawbone that contains the cheek teeth.

MDT. *See* **Minimal divergence time.**

Megaflora. The visible remains of plants, especially leaves.

Mesotarsal. A linear type of ankle joint in which hinge motion in a fore–aft direction occurs between the upper ankle bones (the astragalus and calcaneum) and the rest of the foot.

Mesozoic. That interval of time from 245 Ma to 65 Ma.

Metabolism. The sum of the physical and chemical processes in an organism.

Metacarpal. Bone in the palm of the hand.

Metapodial. A general name for metacarpals and metatarsals.

Metatarsal. Bone in the sole of the foot.

Microtektite. A small, droplet-shaped blob of silica-rich glass thought to have crystallized from impact ejecta.

Minimal divergence time (MDT). The minimal amount of time missing between the two descendent species and their common ancestor; calculated by comparing phylogeny and age of fossils.

Mold. Ichnofossil that consists of the impression of an original fossil.

Monophyletic group. A group of organisms that has a single ancestor and contains all of the descendants of this unique ancestor (synonymous with clade and natural group).

Morphology. The study of shape.

Mosasaurs. Late Cretaceous marine-adapted lizards.

Motile. Moving.

Nares. Openings in the skull for the nostrils.

Natural group. *See* **monophyletic group.**

Neutron. Electrically neutral subatomic particle that resides in the nucleus of the atom.

Node. A bifurcation or two-way split point in a phylogenetic diagram (cladogram).

Non-avian dinosaurs. All dinosaurs except birds.

Norian. The last stage (time subdivision) of the Triassic.

Notochord. An internal rod of cellular material that, primitively at least, ran longitudinally down the backs of all chordates. May be thought of as a precursor to the vertebral column.

Nucleus. Central core of the atom.

Obligate biped. Tetrapod that must walk or run on its hind legs.

Obturator foramen. A large hole in the pubis near the hip socket.

Obturator process. A flange down the shaft of the ischium.

Occipital condyle. A knob of bone at the back of the skull with which the vertebral column articulates.

Occlusion. Contact between upper and lower teeth; necessary for chewing.

Ontogenetic. Pertaining to ontogeny.

Ontogeny. Biological development of the individual; the growth trajectory from embryo to adult.

Opisthopubic. The condition in which at least part of the pubis has rotated backward to lie close to, and parallel with, the ischium.

Orbit. Eye socket.

Organelle. Special structure within a eukaryotic cell.

Ornithischia. One of the two monophyletic groups composing Dinosauria.

Ornithodira. The common ancestor of pterosaurs and dinosaurs, and all its descendants.

Ossify. To turn into bone.

Osteichthyes. Bony fishes that include ray-finned and lobe-finned gnathostomes.

Osteoderm. Bone within the skin; may be small nodule, plate, or a pavement of bony dermal armor.

Osteology. The study of bones.

Oxidation. Bonding of oxygen.

Palate. The part of the skull that separates the nasal cavity (for breathing) from the oral cavity (for eating); usually strengthened by a paired series of bones.

Palatine. One of the bones of the palate.

Paleoclimate. Ancient climate.

Paleoecology. The study of ancient interactions among organisms.

Paleosol. Ancient soil profile.

Paleozoic. That interval of time from 543 Ma to 245 Ma.

Palpebral. A rod-like bone that crosses the upper part of the eye socket.

Palynoflora. Spores and pollen.

Pangaea. The mother of all supercontinents, formed from the union of all present-day continents.

Parasagittal process. A flange of bone, lateral to the midline of the skull, that helps to subdivide the nasal cavity.

Parasagittal stance. Stance in which the legs are held under the body.

Parascapular spine. An enlarged spine over the shoulder.

Parsimony. A principle that states that the simplest explanation that explains the greatest number of observations is preferred to more complex explanations.

Pectoral girdle. The bones of the shoulder; the attachment site of the forelimbs.

Pedestal. In paleontology, a pillar of matrix underneath the fossil.

Pelvic girdle. The bones of the hips; the attachment site of the hindlimbs.

Perforate acetabulum. A hole in the hip socket.

Period. Subdivision of an era, consisting of tens of millions of years.

Permineralization. The geological process in which the spaces in fossil bones become filled with a mineral.

Phalanx (pl. phalanges). Small bones of the fingers and toes that allow flexibility.

Phanerozoic. That interval of time from 543 Ma to the present; it also refers to the time in earth history during which shelled organisms have existed.

Photosynthesis. The process by which organisms use energy from the sun to produce complex molecules for nutrition.

Phylogenetic. Pertaining to phylogeny.

Phylogeny. The study of the fundamental genealogical connections among organisms.

Phylum. A grouping of organisms whose make-up is supposed to connote a very significant level of organization shared by all of its members.

Phytosaur. Long-snouted, aquatic, fish-eating archosaur.

Pineal. The so-called "third eye," a light sensitive window to the braincase that has been lost in mammals and birds.

Planktonic (or planktic). Living in the water column.

Plate. Large, mobile block of the crust of the earth.

Plate tectonics. The concept that the earth's surface is organized into large, mobile blocks of crustal material.

Plesiosaurs. Long-necked fish-eating reptiles with large flippers that inhabited Mesozoic seas.

Pleurocoel. A well-marked excavation on the sides of a vertebra.

Pleurokinesis. Mobility of the upper jaw.

Pneumatic. Having air sacs or sinuses.

Pneumatic foramina. Openings for air sacs to enter the internal bone cavities.

Poikilotherm. Organism whose core temperature fluctuates.

Postacetabular process. The part of the ilium behind the hip socket.

Post-temporal opening. Opening along the back of the skull that transmits the dorsal head vein out of the brain cavity.

Preacetabular process. The part of the ilium in front of the hip socket.

Precipitation. Rain or snow.

Precocial. The condition in which the young are rather adult-like in their behavior.

Predentary. The bone that caps the front of the lower jaws in all ornithischians.

Prepare. To clean a fossil; to get it ready for viewing by freeing it from its surrounding matrix.

Prepubic process. A flange of the pubis that points toward the head of the animal.

Primary bone. Bone tissue that was deposited or laid down first.

Primary productivity. The sum total of organic matter synthesized by organisms from inorganic materials and sunlight.

Primitive. *See* **ancestral.**

Process. Part of bone that is commonly ridge-, knob-, or blade-shaped and sticks out from the main body of the bone.

Productivity. The amount of biological activity in an ecosystem.

Prokaryotic. Small cells with no nucleus or other internal cell partitions.

Propalinal jaw movement. Fore and aft movement of the jaws.

Prospect. To hunt for fossils.

Proton. Electronically charged (+1) subatomic particle that resides in the nucleus of the atom.

Pubis. One of the three bones that make up the pelvic girdle.

"Pull of the Recent." The inescapable fact that as we get closer and closer to the Recent, fossil biotas become better and better known.

Pygostyle. A small, compact, pointed structure made of fused tail bones in birds.

r-strategy. The evolutionary strategy where organisms have lots of offspring and no parental care. The symbol "r" stands for unrestricted.

Radiometric. The dating method to determine unstable isotopic age estimations.

Radius. One of the two lower arm bones; the other is the ulna.

Recombination. The production of new combinations of DNA in the offspring with each reproduction event.

Recrystallization. The process whereby the original mineral is dissolved and reprecipitated, commonly retaining the exact original form of the original mineral.

Red bed. Rock of orange-red color due to an abundance of iron oxides.

Regression. Retreating of seas due to lowering of sea level.

Relative dating. The type of geological dating that, although not providing ages in years before present, provides ages relative to other strata or assemblages of organisms.

Remodel. In bone histology, to resorb or dissolve primary bone and deposit secondary bone.

Replace. To exchange the original mineral with another mineral.

Reptilia. The old Linnaean category for turtles, the tuatara lizards, snakes, and crocodiles. Reptilia as formulated by Linnaeus and as commonly used is not monophyletic; only the addition of birds to these four groups constitutes a monophyletic group.

Resolution. The degree to which one can distinguish detail.

Respiratory turbinate. A thin, convolute or complexly folded sheet of bone located in the nasal cavities of living endothermic vertebrates.

Retroarticular process. A very short projection of the lower jaw beyond the jaw joint.

Rework. To actively erode sediment and fossils from wherever they were originally deposited and redeposit them somewhere else.

Rhamphotheca. Cornified covering on the upper and lower jaws (e.g., a beak).

Robust. (1) In the context of hypothesis testing, a hypothesis is said to be robust when it has survived repeated tests; that is, despite meaningful attempts, it has failed to be falsified. (2) In anatomy, strong and stout.

Rock. A aggregate of minerals.

Rostral. Referring to the rostrum, or snout region of the skull.

Rostral bone. A unique bone on the front of the snout of ceratopsians, giving the upper jaws of these dinosaurs a parrot-like profile.

Sacrum. The part of the backbone where the hip bones attach.

Sarcopterygii. Lobe-finned fish.

Saurischia. One of the two monophyletic groups comprising Dinosauria; the other is Ornithischia.

Scapula (pl. scapulae). The shoulder blade.

Scenario. A story; in evolutionary terms, the combination of phylogenetic patterns and evolutionary explanations.

Sclerotic ring. A ring of bony plates that support the eyeball within the skull.

Seasonality. Highly marked seasons.

Secondarily evolve. To revolve a feature.

Secondary bone. Bone deposited in the form of Haversian canals.

Sedimentary rock. A rock which that generally represents the lithification of sediment.

Segmentation. The division of the body into repeating units.

Selectivity. With reference to extinctions, those who survived and those who did not.

Semi-erect stance. The stance in which the upper parts of both the arms and the legs are directed at about 45° away from the body.

Semi-lunate carpel. A distinctive, half-moon shaped bone in the wrist.

Sessile. Stationary.

Sexual dimorphism. Size, shape, and behavioral differences between sexes.

Sexual selection. Selection not between all of the individuals within a species, but between members of a single gender.

Shaft. (1) The hollow main vane of a feather. (2) The title and eponymous lead male character in the first and most famous of the 1970s "blacksploitation" films.

Shocked quartz. Quartz that has been placed under such pressure that the crystal lattice becomes compressed and distorted; correctly termed "impact metamorphism."

Sigmoidal. Having an "S" shape.

Sinus. A cavity.

Skeleton. The supporting part of any organism. In vertebrates, the skeleton is internal and consists of tissue hardened by mineral deposits (sodium apatite). Such tissue is called "bone."

Skull. That part of the vertebrate skeleton that houses the brain, special sense organs, nasal cavity, and oral cavity.

Skull roof. The bones that cover the top of the braincase.

Soft tissue. In vertebrates, all of the body parts except bones, teeth, beaks, and claws.

Specific. Diagnostic of a monophyletic group; uniquely evolved.

Sphenopsid. A primitive type of vascular plant.

Sprawling stance. Stance in which the upper parts of the arms and legs splay out approximately horizontally from the body.

Stable isotope. An isotope that does not spontaneously decay.

Stapes. The middle-ear bone that transmits sound (vibrations) from the tympanic membrane to a hole in the side of the braincase (allowing auditory nerves of the brain to sense vibration).

Sternum. The breastbone.

Stratigraphy. The study of the relationships of strata and the fossils they contain.

Subatomic. Smaller than atom-sized.

Subnarial foramen. An opening in the skull beneath the nostril area.

Superposition. The geological principle in which the oldest rocks are found at the bottom of a stack of strata and the youngest rocks are found at the top.

Survivorship. The pattern of survival measured against extinction.

Synapsida. The large clade of amniotes, including mammals, diagnosed by a single temporal opening.

Sympatric. Living in the same place at the same time.

Synsacrum. A single, locked unit consisting of the sacral vertebrae.

Taphonomy. The study of all of what happens to organisms after death.

Tarsal. Ankle bone.

Tarsometatarsus. The name for the three metatarsals fused together with some of the ankle bones.

Taxon (pl. taxa). A group of organisms, designated by a name, of any rank within the biotic hierarchy.

Temporal. Referring to time.

Testable hypothesis. A hypothesis that makes predictions that can be compared and assessed by observations in the natural world.

Tetrapoda. A monophyletic group of vertebrates primitively bearing four limbs.

Thecodontia. A paraphyletic taxon that at one time was used to unite the separate ancestors of crocodilians, pterosaurs, dinosaurs, and birds.

Therapsida. The clade of synapsids that includes mammals, some of their close relatives, and all of their most recent common ancestors.

Thorax. In vertebrates, the part of the body between the neck and abdomen.

Thyreophora. Armor-bearing ornithischians; stegosaurs, ankylosaurs, and their close relatives.

Tibia. One of the two lower bones in the tetrapod hindlimb; the other is the fibula.

Tibiotarsus. The fused unit of the tibia and the upper ankle bones.

Torpor. A trance-like state in which the temperature of an animal is considerably lowered.

Trace fossil. Impressions in sediment left by an organism.

Trachea. The windpipe.

Trackway. Group of aligned footprints left as an organism walks.

Transgression. Advancing of seas due to raising sea level.

Turn (a fossil). To separate the fossil from the surrounding rock at the base of the pedestal and to rotate it 180°.

Tympanic membrane. Eardrum.

Type section. The outcrop of rock where a stratigraphic unit was originally described. For example the type section of the Hell Creek Formation is an outcrop of rock found in the Hell Creek Recreation Area, Garfield Co., Montana, USA.

Ulna. One of the two lower arm bones; the other is the radius.

Unaltered. When original mineralogy is unchanged.

Ungual phalange. An outermost bone of the fingers and toes.

Unstable isotope. An isotope that spontaneously decays from an energy configuration that is not stable to one that is more stable.

Upper temporal fenestra. The opening in the skull roof above the lower temporal fenestra.

Vane. The sheet of feather material that extends away from the shaft.

Vertebrae. The repeated structures that compose the backbone and that, along with the limbs, support the rest of the body.

Wedge. The evolutionary pattern of waxing and waning dominance among groups of organisms.

Zygapophysis. A fore-and-aft projection from the neural arches (of vertebrae).

Subject index

Bold type indicates figures.

Generic index

Bold type indicates figures.

Author index